CONTENTS

CREATING THE DEPARTMENT OF HOMELAND SECURITY: CONSIDERATION OF THE ADMINISTRATION'S PROPOSAL

TUESDAY, JUNE 25, 2002

House of Representatives,
Committee on Energy and Commerce,
Subcommittee on Oversight and Investigations,
Washington, DC.

The subcommittee met, pursuant to notice, at 9:10 a.m., in room 2123, Rayburn House Office Building, Hon. James C. Greenwood (chairman) presiding.

Members present: Representatives Greenwood, Stearns, Gillmor, Burr, Whitfield, Bass, Fletcher, Tauzin (ex officio), Deutsch, Stupak, Strickland, and DeGette.

Also present: Representatives Deal, Cubin, Waxman, Markey, Sawyer, Capps, and Harman.

Staff present: Tom DiLenge, majority counsel; Amit Sachdev, majority counsel; Ray Shepherd, majority counsel; Nandan Kenkeremath, majority counsel; Edith Holleman, minority counsel; and Chris Knauer, minority investigator.

Mr. Greenwood. The subcommittee will come to order. The Chair would announce before the commencement of opening statements that, pursuant to the rules, the chairman of the subcommittee and the ranking member and the chairman of the full committee and the ranking member of the full committee will be accorded 5 minutes for opening statements; other members of the subcommittee shall be accorded 3 minutes apiece.

We welcome the participation of other members of the full committee who are not members of the subcommittee, and should they arrive and wish to make opening statements, we will grant them time—yield them time, the amount of time being dependent upon how many of them there are.

And the Chair welcomes Governor Ridge, my friend—good to have you with us—and yields himself 5 minutes for the purpose of an opening statement.

Good morning. Today the subcommittee will hold the first day of a multipart hearing to examine how the Bush Administration's proposal to establish a Department of Homeland Security will affect the agencies and the operations over which this committee now exercises principal jurisdiction. Our first witness is the current Director of the Office of Homeland Security and our former colleague, Governor Tom Ridge, who is appearing today in his capacity as the chief of the transition team for this new department.

The President could have made no finer choice in responding to the disaster of September 11 than by appointing Tom Ridge to be Director of the Office of Homeland Security. The challenge before him is daunting, but those of us who know Tom also know that he has always heeded his country's call.

In 1968, while still in law school, Tom Ridge was drafted into the U.S. Army. He fought in Vietnam as an infantry sergeant and was awarded the Bronze Star. He was the first enlisted Vietnam veteran elected to Congress.

Now he has been enlisted in a new struggle. True to form, he has labored tirelessly since last September to help improve the security of our homeland and our fellow citizens.

The President's proposal is a bold one. It envisions a department whose mission includes border and transportation security; emergency preparedness and response; chemical, biological, radiological, and nuclear countermeasures; information analysis and infrastructure protection. If approved as now proposed, only the Department of Defense and the Department of Veterans' Affairs would have more employees than the almost-170,000 workers proposed for the Department of Homeland Defense.

Few would dispute the need for consolidation and coordination of the nearly 100 agencies that now share responsibility for these critical tasks. This subcommittee's oversight over the past 2 years also has demonstrated the need for a single agency to take charge of the responsibility to enhance the protection of our Nation's critical infrastructure and key terrorist targets, both in the public and private sector. The latter includes several industry sectors over which this committee has principal jurisdiction, including the electricity and telecommunications grids and our Nation's drinking water systems.

As our hearing last April demonstrated, precious little has been done since 1997 when a Presidential blue ribbon panel urged the establishment of a robust public-private partnership to identify critical assets, assess their interdependencies and vulnerabilities, and take steps to mitigate our risks.

Moreover, this subcommittee's oversight with respect to Federal counterterrorism R&D programs has raised many of the same concerns. As the General Accounting Office reported to this subcommittee last September, just prior to the anthrax attacks on this city, our Federal bioterrorism research programs, scattered throughout a dozen or more agencies, are poorly coordinated and lack a clear sense of priority and focus. The same is true for the myriad of Federal programs aimed at improving the preparedness of Federal, State and local governments and emergency response providers to deal with major disasters, terrorist attacks and other public health emergencies. In fact, there were so many such programs within the Department of Health and Human Services itself that in the bioterrorism bill this committee recently shepherded through the Congress, we created a new Assistant Secretary at HHS just to coordinate all these emergency preparedness and response functions.

And this is just one department. Can there be any doubt why every serious study of this issue has ended in a call for some form of centralization, or focal point of coordination in the executive branch? The President's proposal moves us firmly in that direction.

The focus of today's hearing is on the critical aspect of emergency preparedness and response and how the President proposes to improve our national efforts in this area. We cannot move too soon. Yesterday, for example, CNN reported on the new threats being made by a spokesman for al Qaeda who, in a sickening and warped reference to September 11, told Americans they should, quote, fasten their safety belts and then spoke of the death of up to 4 million Americans including 1 million children through the use of chemical and biological weapons.

Although Governor Ridge will testify today on all aspects of the President's proposal, the remainder of our panels and witnesses will focus on the emergency preparedness and response issue, namely Title V of the administration's proposal. With respect to those functions or programs that are proposed for transfer from any agency to the new department, two questions seem in order: First, how do these programs operate currently; and second, what are the potential advantages or disadvantages to the proposed transfer?

In our case, while the President's bill is a useful blueprint, many important questions remain to be resolved. For example, what is the scope of the new secretary's authority over HHS's public health preparedness programs and how might it alter the current focus on important dual-use programs? Why are some of the agencies' preparedness and response programs transferred completely, others transferred partially and others left unchanged in their respective departments? And for those assets or functions not fully transferred to the new Secretary, but under his authority, how does the administration plan to ensure a workable model with one Secretary directing the assets or programs of another?

As I said at the outset, the task before the President, the Congress and today's chief witness is daunting, but whatever the challenge, we must meet it. In the midst of the battle of Bunker Hill, Abigail Adams wrote these words to her husband in Philadelphia:

"Dearest friend, the day, perhaps the decisive day, has come on which the fate of America depends. Now the fate of America rests with us, and of one thing I am certain. Unless a spirit of cooperation and trust informs all of our efforts, we are unlikely to succeed. And to be successful, we have a duty to speak plainly to the American people about the clear and present dangers that lead us to this enormous investment in this massive undertaking."

Again, I want to thank Governor Ridge and all of our witnesses for agreeing to appear before us today, many on short notice.

I will recognize the ranking member, the gentleman from Florida, Mr. Deutsch, for an opening statement.

Mr. DEUTSCH. Thank you, Mr. Chairman. And thank you, Governor Ridge.

This is an issue where I think it is accurately described that there is no light between any of us in the Congress, the 435 Members of the House and the 100 Members of the Senate. And I think that we stand completely with the President on the creation of this department, which is an integral part of the war on terrorism.

I think if we have learned anything post-September 11, it is reminding us that the most fundamental thing we can do as a government and as elected officials is the security of our constituents. And, in fact, I think we understand that unfortunately, prior to September 11, we were not looking at it quite the way we should. And specifically, I think, we acknowledge at this point that terrorists' or terrorist states' particularly weapons of mass destruction are an existential threat to the United States and to our people.

And, Governor Ridge, I have read your comments and I would completely agree with basically all of them, but one I want to focus on which I think is the—in a sense, the essence for the creation of the department is that, at the present time, there really is no one who is responsible or no agency that is responsible, but—you are in your position, but no agency that is responsible for homeland security.

And my experience in life—and I think for most of us if we think about our experience in life—is, something never gets done correctly unless someone is responsible and in charge. And I think that is the essence of, the purpose of this agency where I think the goal, the need, is absolutely imperative.

I also think the facts of, again, what you have put together and what others have put together at this point specifically show the sort of ad hoc dispersed nature of some of these responsibilities. I think as we move forward—and I think this is one of these issues where we really are working hand-in-hand—in a very bipartisan tradition in this committee, although we have many disagreements, we have many agreements as well.

We will disagree, as we did last week on prescription drugs, but on this, I think there are no disagreements. And I think what we are really looking for is working with you, working with each other, just really trying to make it as good as possible.

And I think we are at the level of details. I don't think that this is a case where the devil is in the details. I really don't. I think it is the details of working with you to really try to structure a department that will maximize the imperative that we are successful.

One of the analogies that I have used in talking about post-September 11 and I would add to this creation of this department, I think there are several World War II analogies—two, really, I think, at least for me, and when I have spoken about this, they have been very on point.

One is clearly, obviously, Pearl Harbor where the United States wasn't prepared; and if we look historically, the Japanese might have seen it as a short-term victory. But I think historically, obviously it was an incredible disaster for them. Had the United States entered the war in the Pacific, which is unclear whether we would not have—would have, and I think it was overdetermined once we entered the war that we would be successful.

The other analogy is the Manhattan Project. And when it was started it was not overdetermined that we would be successful in that effort. But if we were not successful, obviously history would be a lot different.

Governor, I speak to you, and I know your commitment is total on this; and I speak to ourselves about this, that I think that just as we had no choice but to be successful with the Manhattan

Project, we have no choice but to be successful with what we are doing to prevent weapons of mass destruction attacking the United States. And I believe the creation of this department is a critical component of that.

So I look forward to working with you and with my colleagues on both sides of the aisle over the next, really, hopefully, just several months. I think setting the date of September 11 to try to get it resolved by is doable. As you well know as a former Member, we can always argue about things. We will have enough things to argue about between now and January 20 if we want to. Hopefully, we won't.

Hopefully, we will put deadlines on ourselves and force us with the minutia of details, with the minutia of jurisdiction. Hopefully, we will get over that and understand that we are all working together for one goal.

So I yield back the balance of my time.

Mr. GREENWOOD. The Chair thanks the gentleman and yields 5 minutes for an opening to the chairman of the full committee, the gentleman, Mr. Tauzin.

Chairman TAUZIN. Thank you, Chairman Greenwood, and I am pleased to join you in welcoming Governor Ridge to testify on President Bush's historic proposal for the creation of the new Cabinet-level Department of Homeland Security.

Governor Ridge, I think you and we, too, understand that we are going to play some important roles here. But the truth is that bureaucrats and legislators and even Cabinet-level officials really play a second-place role when it comes to defending the country in this very important time. It is the men and women of the military, the National Guard or the fire and emergency response teams and the incredible heart and courage of the people of America who are on the front line, the eyes and ears of our country, the first responders who really have this task at hand; and our job is to help arm them and properly coordinate them.

And I, first of all, want to thank you because the other side of that coin is that we have learned since September 11 that there can be a lot of finger-pointing in this country when things go wrong, and there can be a lot of people trying to put the blame on someone else for not sharing information or coordinating properly.

You, however, left your job as Governor of the great State of Pennsyvania at the summoning of our President, and you decided to be the person where the buck stops in coordinating and making sure this awful finger-pointing exercise doesn't happen again. And this is the next, obviously, important step in that process, to make sure there is someone at a Cabinet level for whom the final responsibility rests in coordination.

That is an awesome responsibility, sir, and I commend you for taking it on in this temporary position. And frankly, I would hope that the President has the good sense, when we are through with this work, to continue you in a permanent position if you are willing to undertake it.

I wanted to talk briefly with you this morning about some of our roles in connection with your role in the establishment of this new department. First, our committee has jurisdiction, and we will continue to have jurisdiction, obviously, over many of the programs that the Department of Energy and the national labs, the Department of Health and Human Services, all of which serve vital roles in preparing and responding to chemical, biological, radiological and nuclear attacks. All areas where—if this spokesman for al Qaeda is real and his statements are believable, all areas of vulnerability these people hope to exploit in these programs, such as the nuclear emergency support teams that identify and respond to radiological and nuclear threats as well as public health programs; such as the strategic national stockpile of drugs and vaccines that must be stocked and rapidly deployed, this new department will now play an important role.

Title V of the President's proposal contains a plan for consolidating and coordinating these functions. Well, obviously we have to help you make sure that that is done properly. It is a critical function as we face new threats.

Second, our committee has jurisdiction and will continue to have jurisdiction over research and development programs for chemical, biological, radiological and nuclear countermeasures. Programs that the Health and Human Services Department, DOE and national labs in which the country's top scientists are currently working on new methods for detecting and detecting terrorist attacks. For example, there are improved sensors to detect radiological devices, new scanners to screen luggage and cargo, new technologies to detect and neutralize biological hazards.

Title III of the President's plan would transfer many of these programs, and it is important, I think, as we handle this transfer, to see what we can do about somehow coordinating the very diverse efforts that are going on in as many as four different labs on the same subject, and to make sure we get the best in new, innovative technologies out there to protect our borders and to make travel in this country as safe as we can make it.

And a third of the department's jurisdiction will continue to have jurisdiction over the regulation of many of the Nation's most critical infrastructure and assets, including both publicly and privately owned assets in telecommunications and energy and safe food and drinking water, as well as many manufacturing facilities in the country that could be targets.

Governor Ridge, I want to thank you for something else: for being accessible to this committee without subpoena, voluntarily meeting with us, counseling with us, as we went through the process post-9/11 of examining all the agencies under our jurisdiction and all these critical assets, and where the vulnerabilities might be and what we might do to encourage the agency heads to begin developing protection and countermeasures to make sure these assets are protected.

The key is to recognize that most of the critical, important infrastructures are privately owned, privately operated. And the only way to succeed is going to be creating the strong public-private partnerships for national security. It doesn't create new regulatory regimes in this country, new bureaucracies that are going to make the economy worse off, but literally relies upon the strength of those private-sector-owned and -operated entities to work with us in a partnership to make sure they are protected properly.

We want to point out one more thing, and I will be asking you a couple of questions about it. In the meetings we had post-9/11, we were shocked to find out how many of the vulnerability assessments that exist in this country, how many of the detailed plans and drawings and important critical assets in this country are on the Internet, were available under the Freedom of Information for anybody to obtain. And this committee is vitally concerned, as we create this new department, that there are some common standards for vulnerability assessments and there are some real strong amendments, the Freedom of Information Act and other acts that would unfortunately allow some of this critical information to be available to people who might use it as a road map for terror in the future.

We have to cut a delicate balance here because we are a free society, and we want people to know what our Government is doing; but there is a line we have to draw when it comes to providing free to anybody who wants it a road map of how to get into a nuclear plant or how to find a critical telecommunications infrastructure, and doing something with it.

Finally, Governor Ridge, we just passed the Bioterrorism Act. This committee was primarily responsible for its development, as you know. There are some conflicts now in the new proposals. We are really beginning to assess, to coordinate the act we just passed with the new proposal the President just made. We are going to need your help in doing that. We don't want to leave some of the good work we did on bioterrorism undone because we are now changing the structure of things.

Finally, I want to thank the chairman for also calling today Deputy Secretary Claude Allen and General Gordon, who are also going to assist us in this inquiry.

Let me say, Mr. Chairman, yesterday I spent some time with Leader Armey, and I want to inform the committee and the Governor that we are sticking firmly to the July 12 timetable. We are going to get this work done quickly. And we in the House are going to finish the work on this critical national proposal, and we are going to do it well; and I am going to thank you for helping us do it right.

[The prepared statement of Hon. W.J. "Billy" Tauzin follows:]

PREPARED STATEMENT OF W.J. "BILLY" TAUZIN, CHAIRMAN, COMMITTEE ON ENERGY AND COMMERCE

Thank you Chairman Greenwood, I am pleased to join you today in welcoming Governor Tom Ridge to testify on behalf of President Bush's historic proposal for the creation of a new Cabinet-level Department of Homeland Security.

Governor Ridge, let me thank you for the job that you have been doing—tirelessly and without complaint—to defend our borders and keep the citizens of this great country safe and secure, in our cities, our communities, and our homes. After the terrorist attacks last fall, President Bush asked you to accept perhaps the single most important, and certainly the most difficult, job in the Nation. And you have risen to the challenge.

We in the Congress appreciate the job you are doing, and we will continue to do our part for this cause—a cause that requires us to make absolutely sure that the men and women who are fighting this war against terrorism on our behalf, including our military, our Reservists, the National Guard, and Federal, State, and local law enforcement personnel, have the tools, the resources, and the support they need to keep us safe from the harm our enemies seek to bring to our shores.

With regard to the President's proposal, I support creating a Cabinet-level department—one that will not only pick up the role of homeland security coordinator, but a new Department with an empowered Secretary who has the authority and resources needed to protect our country from the threats of terrorism.

The Committee on Energy and Commerce has an important responsibility to assist the Administration with this proposal. First, we have jurisdiction—and will continue to have jurisdiction—over many of the programs at the Department of Energy (DOE), the National Labs, and the Department of Health and Human Services (HHS) that serve vital roles in preparing for and responding to chemical, biological, radiological and nuclear attacks. These include energy programs such as the nuclear emergency support teams that identify and respond to radiological and nuclear threats, as well as public health programs, such as the Strategic National Stockpile of drugs and vaccines that must be stocked and rapidly deployed in the event of a chemical or biological attack. Title 5 of the President's proposal contains a plan for consolidating and coordinating these functions in the new Department of Homeland Security. We must ensure that this is done properly and that these programs are integrated in a manner that allows them to respond promptly in the event of a future attack.

Second, this Committee has jurisdiction—and will continue to have jurisdiction—over research and development programs for chemical, biological, radiological and nuclear countermeasures. These are programs at HHS, DOE and the National Labs in which our country's top scientists are working to develop new methods for detecting and preventing terrorists attacks—such as improved sensors to detect radiological devices, new scanners to screen luggage and cargo, and new technologies to detect and neutralize biological hazards. Title 3 of the President's proposal contains a plan for transferring many of these programs to the new Department. It is important for us to remember that new and improved technologies and American ingenuity and innovation are among the greatest advantages we have in fighting terrorism, second only to the heart and conviction of the people of this country.

Third, this Committee has jurisdiction—and will continue to have jurisdiction—over the regulation of many of our Nation's most critical infrastructures and assets, including both publicly and privately owned assets that are integral to the delivery of telecommunications and information technology services, the production and distribution of energy, and the delivery of safe food and drinking water, as well as manufacturing facilities that may be targets of potential terrorist actions. Title 2 of the President's proposal would add to the mission of the new Department the responsibility to analyze vulnerabilities and improve protection for these critical assets and infrastructures. The key to our success in this area is to recognize that many of the most important critical infrastructures are privately owned and operated, and the only way to succeed in assuring their protection is through a strong and effective public-private partnership for national security.

After the September 11th attacks, I and other senior Members of this Committee on a bipartisan basis met with high-ranking private sector officials to encourage them to work together in a public-private partnership to ensure that our critical infrastructures are adequately protected against potential terrorist attacks. Not only must potential targets of terrorism be adequately protected, but we also must ensure that sensitive information about these assets, such as vulnerability assessments, are never allowed to be used as roadmaps for terrorist action. I believe that the new Department should develop a comprehensive framework across the critical infrastructure sectors, including common standards for vulnerability assessments, and that we in Congress must provide additional legal protections to protect such sensitive information from improper public disclosure.

Finally, it is worth noting that, just this month, the President signed a sweeping $4.6 billion dollar bioterrorism preparedness bill into law, which was shepherded through Congress by Members of this Committee on a bipartisan basis. Many of the issues that we dealt with in crafting that new law, and many of the proposals to combat bioterrorism, will need to be evaluated in the context of the new Department of Homeland Security. Sorting out roles and responsibilities for the new Department and the other Federal agencies already tasked with many of these functions will be a significant challenge that we must complete quickly.

I commend the President for his proposal. It reflects a sound framework to get this job done, and I believe credit is due not only to the President for taking this bold step, but also to those, such as former Senators Warren Rudman and Gary Hart, who have for sometime recognized this need and whose foresight and ideas are undoubtedly reflected here.

Again, I want to thank Governor Ridge, and each of our other witnesses, including Deputy Secretary Claude Allen from the Department of Health and Human Services, and General John Gordon, Administrator of the National Nuclear Security Administration at the Department of Energy, for coming here today. I look forward to today's testimony and to working with the Administration and my colleagues on both sides of the aisle to craft legislation that creates a Cabinet-level Department of Homeland Security worthy of the people who work tirelessly everyday to protect us. Mr. Chairman, I yield back the balance of my time.

Mr. GREENWOOD. The Chair thanks the chairman of the committee, and recognizes the gentleman from Michigan, Mr. Stupak, for 3 minutes for an opening statement.

Mr. STUPAK. Thank you Mr. Chairman. I look forward to today's hearing and welcome Governor Ridge.

We have spent a lot of time since September 11, and I am sure we will do more in the future. Let me say right away that I accept the principle that homeland security is so important that it demands a Cabinet-level position. In fact, as one of the early cosponsors of some of the proposals put forth by the Democratic Caucus, it is not whether what caucus put it forward, but the idea and the principle that we do need a Cabinet-level position for homeland security.

As such, the Secretary serving as the head of this department should have the information, the authority and resources to carry out the task of protecting our citizens and our domestic resources and infrastructure.

That said, however, I believe that Members of Congress of both parties want to see a homeland security proposal from the administration that is more than just a mere shuffling of the chairs at the table. If the chain of command for organizations like the Coast Guard and FEMA, the Federal Emergency Management Agency, are reorganized, we want to be able to ask about the missions and the staffing and the cost of the change.

If information-gathering is reorganized—if information-gathering is reorganized, we want to know what intelligence will be collected, how it will be distributed and whether the net change puts information in fewer hands or more hands, and whether it speeds distribution of intelligence, or does it encumber it?

Reorganization will come, and the public needs to stay involved; and it needs to make the President, the Republican leadership in the House and the Democratic leadership in the Senate aware of its concerns. And with the chairman putting forth that July 12 deadline, it is even more important that those concerns are expressed immediately. Whether reorganization winds up being merely changed for the sake of change or a real improvement in protection of our Nation will depend on the questions that are asked, the debates that are held and the attention paid to the details of the President's proposal.

Again, welcome, Governor Ridge; and I look forward to hearing from you and other witnesses today.

And, Mr. Chairman, with that, I will yield back the balance of my time.

Mr. GREENWOOD. The Chair thanks the gentleman and recognizes the gentleman from Kentucky, Mr. Whitfield, for 3 minutes for an opening statement.

Mr. WHITFIELD. Thank you, Mr. Chairman.

And, Governor Ridge, we welcome you to the committee today and look forward to your testimony on what President Bush has described as the biggest restructuring of the Federal Government in 40 years. We also look forward to the testimony of the witnesses on the other three panels.

I think all of us understand and recognize that this is a complex piece of legislation, and it will be interesting to determine exactly how the new Department of Homeland Security will interact with the existing agencies in working out the areas of responsibility, and who has direct authority.

So I am looking forward to the testimony today as we embark on this very important legislation, and thank you for being here.

Mr. GREENWOOD. Chair thanks the gentleman, and the Chair notes the presence of the two gentleladies from California who are members of the full committee, but not members of the subcommittee. We welcome your participation.

The Chair recognizes the presence of the gentleman from California. The Chair will grant each of you 3 minutes for an opening statement, beginning with Mrs. Capps.

Mrs. CAPPS. Thank you, Mr. Chairman, and I thank you for holding this hearing.

And welcome and thank you, Governor Ridge, for yet again coming before us with information and insight into what is happening to this restructuring effort.

I don't have formal remarks; I am very eager to get into the conversation. I came to Congress after putting in a couple decades' work in public health in my community. I am very eager to hear how this legislation, which I helped craft—the bioterrorism preparedness bill—to ensure those resources get in the hands of the first responders.

Each time I go back to my district, the safety and health people there are wondering and asking about this. And I am very concerned that we do this with all haste. While this restructuring is very preoccupying, and I can understand that, we can't forget that our mission really is in the local communities, because that is where this battle needs to be waged.

So I will be yielding back my time and looking forward to the hearing. Thank you very much.

Mr. GREENWOOD. The Chair thanks the gentlelady and yields 3 minutes to the other gentlelady from California, Ms. Harman.

Ms. HARMAN. Thank you, Mr. Chairman. I also would like to thank the chairman of the full committee for personally inviting me to participate.

Good morning, Governor Ridge. I would hope that once we set up this Department of Homeland Security, you would not have to spend the entire summer testifying before Congress ever again.

I think this concept is very important. As you know, many pieces of it were borrowed from legislation some of us introduced on a bipartisan basis up here. You have put them in a different order, but I am proud to support your proposal and am one of the original cosponsors of the Armey bill that was introduced yesterday.

I think that we, up here, can contribute a few refinements that would help the legislation be more successful. And I just want to address one area right this minute in my remaining few seconds, which is public-private partnerships.

You were nice enough to participate last week in a really spectacular meeting that 12 members cohosted on a bipartisan basis called

Technology and Terrorism. We had 120 CEOs up here, and they were talking about their frustration with connecting their technologies into our homeland security effort.

The mechanism for doing this needs to be refined in this new department. H.R. 4629, introduced by Congressman Tom Davis, has some very good ideas in it, but I would hope, as we proceed, that we do refine this procurement process.

Second, I said public-private partnerships. On the partnership point, the government at the Federal, State and local levels must work more closely with private entities to ensure homeland security. The Government is responsible for providing security for citizens, but the private sector shares the responsibility to protect against attack or disruption, and it controls many of the assets needed to do so.

When we have questions, I will ask you more about this, but let us as a committee, especially one focused on commerce, lend our expertise, working with your office to make the public-private partnership piece of this legislation more effective.

I yield back the balance of my time.

Mr. GREENWOOD. Chair thanks the gentlelady and recognizes 3 minutes for purposes of an opening statement to the gentleman from Florida, Mr. Stearns.

Mr. STEARNS. Good morning and thank you, Mr. Chairman.

Governor Ridge, you are going to have a lot of patience in life. Governor of Pennsylvania might look pretty good to you after this process. They think you are doing a great job, and we are here to support you in any way we can and we're just glad, as a U.S. citizen, you're willing to tackle this.

Most of my speech, Mr. Chairman, I will make part of the record by unanimous consent.

Mr. GREENWOOD. Without objection.

Mr. STEARNS. I think it's already been pointed out, not since—the creation of such an enormous department, like this, encompassing a vast organization of Government resources has not been attempted since the National Security Act of 1947.

I think one of the concerns some of us have, Governor Ridge, is that while we take all this organization and move all these departments together, what about the intelligence failures and what are we doing to streamline within a department—if you just take all these departments and put them together and do nothing to change the individual departments and streamline them and give them more high tech equipment and make sure that these departments are talking to each other—you know, that would be the question: Is the President's proposal adequate in that respect?

Two FBI units, a national domestic preparedness office and the National Infrastructure Protection Center would be transferred to the department under the President's plan. What about reform or transformation of the FBI, the CIA, related to counterterrorism? You know, in light of what we learn and see in time and U.S. News report, there has got to be something done there, and I think it would be a false assumption for Americans to think just making this new Homeland Security is going to solve all the problems.

We on the Energy and Commerce Committee are very concerned about some of our jurisdiction and how that is going to work, because once we have a department getting its funds through you, yet the department remains in one agency, how is that going to work?

So you have a daunting task ahead of you, and I want to commend you. And I assume you are part of the wellness preparedness program the President has in running every day and making sure you are not stressed out here. Godspeed to you and thank you for testifying.

Mr. GREENWOOD. The Chair thanks the gentleman, and the Chair recognizes for 3 minutes the gentleman from California, Mr. Waxman.

Mr. WAXMAN. Thank you very much, Mr. Chairman. And welcome, Governor Ridge, to this hearing. I am very glad to have this opportunity to further examine the Bush Administration's proposal for the Department of Homeland Security.

The proposal raises many questions of importance to this committee as well as other committees. I am very concerned about the proposed transfer of important public health functions of the Department of Health and Human Services. I believe that the transfer of these functions may undermine the rebuilding of core public health capacities that is now under way. If our public health system is structured and viewed exclusively through the lens of fighting terrorism, it may seriously weaken our ability to respond to other threats to the health of the American people.

It appears that several HHS offices are to be transferred. These include Office of Emergency Preparedness, the National Disaster Medical System and the Metropolitan Medical Response System. With these offices may go significant authority to oversee our Nation's response to public health emergencies.

Such a transfer may also shift to the Department of Homeland Security the power to make bioterrorism and emergency preparedness grants to State and local public health systems. These grants were the cornerstone of the recently enacted Public Health Security and Bioterrorism Response Act. Their purpose was not only to fund specific preparations for bioterrorism. Just as critically, the grants were intended to turn around decades of neglect of our Nation's public health infrastructure.

It is beyond argument that our public health system is in disrepair, and we cannot protect our citizens from bioterrorist attacks if our public health system is not working. Detecting and responding to a bioterrorist attack is just like detecting and responding to other emerging epidemics. It requires fully functioning and coordinated public health systems at the local, State and Federal levels.

For this reason, the bioterrorism bill directed HHS to coordinate the repair of Federal, State and local public health systems as part of bioterrorism and emergency preparedness. The expertise to establish priorities and coordinate this effort lies with the public health experts and scientists at HHS and CDC. If priority-setting, coordination and/or grant-making functions are transferred to a new department, focused on terrorism, I am very concerned that the necessary rebuilding and upgrading of our public health response system will take a back seat.

If we attempt to protect ourselves against terrorist attacks at the expense of our Nation's public health system, we may find that we have undermined rather than enhanced our Nation's true security.

And I thank you for this opportunity for an opening statement, and I look forward to working with you, Governor Ridge, on this very important issue.

Mr. GREENWOOD. The Chair thanks the gentleman.

[Additional statements submitted for the record follow:]

PREPARED STATEMENT OF HON. TED STRICKLAND, A REPRESENTATIVE IN CONGRESS FROM THE STATE OF OHIO

Let me start by thanking Chairman Greenwood and Ranking Member Deutsch for holding this hearing today. All Americans are aware of the need to rethink how we defend our country, and so I thank Governor Ridge, as well as the witnesses who will follow him, for being here to answer our questions about the president's proposed Department of Homeland Security. I am pleased that the Administration has attempted to put together all the ideas for increased domestic security that have been raised during the past eight months, many of which have been discussed in hearings like this. Now Congress must fulfill its role to balance the power of the Executive Branch and question the president's proposal. It's our responsibility on this panel today to ask questions of our witnesses that will allow us to flesh out the skeletal suggestion put forth by the president as well as to create a new department that will best serve the constituents whom we represent here in Washington.

When we talk about protecting America, we should be thinking in terms of what's proactive and preventative instead of only what's reactive and responsive. While we all understand the need to formulate "countermeasures" and to devise plans for "emergency preparedness and response," I am concerned that the president's proposal may not give the secretary of the new department enough authority to prevent disaster. We have learned from the news media in recent weeks that we might have averted the terrorist attacks on September 11th if our federal agencies had been configured differently or had communicated with each other more effectively. In other words, we might have been able to prevent disaster.

In my view, we have two main strategies at our disposal: we can deter future attacks with our brawn, or we can halt them with our brain—with our intelligence capabilities. We can spend hundreds of millions of dollars on star wars, or we can spend a couple hundred dollars on language courses so that we have linguists who can translate the mountains of raw intelligence data that we collect but never analyze. But, even if all the data are analyzed and packaged in a form that is presentable to the secretary of the new department, what assurances do we have that one intelligence gathering agency, be it the CIA, the NSA, or the FBI with its new powers, would share its reports with the others? Will the new secretary have any authority to ensure that information is shared and that our intelligence operations are working together to prevent disaster? These questions are among many that we will be seeking answers to in the coming weeks.

In particular, last Fall I wrote to Secretary Abraham to express my concern for the safeguarding of our federal nuclear facilities and the nuclear materials stored at these sites. Substantial quantities of nuclear materials, including highly enriched uranium and plutonium, are stored in chemically and physically unstable forms across the Department of Energy complex. Some of these nuclear materials are stored in outdated containers that often sit in deteriorated facilities or even outside, exposed to the elements. In either case these storage facilities were not built with the intention of protecting nuclear materials from terrorist attacks. At the DOE facility in Piketon, Ohio, for example, the majority of the 16,000 depleted uranium hexafluoride canisters stored onsite are out in the open.

I think it is tremendously important that we have an understanding of how the Department of Homeland Security will protect America and its citizens from acts of malice against the physical structures and containers holding special nuclear materials, by-products, and source materials, especially in those cases where the physical structures may be vulnerable to significant radiological and other consequences.

I anticipate hearing from the witnesses about how such drastic governmental restructuring will affect—good or bad—the ability of the different agencies to fulfill their objectives. I look forward to a thoughtful and candid discussion of the proposals to protect our nuclear assets, in addition to plans for safeguarding Americans if terrorists were to strike at nuclear facilities. I thank the Chair and yield back the remainder of my time.

PREPARED STATEMENT OF HON. JOHN D. DINGELL, A REPRESENTATIVE IN CONGRESS FROM THE STATE OF MICHIGAN

Thank you, Mr. Chairman, for holding this initial hearing on the President's proposed new cabinet agency for homeland security. I have made no secret of my skepticism that mere reorganization can solve the problems we face, or that reorganization would not create significant new problems. That is why this hearing, and others like it across the Congress, are so necessary. They cannot simply be "check the box" exercises.

The Committee on Energy and Commerce will need to address several questions in the coming weeks about the proposed new structure. First, I note we just passed, and the President just signed, a carefully crafted comprehensive bioterrorism measure. It established programs to rebuild our public health infrastructures at the state and local levels, which are where responses to terrorism occur, as well as strengthened the federal capacity to address possible threats. Will the new Department actually increase fragmentation in the largely cohesive federal effort against bioterrorism and other public health emergencies? Will the new Department undermine the state and local public health focus of the new law?

Second, will the Department's security activities undermine the enforcement of existing environmental, health and safety protections, or be otherwise detrimental to such safeguards developed over many years after full and open consideration by the Congress? Will the Department be given broad authority to override existing statutes and regulations? Will the accelerated and superficial treatment accorded thus far to this proposed reorganization provide an opportunity for major mischief?

Third, and more broadly, will this reorganization result in more confusion, more expense, more bureaucracy, more people, more harm to the civil service, more harm to public employee unions—and less work? Will the country actually be more vulnerable during what will likely be a lengthy transition period? Will the Department remain fully accountable to the people, and to the Congress, for its security mission as well as for the non-security functions it may inherit?

Our constituents will expect us to know the answers to these and many other questions before we act. Today's hearing is a small step towards developing the kind of understanding we will need to address this matter responsibly.

Mr. GREENWOOD. Governor, you are aware that the committee is holding an investigative hearing and when doing so has had the practice of taking testimony under oath. It is my understanding that you have no objection to offering your testimony under oath.

Mr. RIDGE. None.

Mr. GREENWOOD. The Chair also advises you that under the rules of the House and the committee, you are entitled to be advised by counsel. My understanding is that you don't feel the need to be advised by counsel.

Mr. RIDGE. That's correct.

Mr. GREENWOOD. If you would stand and raise your right hand.

[Witness sworn.]

Mr. GREENWOOD. Thank you Governor, you are under oath and we look forward to your testimony and please begin.

TESTIMONY OF HON. TOM RIDGE, DIRECTOR OF TRANSITION PLANNING FOR PROPOSED DEPARTMENT OF HOMELAND SECURITY AND ASSISTANT TO THE PRESIDENT FOR HOMELAND SECURITY

Mr. RIDGE. Chairman Greenwood, Ranking Member Deutsch and subcommittee members, I certainly appreciate the opportunity to testify—with the microphone on—in support of the President's historic proposal to unify our homeland security efforts under one Cabinet-level Department of Homeland Security.

Since the terrorist attacks of September 11, all of America has risen to the challenge of improving the security of our homeland. In partnership with Congress, with States and localities, with law enforcement, with the private sector and academia, America has made great progress in securing its borders and preserving its way of life and the security of its citizens.

The President believes our Nation must now take the next critical step by unifying our efforts under a single Department of Homeland Security. Only Congress can create such a department, and I am here today to personally convey the President's deep desire to work with Members to accomplish this goal. The President believes that the creation of a single department with a single, clear line of authority, as quite a few of the members of the committee have discussed, would not only improve our preparedness for future attack, but also strengthen these partnerships, thereby helping to prevent a future attack.

Earlier this month, the President signed an executive order appointing me as Director of the Transition Planning Office for the Department of Homeland Security, to be housed within the Office of Management and Budget. While I will still retain the title of Assistant to the President for Homeland Security, my testimony today will be given as Director of this new entity.

This proposal was the result of a deliberative planning process that really began with an effort led by Vice President Cheney a year ago, in May of 2001, and continued as part of the mission of the Office of Homeland Security when it was created on October 8, 2001, as well.

My staff and I have met with thousands of Government officials at the Federal, State and local levels, with hundreds of experts and many, many more private citizens. Throughout these discussions, we have constantly examined ways to organize the Government better.

The President's proposal also draws from the conclusion of many recent reports on terrorism, reports by blue ribbon commissions such as Hart-Rudman, Bremmer and the Gillmore Commissions, as well as a variety of reports from the many think tanks who have really investigated the issues relating to international terrorism and homeland security over the past several years.

It also draws, admittedly—and proudly I might add—from the legislative proposals of Congressmen and Congresswomen, including Mac Thornberry and Jane Harman, Ellen Tauscher, Jim Gibbons, Saxby Chambliss and others, along with Senators Joe Lieberman and Arlen Specter and Bob Graham.

This historic proposal would be the most significant transformation in the U.S. Government since 1947. The creation of this department would transform the current, rather confusing patchwork of Government activities related to homeland security into a single department whose primary mission—whose primary mission is to protect our homeland.

Responsibility for homeland security is currently dispersed among more than 100 different Government organizations, and the President believes—and I sense that it is a belief shared with many Members of the Congress of the United States, both Chambers, both parties—that we need a single department whose primary mission is to protect our way of life and our citizens; a single department to secure our borders, synthesize and analyze intelligence, combat bioterrorism and direct Federal emergency response activities.

The proposal to create a Department of Homeland Security is one more key step in the President's national strategy for homeland security. Like the national security strategy, ladies and gentlemen, the national strategy for homeland security will form the intellectual underpinnings to guide the decisionmaking of planners, budgeters and policymakers for years to come.

From securing our borders to combatting bioterrorism to protecting the food supply, most of the initiatives of the Federal Government in pursuing—excuse me, the majority of the initiatives the Federal Government is pursuing as part of our strategy to secure the homeland have already been discussed publicly. We will certainly refine them with the national strategy. The strategy will pool together all of the major ongoing activities and new initiatives that the President believes are essential to our long-term effort to secure the secure the homeland.

Now permit me, if you will, just a few comments with regards to details of the President's plan.

Preventing future terrorist attacks must be our No. 1 priority. Because terrorism is a global threat, we must have complete control over who and over what enters the United States. We must prevent foreign terrorists from entering and bringing in instruments of terror, while at the same time facilitate the legal flow of people and goods upon which our economy relies. Protecting our borders and controlling entry to the United States has always been the responsibility of the Federal Government. Yet this responsibility is currently dispersed among more than five major Government organizations in five different departments.

The new department would unify authority over the Coast Guard, Customs Service, Immigration and Naturalization Service and Border Patrol, the Animal and Plant Health Inspection Service of the Department of Agriculture and the recently created Transportation Security Administration. All aspects of border control, including the issuing of visas, would be informed by a central information-sharing clearinghouse and compatible data bases. It will be greatly improved in that process.

The new department would unify government's efforts to secure our borders in the transportation system that move people from our borders to anywhere in this country within just a matter of hours.

Although our top priority is preventing future attacks, Mr. Chairman, we cannot assume that we will always succeed. We cannot assume—it would be perilous to assume we could create a fail-safe system. Therefore, we must also prepare to recover as quickly as possible from attacks that do occur.

The Department of Homeland Security will buildupon the Federal Emergency Management Agency as one of its key components in this effort. The new department would assume authority over Federal grant programs for local and State first responders, such as fire fighters, police and emergency medical personnel, and manage such critical response assets as the nuclear emergency search team and the national pharmaceutical stockpile. It would build a comprehensive national management system that would consolidate existing Federal Government emergency response plans into one genuinely all-hazard plan.

The department would ensure that response personnel have and use equipment and systems that allow them to communicate with one another. As the President made clear in the State of the Union address, the war against terrorism is also a war against the most deadly weapons known to mankind—chemical, biological, radiological and nuclear weapons. If our enemies acquire these weapons, there is no doubt in anyone's mind, I believe, that they will certainly use them. They will use them with consequences potentially far more devastating than those we suffered on September 11.

Currently, efforts to counter the threats of these weapons are too few and too fragmented. The President believes we must launch a systematic national effort against these weapons that is equal in size to the threat that they pose, and the President's proposal, we believe, does just that. The new department would implement a national strategy to prepare for and respond to the full range of terrorist threats involving weapons of mass destruction.

The Department of Homeland Security would set national policy and establish guidelines for State and local governments to plan for the unthinkable, and direct exercises and drills for Federal, State and local weapons of mass destruction response teams. At the very heart of this particular feature of the President's proposal is to develop even stronger partnerships with the State and local first responders. The homeland will be secure when the hometown is secure, and that is why the President believes very strongly that we need to have this relationship with the State and local governments and build in that relationship as the Federal support for the kind of equipment, drills and training essential to build a national capacity to be able to respond to these threats.

The Department of Homeland Security would provide direction and establish priorities for national research and development, for related tests and evaluations and for the development and procurement of new technology and equipment. Additionally, the new department would incorporate and focus the intellectual power of several important scientific institutions including our national labs in this effort.

Finally, preventing future terrorist attacks requires good information in advance. The President's proposal recognizes this and would develop a new organization with the authority and the capacity to generate and provide such critical information. The new department would fuse intelligence, integrate intelligence from multiple sources and other information pertaining to threats to the homeland, including information from the CIA and the FBI, as well as the NSA, INS, Customs and the many other departments and agencies that have an information-gathering, intelligence-sharing capability within this country.

It would also comprehensively evaluate the vulnerabilities of America's critical infrastructure to which many of the Members alluded and note the pertinent intelligence against those vulnerabilities for the purpose of identifying protective priorities and supporting protective steps being taken either by the department, other Federal departments and agencies, State and local agencies and the private sector.

The individuals that work for the organizations tapped by President Bush for the new department are among the most talented and certainly the most capable patriots in our Government. We are proud of what they are doing to secure our homeland, and we call upon them to continue their crucial work while the new department is created. This consolidation of the government's homeland security efforts can achieve great efficiencies and free up additional resources over time for the fight against terrorism. They should rest assure that their efforts will only be improved by the Government reorganization proposal made by President Bush.

To achieve these efficiencies, the new Secretary will require considerable flexibility in procurement, integration of information technology systems and personnel issues.

Even with the creation of a new department, ladies and gentlemen, there will remain a strong need for a White House Office of Homeland Security. Homeland security will remain a multidepartmental issue and will continue to require interdepartmental collaboration and coordination. Additionally, the President will continue to require the confidential advice of a close assistant. Therefore the President's proposal intends for the existing Office of Homeland Security to maintain a strong role. The President believes this will be critical for the future success for the new office itself.

During the transition period, Mr. Chairman, the Office of Homeland Security will maintain vigilance and continue to coordinate the other Federal agencies involved in homeland security.

The President appreciates the enthusiastic, bipartisan response from Congress and is gratified by the expressions of optimism about how quickly this bill might be passed. Until the Department of Homeland Security becomes fully operational, the proposed department's designated components will continue their mandate to help ensure the security of this country.

During his June 6 address to the Nation, the President asked Congress to join him in establishing a single, permanent department with an overriding and urgent mission, securing the homeland of America and protecting the American people. Extraordinary times call for extraordinary measures. We know that the threats are real and the need is urgent. In working together, we all know we must succeed in this mutual endeavor.

President Truman did not live to see the end of the cold war, but the war did end, and historians agree that the consolidation of Federal resources was critical to our ultimate success.

Ladies and gentlemen, my colleagues in this effort, we, too, have that opportunity for leadership and to create a legacy that will benefit future generations as well. I thank you for the attention you have given my remarks and your public expressions of both desire and will to work together to achieve our mutual goal that is reorganizing Government to enhance our ability to protect our fellow citizens and our way of life; and I thank you very much.

[The prepared statement of Hon. Tom Ridge follows:]

PREPARED STATEMENT OF HON. TOM RIDGE

Introduction

Chairman Greenwood, Congressman Deutsch, Subcommittee Members, I appreciate the opportunity to testify today in support of the President's historic proposal to unify our homeland security efforts under one Cabinet-level Department of Homeland Security.

Since the terrorist attacks of 9-11, all of America has risen to the challenge of improving the security of our homeland. In partnership with Congress, with states and localities, and with the private sector and academia, we have worked to map and protect our critical, infrastructure, including nuclear power plants; to seal our borders from terrorists and their deadly cargo; to strengthen enforcement of our immigration laws; and to prepare for and prevent attacks involving weapons of mass destruction.

The President believes our nation must now take the next critical step by unifying our efforts under a single Department of Homeland Security. Only Congress can create such a Department, and I am here today to personally convey the President's deep desire to work with Members to accomplish this goal. He believes the creation of a single Department with a single, clear line of authority would not only improve our preparedness for a future attack, but also strengthen these partnerships, thereby helping to prevent a future attack.Earlier this month, the President signed an Executive Order appointing me as Director of the Transition Planning Office for the Department of Homeland Security, to be housed within the Office of Management and Budget. While I will still retain the title of Assistant to the President and Homeland Security Advisor, my testimony today will be given as the Director of this new entity. I look forward to responding to your questions after providing a short statement on the proposed legislation and how it would make Americans safer.

The President's Proposal

On June 6, 2002, President Bush addressed the nation and put forth his vision to create a permanent Cabinet-level Department of Homeland Security. Two days ago, on June 18, 2002, I delivered to the Congress the President's proposed legislation for establishing the new Department. This is an historic proposal. It would be the most significant transformation of the U.S. government in over a half-century. It would transform and largely realign the government's confusing patchwork of homeland security activities into a single department whose primary mission is to protect our homeland. The proposal to create a Department of Homeland Security is one more key step in the President's national strategy for homeland security.

It is crucial that we take this historic step. At the beginning of the Cold War, President Truman recognized the need to reorganize our national security institutions to meet the Soviet threat. We emerged victorious from that dangerous period thanks in part to President Truman's initiative. Today we are fighting a new war against a new enemy. President Bush recognizes that the threat we face from terrorism requires a reorganization of government similar in scale and urgency to the unification of the Defense Department and creation of the CIA and NSC.

Currently, no federal government department has homeland security as its primary mission. In fact, responsibilities for homeland security are dispersed among more than 100 different government organizations. Creating a unified homeland security structure will align the efforts of many of these organizations and ensure that this crucial mission—protecting our homeland—is the top priority and responsibility of one department and one Cabinet secretary.

Immediately after last fall's attack, the President took decisive steps to protect America—from hardening cockpits and stockpiling vaccines to tightening our borders. The President used his legal authority to establish the White House Office of Homeland Security and the Homeland Security Council to ensure that our federal response and protection efforts were coordinated and effective. The President also directed me, as Homeland Security Advisor, to study the federal government as a whole to determine if the current structure allows us to meet the threats of today while anticipating the unknown threats of tomorrow. After careful study of the current structure—coupled with the experience gained since September 11 and new information we have learned about our enemies while fighting a war—the President concluded that our nation needs a more unified homeland security structure.

The Department of Homeland Security

The creation of the Department of Homeland Security would empower a single Cabinet official whose primary mission is to protect the American homeland from terrorism. The mission of the Department would be to:

- Prevent terrorist attacks within the United States;
- Reduce America's vulnerability to terrorism; and
- Minimize the damage and recover from attacks that do occur.

The Department of Homeland Security would mobilize and focus the resources of the federal government, state and local governments, the private sector, and the American people to accomplish its mission. It would have a clear, efficient organizational structure with four divisions.

- Information Analysis and Infrastructure Protection

- Chemical, Biological, Radiological, and Nuclear Countermeasures
- Border and Transportation Security
- Emergency Preparedness and Response

Information Analysis and Infrastructure Protection

The Information Analysis and Infrastructure Protection section of the Department of Homeland Security would complement the reforms on intelligence and information-sharing already underway at the FBI and the CIA. The Department would analyze information and intelligence for the purpose of understanding the terrorist threat to the American homeland and foreseeing potential terrorist threats against the homeland.

Furthermore, the Department would comprehensively assess the vulnerability of America's key assets and critical infrastructures, including food and water systems, agriculture, health systems and emergency services, information and telecommunications, banking and finance, energy (electrical, nuclear, gas and oil, dams), transportation (air, road, rail, ports, waterways), the chemical and defense industries, postal and shipping entities, and national monuments and icons. Critically, the Department would integrate its own and others' threat analyses with its comprehensive vulnerability assessment for the purpose of identifying protective priorities and supporting protective steps to be taken by the Department, other federal departments and agencies, state and local agencies, and the private sector. Working closely with state and local officials, other federal agencies, and the private sector, the Department would help ensure that proper steps are taken to protect high-risk potential targets.

In short, the Department would for the first time merge under one roof the capability to identify and assess threats to the homeland, map those threats against our vulnerabilities, issue timely warnings, and organize preventive or protective action to secure the homeland.

Chemical, Biological, Radiological and Nuclear Countermeasures

The war against terrorism is also a war against the most deadly weapons known to mankind—chemical, biological, radiological and nuclear weapons. If the terrorists acquire these weapons, they will use them with consequences that could be far more devastating than those we suffered on September 11th. Currently, our efforts to counter the threat of these weapons to the homeland are too few and too fragmented. We must launch a systematic national effort against these weapons that is equal to the threat they pose.

The President's proposed legislation would accomplish this goal. It would authorize the Department of Homeland Security to lead the federal government's efforts in preparing for and responding to the full range of terrorist threats involving weapons of mass destruction. To do this, the Department would set national policy and establish guidelines for state and local governments. It would direct exercises and drills for federal, state, and local chemical, biological, radiological, and nuclear (CBRN) attack response teams and plans. The result of this effort would be to consolidate and synchronize the disparate efforts of multiple federal agencies currently scattered across several departments. This would create a single office whose primary mission is the critical task of protecting the United States from catastrophic terrorism.

The Department would serve as a focal point for America's premier centers of excellence in the field. It would manage national efforts to develop diagnostics, vaccines, antibodies, antidotes, and other countermeasures. It would consolidate and prioritize the disparate homeland security related research and development programs currently scattered throughout the Executive Branch. It would also assist state and local public safety agencies by evaluating equipment and setting standards.

Border and Transportation Security

Our number one priority is preventing future terrorist attacks. Because terrorism is a global threat, we must attain complete control over whom and what enters the United States in order to achieve this priority. We must prevent foreign terrorists from entering our country and bringing in instruments of terror. At the same time, we must expedite the legal flow of people and goods on which our economy depends.

Protecting our borders and controlling entry to the United States has always been the responsibility of the Federal government. Yet, this responsibility is currently dispersed among more than five major government organizations in five different departments. Therefore, under the President's proposed legislation, the Department of Homeland Security would for the first time unify authority over major federal security operations related to our borders, territorial waters, and transportation systems.

The Department would assume responsibility for operational assets of the United States Coast Guard, the United States Customs Service, the Immigration and Naturalization Service (including the Border Patrol), the Animal and Plant Health Inspection Service, and the Transportation Security Administration. The Secretary of Homeland Security would have the authority to administer and enforce all immigration and nationality laws, including, through the Secretary of State, the visa issuance functions of consular officers. As a result, the Department would have sole responsibility for managing entry into the United States and protecting our transportation infrastructure. It would ensure that all aspects of border control, including the issuing of visas, are informed by a central information-sharing clearinghouse and compatible databases.

Emergency Preparedness and Response

Although our top priority is preventing future attacks, we cannot assume that we will always succeed. Therefore, we must also prepare to minimize the damage and recover from attacks that do occur. The President's proposed legislation would require the Department of Homeland Security to ensure the preparedness of our nation's emergency response professionals, provide the federal government's emergency response to terrorist attacks and natural disasters, and aid America's recovery.

To fulfill these missions, the Department would oversee federal government assistance in the domestic disaster preparedness training of first responders and would coordinate the government's disaster response efforts. The Federal Emergency Management Agency (FEMA) would become a central component of the Department of Homeland Security, and the new Department would administer the grant programs for firefighters, police, emergency personnel, and citizen volunteers currently managed by FEMA, the Department of Justice, and the Department of Health and Human Services. The Department would manage certain crucial elements of the federal government's emergency response assets, such as the Strategic National Stockpile. In the case of an actual or threatened terrorist attack, major disaster, or other emergency, the Secretary of Homeland Security would have the authority to call on other response assets, including Energy's and the EPA's Nuclear Incident Response teams, as organizational units of the Department. Finally, the Department would integrate the federal interagency emergency response plans into a single, comprehensive, government-wide plan, and ensure that all response personnel have the equipment and capability to communicate with each other as necessary.

State/Local Government & Private Sector Coordination

The Department of Homeland Security would consolidate and streamline relations on homeland security issues with the federal government for America's state and local governments, as well as the private sector. It would contain an intergovernmental affairs office to coordinate federal homeland security programs with state and local officials. It would give state and local officials one primary contact instead of many when it comes to matters related to training, equipment, planning, and other critical needs such as emergency response.

Secret Service

The Department of Homeland Security would incorporate the Secret Service, which would report directly to the Secretary. The Secret Service would remain intact and its primary mission will remain the protection of the President and other government leaders. The Secret Service would also continue to provide security for designated national events, as it did for the recent Olympics and the Super Bowl.

Non-Homeland Security Functions

The Department of Homeland Security would have a number of functions that are not directly related to securing the homeland against terrorism. For instance, through FEMA, it would be responsible for mitigating the effects of natural disasters. Through the Coast Guard, it would be responsible for search and rescue, navigation, and other maritime functions. Several other border functions, such as drug interdiction operations and naturalization, and would also be performed by the new Department.

White House Office of Homeland Security and Homeland Security Council

The President intends for the White House Office of Homeland Security and the Homeland Security Council to continue to play a key role, advising the President and coordinating a vastly simplified interagency process.

Making Americans Safer

The Department of Homeland Security would make Americans safer because our nation would have:

- One department whose primary mission is to protect the American homeland;
- One department to secure our borders, transportation sector, ports, and critical infrastructure;
- One department to integrate threat analyses and vulnerability assessments;
- One department to coordinate communications with state and local governments, private industry, and the American people about threats and preparedness;
- One department to coordinate our efforts to protect the American people against bioterrorism and other weapons of mass destruction;
- One department to help train and equip for first responders;
- One department to manage federal emergency response activities; and
- More security officers in the field working to stop terrorists and fewer resources in Washington managing duplicative and redundant activities that drain critical homeland security resources.

The New Department Would Improve Security Without Growing Government

The Department of Homeland Security must be an agile, fast-paced, and responsive organization that takes advantage of 21st-century technology and management techniques to meet a 21st-century threat.

The creation of a Department of Homeland Security would not "grow" government. The new Department would be funded within the total monies requested by the President in his FY 2003 budget already before Congress for the existing components. In fact, the President's FY 2003 budget will increase the resources for the component parts by $14 billion over the FY 2002 budget. We expect that the cost of the new elements (such as the threat analysis unit and the state, local, and private sector coordination functions), as well as department-wide management and administration units, can be funded from savings achieved by eliminating redundancies inherent in the current structure.

In order to respond to rapidly changing conditions, the Secretary would need to have great latitude in re-deploying resources, both human and financial. The Secretary should have broad reorganizational authority in order to enhance operational effectiveness, as needed. Moreover, the President will request for the Department significant flexibility in hiring processes, compensation systems and practices, and performance management to recruit, retain, and develop a motivated, high-performance and accountable workforce. Finally, the new Department should have flexible procurement policies to encourage innovation and rapid development and operation of critical technologies vital to securing the homeland.

Working Together to Create the Department of Homeland Security

President Bush recognizes that only the Congress can create a new department of government. During his June 6th address to the nation, the President asked Congress to join him in establishing a single, permanent department with an overriding and urgent mission: securing the homeland of America, and protecting the American people. I am here to ask, as the President did, that we move quickly. The need is urgent. Therefore, the President has asked Congress to pass his proposal this year, before the end of the congressional session.

Preliminary planning for the new Department has already begun. The formal transition would begin once Congress acts on the President's proposed legislation and the President signs it into law. Under the President's plan, the new Department would be established by January 1, 2003, with integration of some components occurring over a longer period of time. To avoid gaps in leadership coverage, the President's proposal contemplates that appointees who have already been confirmed by the Senate would be able to transfer to new positions without a second confirmation process.

During this transition period, the Office of Homeland Security will maintain vigilance and continue to coordinate the other federal agencies involved in homeland security. Until the Department of Homeland Security becomes fully operational, the proposed Department's designated components will continue to operate under existing chains of command.

Mr. GREENWOOD. Thank you, Governor; thank you very much.

The Chair recognizes himself for 5 minutes for purposes of questions.

Governor, as you know, this committee worked hard to pass the Public Health Security and Bioterrorism Preparedness and Response Act of 2002; and the title of that act, Public Health Security and Bioterrorism Preparedness and Response was meant to underline the dual-use nature of the programs and the grants that we wanted to create.

We directed the Secretary of Health and Human Services in that statute to award grants to States, cities and hospitals and other health care facilities and providers to enhance education, training, supplies and equipment at the local level for bioterrorist attacks and other public health care emergencies, many of them naturally occurring.

The—we noticed in the bill, DOJ—we did that because we know that DOJ and FEMA were geared toward more traditional first responders, such as fire and police, and we wanted to get these grants out to the health care providers.

In the President's homeland security proposal, these bioterrorism programs would be continued to run through HHS, but the Secretary of the Department of Homeland Security could essentially control the HHS programs by establishing its parameters and setting its priorities. The question is, how do we make sure that these resources are there to prepare for an assault by West Nile virus or a new strain of influenza, so we have preparedness for the naturally occurring disasters and still are prepared for possible terroristic—bioterroristic attacks and how do you see the Secretary coordinating those concerns?

Mr. RIDGE. Mr. Chairman, first of all, you and your ranking member on the committee need to be congratulated once again for the extraordinary effort on the bioterrorism measure. It went a long way in helping focus the departments and the Government and on the critical need not only now, but in the future in dealing with this issue.

You raise a very important issue that hopefully is dealt with to your satisfaction within the legislation. You note very appropriately that the public health system really is a dual infrastructure. Whether the microbes of an infectious disease are brought to us in an envelope from a terrorist or as a result of Mother Nature, it is still problematic to citizens and communities.

The Health and Human Services will continue to have an independent funding stream to direct the resources to the dual infrastructure, the CDC and NIH and other laboratories and research facilities as well. But by specific legislative language included in this proposal the President submits to you, there is a direct responsibility for the new Cabinet Secretary to cooperate and coordinate and establish priorities in conjunction with the Secretary of Health and Human Services.

It, incidentally, is a partnership that predated the legislative proposal. Secretary Thompson has worked very, very closely with the Office of Homeland Security and the White House, and in fact, Secretary Thompson and his people worked closely with us on the language of this legislation.

So your interests are appropriate in ensuring that the collaboration that preexisted, that this proposal continues to exist; and we believe that the language in the President's initiative ensures that.

Mr. GREENWOOD. Kind of a day-to-day basis, I mean, what happens if the Secretary of Homeland Security calls up the Secretary of HHS and says, I am concerned about some intelligence that we are gathering about the potentiality of a bioterrorist assault in a particular part of the country, and I would like to marshal some CDC forces out there, and the Secretary of Health and Human Services says, I don't think we can spare that right now, I am worried about an outbreak of a pathogen naturally occurring that the CDC has been monitoring in another part of the country; and the two Secretaries become less than congenial in their cooperation?

How do you see that being resolved?

Mr. RIDGE. I think there probably would be a two-step process. First of all, since the President seeks to retain the Office of Homeland Security within the White House, we will continue to have a coordination role. The matter may be resolved by the intervention of the Assistant to the President, bringing the parties together.

It is a process that we have used on several occasions internally, and I suspect that would be used again. I believe that is at the heart of the President's decision to keep that Assistant to the President for Homeland Security operational within the White House.

But, second, obviously if there is a disagreement between Cabinet members or among Cabinet members, the ultimate tie breaker is the President of the United States.

Mr. GREENWOOD. So it is your understanding that the Secretary of Homeland Security would not be able to say to the Secretary of Health and Human Services, I have listened to what you have said, appreciate your concerns, now do what I tell you—wouldn't have the power to override unilaterally?

Mr. RIDGE. I believe the President preserves the autonomy of both Cabinet Secretaries.

Clearly, the intelligence information that would be available to the Secretary of Homeland Security would also be available to the Secretary of Health and Human Services; and based upon that information, based upon vulnerability assessments that are available to both, it would hopefully result in an agreement on joint action.

But in the possible event that a difference of opinion would arise, there are tie breakers to move quickly.

Mr. GREENWOOD. The Chair recognizes the gentleman from Florida, Mr. Deutsch, for 5 minutes.

Mr. DEUTSCH. Thank you, Mr. Chairman. And I guess my focus is a follow-up on what the chairman mentioned.

We are getting into some of the details. I think there is a concern, just trying to flesh out this issue, of how we envision—because we actually think we have done a good job and are doing a good job and continue to make strides in the public health area that—you know, taking public health into—or what would be left.

What is your vision of what would be left in HHS of public health issues after the Department of Homeland Security takes out the significant component?

Mr. RIDGE. One of the most critical pieces, I believe, is our public health infrastructure. NIH and CDC remain an integral and robust part of the Health and Human Services research effort, outreach effort and response effort.

So I think the point of the legislation is to create an environment and a means by which the Secretary of Homeland Security, working in collaboration with the Secretary of Health and Human Services and understanding that the research infrastructure preexisted the Department of Homeland Security and has a longstanding relationship with Health and Human Services, CDC, NIH and the other laboratories to which they may refer research—that infrastructure continues to exist.

And Health and Human Services will obviously have the opportunity to come up and work with Congress on public-health-related issues specifically. But as they work on health-related issues, bioterrorism issues, there will be that collaborative relationship between the two.

And when it comes to local preparedness, that grant program that heretofore had been in Health and Human Services, will be shifted to the Department of Homeland Security. It will be in everyone's best interest, however, recognizing the dual nature of the infrastructure that exists out there in the public health system, that the work is done in collaboration; and that is the specific reason that the Secretary of Health and Human Services is mentioned in this legislation—in Title III, I believe.

Mr. DEUTSCH. Again, this is not really in any way a critique, but the best result.

Mr. RIDGE. We are trying to work to refine it.

Mr. DEUTSCH. Focusing on this issue specifically—and you just mentioned it, and that is—our understanding is that the public health funding mechanism that HHS does, the department will take over all of that. And theoretically—again within your mission, or not your person, but the mission of the new department, this is again—I guess where the concern lies is that in my opening statement, I talked about the fact that I think people are doing a much better job. In fact, it is a necessary condition that they have responsibility, that they have goals and that they achieve those goals.

Unfortunately, a lot of the things related to public health are not what we, I think, really envision as your goal as a new department. And I guess the concern I have, and I think many of the members of this subcommittee and committee share, is that, if anything, we need to be pushing forward on all sorts of public health issues that are really not a component of—as you said in your answers previous to this, are not really a component of bioterrorism or chemical, you know, potential weapons of mass destruction against the United States.

So how do we—I mean, I understand what you are saying. But as we are structuring an agency, how do we deal with this concern, I think, is a very real question. And I know you responded——

Mr. RIDGE. I think you raised a very important point and you have offered, as all the committee members have, to work with us on refining the language so that it continues to meet the goals of the President as well as the committee's goal of continuing to build-up a public health infrastructure that has been—that has deteriorated over the past decade or so for lack of funding; and that refinement we'll just have to work with you on as we go about moving this legislation forward.

But it is clear that the public health infrastructure, any investment from—either directly from Health and Human Services or Homeland Security will end up having dual value, one in combating terrorism, another just making our public health system more robust and, frankly, long-term, improving the overall health of the country generally.

So working out that refinement with you in the language to make sure that we meet both objectives is certainly something we want to do.

Mr. DEUTSCH. I see my time is running out. I would like to ask one much more general question, which is, what lessons have we learned and going forward at this point in the creation? Obviously we talked about what happened post-World War II and the creation of the National Security apparatus. But really, the more recent agencies, the Department of Energy, other agencies in terms of their creation. And I've read a number of press accounts of just historically your interviews with people that the creation of a new department almost by definition has inherent bureaucratic problems in terms of staffing issues, in terms of other issues.

I mean, how are you approaching the just systemic problems of, you know, creating that large of a bureaucracy, and what's the apparatus that you have in place at this point in time to deal with some of those acknowledged issues that you will face?

Mr. RIDGE. Congressman, first of all, the legislation provides from the effective date a year transition period, because clearly your ability to aggregate all these people and all these departments and the infrastructure is certainly going to take some time. And so there is a year transition process. And you and I can well imagine that it will probably take even longer than that to get the kind of specific changes and refinements we need to maximize the effectiveness of this organization. But we have got a good period of time, a year transition.

Second, the President has asked in his proposal that the new Secretary be given more flexibility and greater agility in order to deal with issues such as the information system integration procurement and, for that matter, personnel. And depending on the wish and will of the Congress of the United States investing in the new Secretary the ability—the flexibility to deal with some of these issues I think would depend how quickly we can get the system operating to maximum effectiveness.

Mr. GREENWOOD. The Chair thanks the gentleman, and recognizes the Chairman of the full committee, Mr. Tauzin.

Chairman TAUZIN. Thank you, Mr. Chairman.

Governor Ridge, I hope you will give me a minute just to get something off my chest. There is a lot of work in this bill and a lot of work that I know you are doing in terms of securing our borders, and they need to be secured, but there's three points I want to quickly make.

One is that the instruments for terrorists to use against our people are here. The jet fuel that was exploded at the World Trade Center and here at the Pentagon was made in America. The airplanes were built in America. And the fuel trucks and the ambulances that a couple of people in New Jersey were trying to buy this week were made in America. And I suspect that we haven't paid enough attention to that. We had better, that someone with an evil intent against our people doesn't have to bring a doggone thing in through our borders. We have got a lot of stuff right here in America that they can turn against us if they are evil enough and intentional enough to do it.

Second, the terrorists are here. They are not in Afghanistan. If anybody has not seen Jihad in America, pick it up from PBS. The cells are operating not just in New York and Washington, but in little communities all over this country, in St. Louis, in New Orleans, in Kansas City and communities all over this country. They are here, they are operating, and they have come in under student visas. And in the 1990's, I started an effort to try to do something about students, and could not get any attraction to the issue. But we have let people in under student visas and left it entirely up to the school to track their movements. Some of them never registered to go to school; if they did register in school in English, they could switch to chemical engineering or nuclear engineering, for all we know, and nobody ever notified the State Department. And if they graduated or if they left, nobody notified the State Department, and they have settled in in communities all over this country. And we need to face that fact. We have let them in and they are here, and they are waiting for new instructions. And we had better face that fact. And the information they need to do is harm is so readily available in a free society. We really have to be careful.

In the 1960's, 1970's, in the State legislature in Louisiana, I tried to require a—pass a bill to require the desensitization of something as common as ammonium nitrate fertilizer and make sure you wouldn't mix it with fuel oil and make a bomb. Couldn't get any traction on it. This committee held hearings on this issue. But a guy named McVeigh simply had to go in an agriculture center and buy some fertilizer and go to a hardware store and buy a few canisters of butane gas, and he built a bomb that took down a Federal building.

We predicted that in the 1970's when we were debating whether we should desensitize ammonium nitrate fertilizer before it's sold in the markets. Information about how to do that is on the Internet. Information about how to use thousands of available chemicals and products we make in America to turn them into weapons of destruction, here in America, not imported, not bringing a doggone thing in through a ship or a plane, but right in this country, the information on how to use those things, readily available.

You have got an awesome task; we have an awesome task. But we have to face the facts: We have let the enemy in; he resides among us; and he is prepared to use the things, the common things in our lives to turn them against us, to do us harm. And a free society, a Nation that prides itself on freedom of information and a free access to goods and supplies and information suddenly is challenged about how to balance all those incredibly important rights that make us special, make this country special, against now the threat that lives at home with us in our own neighborhoods. And, this department is going to be critical.

And I want to ask you a couple of questions about it, but I want to make that statement first, because I hope everyone realizes just how serious this business is, and how creating a department with the absolute buck-stops-here authority to organize and coordinate and to do anything within our legal system to stop these people from harming our citizens here in America now, unlike any threat we have faced in the history of our country, is going to be simply awesome, and we have got to do this thing right.

I noticed in the President's proposal, for example, that the Freedom of Information changes. The changes you recommend being made about providing new protections against public disclosure of some sensitive information is limited only to information that's provided voluntarily, and is non—it is provided by non-Federal entities with respect to critical infrastructure activities. I wonder why that's limited. I wonder why, when the government compels a private entity, such as a safe water drinking facility or an electric generation facility or a manufacturing plant that's manufacturing critical components—when the government complies, they have to submit a vulnerability assessment, and it's under government requirement mandate to do so, why we couldn't protect that information as much as we would protect information that's voluntarily supplied. I hope you look at that.

Mr. RIDGE. We will.

Chairman TAUZIN. I hope you look at whether or not the non-Federal entity limitation is a good one, or whether there are some Federal entities that may supply information to your—to our new Department of Homeland Security that ought not be in the public domain; that may be accessible by the right persons in the government, but nevertheless protected from disclosure on the Internet because it may open the door to some sort of road map for destruction. We need to be careful, very careful about that, as we go down the future.

I notice in the bill, Governor Ridge, that one of the R&D programs, nuclear smuggling, is exempted from complete transfer to the Homeland Security office, that it suggests instead that the DOE jointly operates the program. I wonder if that isn't a better model for a number of R&D programs. And I would—you don't need to respond today, but I would love your office, before we act on this proposal, to explain to us why that model wouldn't work for a number of the other R&D programs which are equally sensitive as nuclear smuggling might be in terms of joint operation, rather than simple pure transfer out of the department.

I want to emphasize the points that Mr. Waxman made about our public health entities, and I believe Chairman Greenman made it, too. When we debated the bioterrorism bill, we were very, very careful not to create a special unit at the CDC that strictly related to terrorist attacks to our public health, because, frankly, when an outbreak of infectious disease hits or something else happens in this country, we don't know at the start how it happened, we just know we have got a problem on our hands. CDC has to respond whether it's a terrorist or whether it's a natural pathogen in our society. And we have to be careful that we don't create a situation where bureaucrats have to first debate where to send the issue before we can respond. And I would hope that as we evolve this new department, we are careful about that.

I would like to point out to the committee again in regards to my opening statement, we discovered just last week that the smallpox—rather, the anthrax bacteria that was sent in the mail was probably cultured here in America, not brought in over borders, again, but cultured here in America and may be cultured again in America.

CDC needs to respond whether it's someone culturing it in a lab and it accidentally gets out, or someone has got an evil intent in sticking it in the mail trying to kill people. They have got to have a clear capacity to respond and not wait for some bureaucrat to say, "Okay. We don't think it's a terrorist attack, so you are in charge instead of us." that's a very, very sensitive decision we have to make.

I want to also mention that in regard to—in regard to the President's proposal, there is a proposal in here to give the new Secretary authority to take, seek—or, seek to effect protective measures to secure critical assets, including those in the private sector. I mentioned this in the opening statement, but I hope you pay an awful lot of attention. I want to look at this very carefully before we complete action on this bill.

The last thing we need is to create another bureaucracy with regulatory authority in this area, and I would hope this is not designed to do that. And we are going to be watching very carefully that this truly represents an effort to coordinate the public/private partnership rather than creating new lines of authority that are going to contradict other regulatory agencies of the government in some of these private sector operations.

Finally, Governor Ridge, I think one of the best pieces of information and advice that came to the President the other day at our meeting with you came from John Dingell of Michigan, the ranking member of our full committee, who pointed out to the President and to you—and I wanted to emphasize his words again—that we have seen in the past creation of Federal agencies cobbled together out of pieces of different other—different agencies, with other different cultures and with other different organizational structures. We have seen the creation of some big messes. He cited the Energy Department as one. I want to second that.

The Energy Department represents one of the most difficult organizations in the government to manage because it was cobbled together, with all sorts of different pieces, some of which contradict one another; there are fiefdoms all over that department that don't cooperate with one another, that the right hand doesn't know what the left hand is doing, and wouldn't want to know if it was told.

The problems inside the Energy Department are not because of the—of any particular leaders, and Mr. Abraham is doing his best, as you know, to manage that department, as other Secretaries have done before him. It was a problem inherent in the way it was constructed.

I would urge you and the President to pay special attention to Mr. Dingell's words here, as we cobble together a new department, one that may be more critical than any we have ever cobbled together in a long, long time. I would hope that you pay special attention to the pieces you put together, and to make sure we don't create another mess like we have created with the Energy Department.

Thank you, Mr. Chairman.

Mr. RIDGE. Thank you very much for your commentary, your observations, and the recommendations and concerns you have expressed. Let me just try to summarize a quick response, noting the many interests and concerns you have with the legislation: That a good organization isn't necessarily a guarantee of success. A flawed organization is guarantee of failure. And that's why we believe that working together with Congress as we refine the ideas and address the concerns, hopefully, we can avoid the pitfalls that have undermined earlier reorganization efforts, and never really led to the unity of command and the kind of effectiveness that I think those who had organized it way back when had intended and had hoped. We need to avoid all those pitfalls as we ramp up this new organization.

Mr. GREENWOOD. The Chair thanks the gentleman, and recognizes for 5 minutes for inquiry the gentleman from Michigan, Mr. Stupak.

Mr. STUPAK. Thank you, Mr. Chairman.

Governor Ridge, you said in your statement that homeland security works when the hometown is secure. I want to talk a little bit about IBETs and some of the intelligence-gathering stuff that we have going on in this country. The IBETs, as you know, are Intelligence Border Enforcement Teams, and there are 13 of them, and after September 11, I think Customs did a pretty good job. But I want to point out where I think there is a gaping hole. I want to see if this is still driven by Customs, or will Homeland Security now take charge.

Thanks to my friend here, Mr. Strickland, here is Michigan. It's just a map of Michigan. You have an IBET down here in the Detroit area, right down here. And that comes right around here, so that's pretty much covered here. But then you don't have another IBET until you get way over here to Thunder Bay, Canada. So all this area in here—and by the way the crow flies, if you did a straight line, it's about 700 miles. But where we have a lot of activity is here in Sault Sainte Marie, Canada.

Mr. RIDGE. Right.

Mr. STUPAK. And that's about 700 miles. When you come over here, the islands are right here by Drummond and then the Channels. It's very easy, St. Mary's River, are very easy to cross. It's a major hole in our IBETs. So my question is, if you are going to do an IBET, who will make that determination now? Customs? Or will Homeland Security?

Mr. RIDGE. Customs will be an integral part of the reorganization effort. Interestingly, you talk about this rather unique alignment of multiple agencies led by Customs. Because when I complete the hearing today, I am going to spend a little time with some of the officials that are running one down in Key West. It's a good model. It's been very effective where it has been deployed. I see no reason why the new Cabinet Secretary would do anything other than to try to continue to enhance and empower its activity.

As you know, the President in his 2003 budget proposal also calls for I think the largest increase in support for the Coast Guard ever. We need additional people and boats and platforms to buildup their capacity, because clearly now border security and port security has taken on an enhanced dimension. So——

Mr. STUPAK. But then who would do an IBET then? Coast Guard or Homeland Security now?

Mr. RIDGE. Well, by definition, if the Coast Guard's doing it under the new department, Homeland Security would be doing it. Again, it is a best practice that I would suspect that the new Secretary would continue to try to deploy. It has proven to be successful.

Mr. STUPAK. Well, at these IBETs, and even—we don't have an IBET here at Sault Sainte Marie, where I think we should. We do have the Sault Area Intelligence Committee, and they are working with the Canadians, and we have 12 Federal agencies working out here trying to help secure the border here. But the problem with that one—that's one of the problems we are concerned about—is, while you have 12 agencies working well with the Canadians and all the local and county law enforcement, no is one is in charge. You have 12 agencies. They are all working together cooperatively—and I don't mean to be critical of what they do. I think they do a great job. But if something happens or if someone has to call a shot, we are going to do this, there is no one there who is in charge. And I think that's one of the problems we have when we start talking about security at our borders and elsewhere. And I would hope the new Homeland Security would have, at least at these area intelligence committees, someone to go to. Who is the go-to person in that local area, is what we sort of need to do.

Mr. RIDGE. You highlight a feature of border security that became evident to me as we put together a team from Customs and Coast Guard and INS and other agencies that deal with border security to develop a 21st century smart border accord with our friends in Canada.

Mr. STUPAK. Sure.

Mr. RIDGE. That's an ongoing process, where we look to critical review of our infrastructure, protective infrastructure, and how we facilitate the flow of people and goods, at the same time enhancing security.

So under the new agency, the coordinating function to a certain extent would be replaced by a command function, because you have Customs in one department, you have INS in one department, you have Border Patrol in another.

Mr. STUPAK. FBI.

Mr. RIDGE. Now, under the President's proposal, they would be all aligned singularly under an under secretary. So I think you will enhance the effectiveness of that kind of program, because you now have a command structure that can direct that it be done. And it is a good practice.

Mr. STUPAK. But if it's the IBET or like the Sault Area Intelligence Committee, I guess what I want to know, so we aren't pointing fingers like we do after September 11, where would I go to get full accountability on the issue? Who or what department—and, as we say—does the buck stop here? And, will the department order Customs to do it, the new department? Who is going to have the accountability? Where does the buck really stop with that new proposal?

Mr. RIDGE. I think it's a very appropriate question that you've asked, because you want the authority to get things done, be aligned with the accountability. And, at the end of the day, that will be determined by the new Cabinet Secretary. But——

Mr. STUPAK. So the Cabinet Secretary would be——

Mr. RIDGE. Clearly, I think that's the primary reason behind the President's reorganization effort aligning authority with accountability.

In here, what you finally have is a consolidation of the many agencies involved in IBET under one command structure. You can do—you can go so far trying to coordinate activity among organizations. I think you can go even further when you can command activity among organizations. And now I think you have a unitary command structure that will enhance the capacity of those multiple agencies to do that kind of job.

Mr. STUPAK. Well, when you see your Florida IBET, I would be interested in seeing your reaction to it, and see if there is one person in charge down there, or are we still all cooperatively.

Mr. GREENWOOD. The time of the gentleman has expired.

Mr. STUPAK. Thank you, Mr. Chairman.

Mr. GREENWOOD. The Chair recognizes the gentleman from Kentucky, Mr. Whitfield, for 5 minutes.

Mr. WHITFIELD. Thank you, Mr. Chairman.

Governor Ridge, Chairman Tauzin touched on a matter that I'm interested in and I'm sure other members of the committee are also, and that was the FOIA protection in the President's proposal being limited to voluntary information supplied by non-Federal entities. And, as he has indicated, EPA and others sometimes require entities to provide vulnerability assessments, which, under the President's legislation, would be subject to a FOIA request. Is that an issue that you all are willing to revisit and determine whether or not his proposal would be subject to change in that area, or not?

Mr. RIDGE. Yes, it is, Congressman. I mean, the legislation as drafted was directed specifically at a problem that has been experienced by a lot of the Cabinet Secretaries, and even during the work of the Office of Homeland Security, and that is, getting an understanding that 80 to 90 percent of the critical infrastructure in this country is owned by for-profit entities. And they are anxious, just as all Americans are, to help. They are anxious to participate. They want to let the government know, for a variety of reasons, where they view themselves as vulnerable. As—the companies are custodians of not only the proprietary interests, but they're neighbors in communities, they're corporate citizens, and have a responsibility to all these different groups. But they are not—our sense is that they would be a lot more forthcoming voluntarily in sharing this kind of information with us if it was part of a limited exemption to the Freedom of Information Act.

So whether or not we expand it is certainly worth consideration, not only in this bill but down the road in the years ahead.

Mr. WHITFIELD. Okay. Well, thank you, Governor. And I notice also that some of the transferred functions that would be coming into the new department relate to DOE's non-proliferation work with certain countries, and particularly Russia. And this is a little bit parochial interest of mine, narrowly focused, and maybe you don't know the answer to it, but recently DOE entered into a new agreement with the United States Enrichment Corporation to be the executive agent for bringing in enriched uranium from Russia as a part of the non-proliferation efforts in that country. Is that the type of an agreement that would be transferred to the new agency, or would that remain with the National Security Council? Or do we know?

Mr. RIDGE. That kind of agreement as presently drafted, I believe, with remains with the National Security——

Mr. WHITFIELD. Okay.

Mr. RIDGE. [continuing] apparatus of this country.

Mr. WHITFIELD. Okay.

Mr. RIDGE. You should know that the agencies and departments and programs that we have drafted into the Department of Homeland Security has been done with very close collaboration with the Department of Energy and others. And because of the complex nature of these programs—you just alluded to one of them—there are international aspects to this that involve issues that are related to homeland security, but also involving the Department of State and the National Security Advisory and the like. So, we have been very careful in drafting these programs. But that would remain with the national security apparatus of this country.

Mr. WHITFIELD. I notice that we have some private companies, like FedEx and the Port of Virginia that are actively testing and pursuing installation of radiation detection devices throughout their systems right now. But there are no Federal standards in this regard for radiation detection devices, and there is no single Federal entity to which the companies can look to guidance—for guidance and support. Will this new Department of Homeland Security be able to assist in providing leadership in that area for these private companies that want to pursue this?

Mr. RIDGE. Congressman, you've raised that question; Congresswoman Harman has raised that question. Literally dozens of your colleagues have done the same thing.

It is the purpose of the creation of the unit within Homeland Security of weapons of mass destruction countermeasures, and to involve a means by which we can establish the kind of standards and the point of access so the companies can work—know, one, the standards that we would like their equipment to meet; and, two, a point of access to get their equipment, their technology tested against those standards.

So, again, this is a work in progress, but develop a center of excellence around the Lawrence Livermore Lab, but using the other national labs and the other research facilities in this country, we would hope to, one, create a point of access for testing and evaluation; and, two, as we develop national strategy, to set national standards.

One of the big challenges we have in setting a—in developing a national strategy over a Federal system is we can't necessarily dictate to State and locals or Federal agencies, for that matter, the kinds of equipment that they must acquire or purchase. But by setting standards, we can go a long way in making sure that the equipment, from whomever the vendor might be, is interoperable with the other equipment that may be needed at the time.

Mr. WHITFIELD. Thank you.

Mr. GREENWOOD. The time of the gentleman has expired. The Chair recognizes the gentleman from Ohio, Mr. Strickland, for 5 minutes.

Mr. STRICKLAND. Thank you. And my friend from Michigan has a quick question here. I do have some questions, but I will yield the time to him temporarily.

Mr. STUPAK. Thanks.

Exactly on that point, on the radiation detection issue. Customs said we are going to do it, and then Customs says we know nothing about it, so they give it to DOE. DOE says we don't know anything about it, so we give it to Lawrence Livermore now. So now we have three ways down the scale. Who is making the decision? Who is going to be ultimately responsible and accountable? This has been going on for some time.

Mr. RIDGE. It has.

Mr. STUPAK. And someone has got to say enough is enough. Let's get the decision done. Let's get it made. Here, you have got Customs in saying, do this. Then they say, well, we really don't know anything about it, so we will give it to DOE. They contract to DOE; DOE says, yeah, good idea. We should do a standard, but we don't know what it is. Let's contract to one of our labs.

Now we are three ways down the ladder and three rungs down the ladder. How is this ever going to get done? We need someone to take the bull by the horns and say get it done.

Mr. RIDGE. Congressman, I think you reflect a challenge not only for Homeland Security in terms of how those three departments operate, but overall the operation of the Federal Government. You know, bringing some kind of a concerted effort to resolve these issues and getting someone to make a decision based upon a national strategy, national priorities, and national vulnerabilities is what needs to be done, and that is at the heart of the President's proposal.

One, the Department of Homeland Security, where this kind of issue can be resolved once a strategy is developed, priorities have been developed based on vulnerabilities and threat assessments, and then targeting the research, the appropriate research dollars to that end.

We have a fairly robust and fairly expansive and expensive series of research and development activities within the Federal Government. It's ad hoc, and at least under the umbrella of the Department of Homeland Security, those kinds of efforts relating to protecting our way of life and our citizens would be given, I think, a strategic focus, long overdue, as you pointed out in your question.

Mr. STUPAK. Right. And in this whole thing, we have entities willing to install the equipment, we have vendors willing to sell the equipment. How do we bring it all together is really sort of the crux. Going back to the accountability issue, we have vendors, again, willing to sell, you have got people willing to install. But what do we install? What's the standard? How do we do it? That's—that's the part we have got to get our hands on, and I'm just looking for more specific proposals in the President's legislation that would put someone in charge to get it done, to get that accountability.

Mr. RIDGE. Well, I think if you take a look at the one unit in there that deals with research and development and science and technology, that is the President's intention, that the centerpiece of the strategic—the strategic direction for homeland security research and development would be here. It would be through the Lawrence Livermore Laboratory. You would have centers of excellence at some of the other laboratories. We have got an extraordinary system of academic research institutions around this country. I mean, we have got plenty of people who are prepared intellectually with the laboratories and the experience to direct their focus once the Federal Government decides where that—where that research should be directed. We have got plenty of people out there that can help us do it, once we give them specific direction. We don't do it now. We just do it on an ad hoc basis.

Mr. STRICKLAND. Mr. Secretary, I just wanted to share some information that was in the Youngstown Vindicator regarding the possible location of the new department, and just to let you know that Youngstown, Ohio would be more than happy to provide a home for your new department. There has been some discussion.

Mr. RIDGE. I thought maybe in Pennsylvania somewhere. But apparently——

Mr. STRICKLAND. That's exactly what I was thinking.

Mr. RIDGE. We can get close to the river.

Mr. STRICKLAND. Sir, so that we both can benefit. But I have been thinking about this new department since the President has made his proposal, and one of the things that has concerned me is the fact that it appears that there was at least some failure to analyze data and to communicate data and so on. And I've been wondering how this new Secretary who is going to be responsible for homeland security is going to be able to do what they need to do—he or she needs to do if they don't have some direct authority over the agencies that are most responsible for intelligence in this country, specifically the FBI. And I'm wondering, how will the new Secretary be able to assure us and the Nation that the failures that have been identified in terms of not only data collection but data analysis and dispensation and the sharing of data and so on, how will the Secretary be able to deal with that problem, if it continues to exist, without having some direct authority over that agency?

Mr. RIDGE. Congressman, your question goes to the heart of the ultimate desire of the President, the Congress of the United States, and the people of this country, must do everything we can to prevent the attacks from occurring in the first place. And at the very heart of that effort is acting on credible intelligence and information, interdicting and preventing the attacks from occurring.

By specific legislative language, the Congress of the United States will empower the new Secretary to secure the reports and the assessments and the analytical work done by the CIA and the FBI, but also be empowered to get the information and intelligence that any other agency generates. This is an historic new capacity within the intelligence community, because within the Department of Homeland Security there will be an integration and fusion function that heretofore has not existed. It will be based upon whether or not that assessment—there are credible assessments with regard to threats, because in the same department you will match that up against potential vulnerabilities. More often than not, private rather than public. But if you have a credible threat targeted to a specific sector, to a specific company, to an area, you will be able to match and take a look at the vulnerabilities that may exist there, and then, again, in the same department have a recommendation of prescriptive or protective measures to be taken in response to the threat based upon the vulnerability.

Let me just say, if I might, that the President believes very, very strongly that the CIA, which obviously gathers from time to time information that is relevant to domestic terrorism, also secures information with regard to terrorism around the world, also is involved on a daily basis with securing information with regard to challenges from sovereign states. Weapon systems, biochemical systems, and the like. So the portfolio of responsibilities for the CIA far, far exceeds just the targeting of domestic terrorist information.

The President also believes very strongly that there is a direct line of authority, the DCI to one person in the executive branch, and that's directly to the President of the United States.

The President also believes that the FBI should remain an integral part of the chief law enforcement agency of this country, the Attorney General's Office.

But again, by specific legislative language, if the Congress adopts the President's proposal, you will create a new capacity of intelligence, integration, fusion, analysis, and then application. Because the reports and the assessments—the Phoenix memo would come to the new agency. Prior to this legislation, the Phoenix memo might have been lost in the department, in the FBI; but as the language is written with regard to the President's new Department of Homeland Security, the Phoenix memo would obviously be shared internally, but also be a piece of the information, the gathering that the FBI has done that would be shared with the new Department of Homeland Security.

Mr. GREENWOOD. The time of the gentleman has expired.

Mr. RIDGE. I'm sorry. It's a long-winded answer to a very appropriate question.

Mr. GREENWOOD. That's what we are here for.

Mr. RIDGE. All right.

Mr. GREENWOOD. The gentleman from Kentucky is recognized for 5 minutes.

Mr. FLETCHER. Thank you, Mr. Chairman.

And, Governor, I want to thank you. I know we all have some questions how this new agency will operate. I think there is very little question as to your capability of leadership and the choice that the President has made in you. So I want to say thank you for your leadership thus far.

As I look over your testimony, and of the three really mission areas of this new agency, to prevent terrorist attacks, reduce America's vulnerability, and minimize the damage and recovery from attacks that do occur, I think I understand a little more clearly the prevention portion and kind of the reducing vulnerability. In the minimizing the damage and more in the response, as I understand it, if there were a major terrorist attack today, of whatever type it might be that the roles and responsibilities of the various Federal agencies that respond to such emergencies are currently well-defined in the Federal response plan.

Mr. RIDGE. Right.

Mr. FLETCHER. The FBI would be the lead for the crisis management portion; FEMA would lead for consequence management; and, if the attack occurred overseas on foreign soil, then the State Department would take the lead. And there are various other scenarios as well where the lead Federal agency may change.

I think we have all been assured that this seemingly convoluted system would work and that everyone would understand the chain of command in it. But under this new plan, let me ask you, would the new Homeland Security Secretary be the lead Federal agency for all events, whether criminal or whether of natural origin, whether domestic or foreign? How would that be sorted out?

Mr. RIDGE. I believe it is the intent of the President that the unit within the new department dealing with the emergency preparedness and response become an all-hazard agency, and that is the Federal Emergency Management Agency. Heretofore, it would be responsible for the consequence management of acts of nature and potentially even horrific accidentally caused acts, such as the fires out in Arizona, but under the President's proposal become the lead agency to respond to both terrorist incidents and natural incidents as well.

Mr. FLETCHER. Well, what—given that, and the FBI—say you went back to an event like 9/11—of course, the Justice Department, there is criminal investigations of the Department of Defense.

Mr. RIDGE. Right.

Mr. FLETCHER. How would you see as far as the leadership role of the Secretary of the agency in responding? What roles would the FBI take? Would they still lead the criminal aspect and FEMA the natural disaster, if it were a different situation? And what would the new Secretary's responsibility—and who would be the lead—who is going to be the boss in some of these decisions?

Mr. RIDGE. First of all, I would share with you that at the time the disaster occurs, I think the lines are—between law enforcement and FEMA are very much blurred, because the natural impulse of the men and women who rush to the scene, whether they are police, firefighters, emergency medical folks, or civilian volunteer, are to save as many lives as possible. And so I think you'll find that the first responders at the scene as you go about trying to save lives as quickly as possible will ultimately have the responsibility. That means as soon as FEMA can get to the scene, they would oversee the response and recovery effort. That is not to exclude, if the circumstances warranted, the FBI from the very beginning trying to preserve whatever evidence there might be at the scene. But as we have discovered in the two horrific—in the multiple horrific events around 9/11, the first impulse is to save lives. And that's exactly what they did. And the information that the FBI has gleaned isn't so much from the scene of the crime, it's from other sources as they patch together the profile of the terrorists and learned what they did and how they did it in preparation of the 9/11 tragedies.

Again, the anthrax is a little bit different situation where you really had to have a collaborative effort at the scene.

So I think it's going to vary from incident to incident. But at the end of the day, I believe you are going to have—you need FEMA to be in charge of the response. Mr. FLETCHER. Then the FBI would still maintain control and the lead of the criminal aspect of it?

Mr. RIDGE. Correct.

Mr. FLETCHER. FEMA, kind of the first response and the humanitarian——

Mr. RIDGE. Right.

Mr. FLETCHER. To make sure to reduce the loss of life, and recovery.

Mr. RIDGE. Correct. Interestingly enough, when I visited Fort McClellan in Alabama where they are preparing first responders to get to the scene, they were training the firemen and the emergency medical technicians and others to be sensitive, depending on the scene and the kind of incident, about the necessity of trying to preserve what might be viewed later as evidence. And, at the same time, they were training the police, the local police, the State police, the auxiliary police, how to respond in a more traditional lifesaving capacity.

So there is a sensitivity within the first responder community to protect each—to support each other in the long-term—with regard to their long-term duties. But the first response when people get to that scene is to save lives, not to gather evidence. But then it sorts itself out down the road.

Mr. FLETCHER. And I think, certainly, as this goes along I think, at least in my mind, it would help to be a little more clear of, you know, who is going to be in charge of what, who's—because one of the problems you have in management is always if you have two or more bosses, it makes it very difficult where the responsibility lies in a lot of these issues.

Mr. RIDGE. Clearly, the law enforcement function related to a terrorist incident, the investigation, the follow-on would vest in the Federal Bureau of Investigation. I mean, hopefully, there is no confusion there. Where there is confusion from time to time is who is in charge as soon as the incident occurs. And the experience that America witnessed and participated in on 9/11, people didn't pay any attention to the authority given to them by virtue of the badges, whether it was law enforcement or first responder. The first impulse is, let's go in and save lives. Then you have a very appropriate delineation of responsibilities. But the investigative, the law enforcement side of this still belongs to the FBI.

Mr. FLETCHER. Thank you. I see my time has expired.

Mr. GREENWOOD. The Chair thanks the gentleman.

The gentlelady from California, Mrs. Capps, is recognized for 5 minutes.

Mrs. CAPPS. Thank you, Mr. Chairman.

And, again, Governor Ridge, I want to pick up on a theme you referred to earlier, that our homeland is secure when the hometown is secure, going back to that local system and systems in place.

I want to concentrate, if I could, on the Center for Disease Control, the CDC, and how that affects our local communities. In the third panel, a representative from the GAO, Janet Heinrich, has made a couple of statements that I want to bring into this and give you a chance to respond to her.

She is expressing “concerns about the proposed transfer of control from HHS, to the new Department for Public Health Assistance programs that have both basic public health and homeland security functions.” And she says “these dual-purpose programs have important synergies that we believe should be maintained.” And she expresses concern “that transferring control over these programs, including priority setting to the new department, has the potential to disrupt some programs that are critical to basic public health responsibilities. We do not believe”—these are her words—“that the President's proposal is sufficiently clear on how both the homeland security and public health objectives would be accomplished.”

And, if I could, again, I was privileged to visit with Congressman Deutsch the Center for Disease Control site not long after 9/11, and to see that CDC was stretched beyond capacity before that date and now have so many additional responsibilities. And acknowledging that when I, in my years of being a school nurse, relied on them very directly for help with ongoing epidemics and issues of, for example “is there enough flu vaccine on hand?” These are the questions that my first responders are asking me. And so can you describe and will you describe how these fears can be allayed?

Mr. RIDGE. Well, first of all, again, you and your colleagues have raised a very important question with regard to the distinction between homeland security, related research and activities of the CDC, and the traditional public health work of the CDC. And we believe there is a very distinct care line here where the Department of Homeland Security would be involved in those issues that had primary—not necessarily exclusive, but primary homeland security dimension. That's not to say that the CDC would not continue to deal with public health issues, maternity care, child care, immunizations. I mean, are they going to continue to have the same programs they have working with the States and the localities on a variety of public health issues, continue to have the programs dealing with the restoration of some of the public health infrastructure, continue to have money for research-related issues of cancer and smoking and things of that sort?

Mrs. CAPPS. Right.

Mr. RIDGE. So I think—I think there is a distinguishable line now. And if we need to further clarify that with language in the legislation, we certainly want to entertain that. But it's also, I think, very important to note that the legislation specifically calls for the two Secretaries to establish the kind of relationship so that both can take advantage of the dual-use infrastructure that has been built up through the extraordinary work of the Health and Human Services and the CDC over the past decades.

Mrs. CAPPS. Let me thank you, and—but push this even further.

Mr. RIDGE. Sure.

Mrs. CAPPS. Because we can talk about charts and flow charts, but it really becomes clear when you talk about dollars. And CDC, many would say, including me, was underfunded before 9/11. How will the dollars flow to do those basic activities?

And let me add on to that an additional challenge that we have faced here in our House subcommittee, what some would say—at least from where I sit in California—a crisis of health care delivery. And the upper payment limit cuts to the State of California, for example, will take $300 million from our public health safety net hospitals. That's going to be difficult if there is no bioterrorism attack. That's going to be a real hardship on a State like ours. And those institutions are exactly where people go when they—when the flu epidemics hit and when if, God forbid, there is a bioterrorist attack. That's exactly where people will go.

If we continue to cut resources to these programs, these hospitals, how can we add on another layer of preparedness?

Mr. RIDGE. Well, I tell you, I think you raise a question that under a new configuration of the executive branch would be appropriately raised with both the new Secretary of Homeland Security and the Secretary of Health and Human Services. The point being is that there will be an identifiable money stream with regard to specific programs that I think that can be identified and can be identified today. Over the years, obviously the Congress of the United States will have opportunity to increase dollars, whether it's through homeland security for those issues and that research relating more particularly to weapons of mass destruction, bioterrorism, chemical attacks and the like, but also work with the Secretary of Health and Human Services to bolster and fund programs related strictly to public health.

I mean, so many of these programs—and again, that will be a balancing act that will require the best efforts of both the executive branch, but working in collaboration with the Congress of the United States that ultimately has the constitutional responsibility and authority to appropriate the monies. So you'll help create that balance.

Mrs. CAPPS. Well, I know my time is up. But, you know, the President has said there are no additional dollars for this effort; and we are saying there weren't enough in the beginning. What shall we do now?

Mr. RIDGE. Well, there are—for the—for 2003, as we ramp up the new Department of Homeland Security, the President has spoken, recognizing what he has in the 2003 budget, which includes about a $14 billion increase for homeland security initiatives over the 2002 budget. What happens in the 2004 and beyond again will depend upon the interaction and the priorities set collectively between the Congress of the United States and the President.

Mr. GREENWOOD. The Chair thanks the gentlelady, and recognizes for 5 minutes the gentleman from Ohio, Mr. Gillmor.

Mr. GILLMOR. Thank you, Mr. Chairman.

And, Governor, one of the things I wanted to inquire about was in the bioterrorism bill, which we just completed, we provided for drinking water systems' vulnerability assessments and recommendations for action be done by EPA. Now under the proposed bill we have, it's my understanding that would take that authority out of EPA and put it under the new department. I guess the question is, does it make a lot of difference? Is it something that you feel really is an improvement in homeland security, or would it just as well be left with EPA where there is at least some body of expertise?

Mr. RIDGE. It would be our hope that the President's initiative could be embraced to include pulling that into the Department of Homeland Security because of the vulnerability assessment requirements that will be imposed upon the new department. If it's the congressional will to keep it at the EPA and mandate that that information be shared and become part of the infrastructure, the information infrastructure upon which the Department of Homeland Security operates, so be it. But it's just a feeling that we—in this new department, we have got, remember, the threat assessment matched against the vulnerabilities. And clearly, the water system, the energy systems, telecommunications, utilities, financial systems and the like are part of our critical infrastructure. So it was consistent with the President's belief that we ought to have that information-gathering capacity with regard to critical infrastructure within this department.

Mr. GILLMOR. I wasn't strongly suggesting that it stay with EPA; I was just trying to feel you out on where you were coming on that.

Mr. RIDGE. We think it would be better to put all this within this—this assessment within the new department.

Mr. GILLMOR. In title 3, the President proposes to transfer certain R&D programs from DOE to the new Secretary. And mostly those are the ones dealing with development of detectors or sensors for nuclear, bio, and chemical agencies.

Now, most of the research is done by DOE's laboratories, which are public and private entities under control of DOE. The labs conduct such research, however, not just for DOD; they do similar research under the work for other programs where the CIA, FBI, State, and the Secret Service can also request their own work.

Now, while it seems to make sense to have a single agency coordinating and prioritizing all the research, I'm not sure that the proposal does that since it only transfers the DOE programs and doesn't touch the rest of them.So why just transfer the DOE programs? Why not also transfer the work for other programs at the labs? Is that an oversight, or is there a reason for that?

Mr. RIDGE. I think we focused, Congressman, on the programs within the Department of Energy because of the very specific focus they have at the national labs and the expertise they have developed. But particularly, the Chem-Bio National Security Program where they have as their mission the development, the demonstration, and delivery of technologies and systems that will help this country prepare for, prevent, and respond to a terrorist attack. And they have been—this is work that they have been doing for years. It deals with bio and chem detectors, it deals with modeling capabilities to predict the effects of a chemical-bio attack. And again, in consultation with the Department of Energy, as we try to pull into the new Department of Homeland Security those programs, if not exclusively, then at least primarily deal with securing the homeland, this was very appropriate.

Mr. GREENWOOD. The time of the gentleman has expired. The gentlelady from California, Ms. Harman, is recognized for 5 minutes.

Ms. HARMAN. Thank you, Governor Ridge, for your testimony. I have been listening carefully, and agree with your testimony and with your answers to questions.

I would like to associate myself with the comments of our Chairman about the urgency of the threat and the fact that it is among us right now. And that prompts me to talk about the urgency I believe there is, not just to pass this legislation, but to implement certain changes which we could do this minute and not even wait for the legislation. One of them is information-sharing across the Federal Government and between the Federal Government and local first responders.

As you know, Governor, H.R. 4598, a bill that Saxby Chambliss and I introduced some months back, has now been reported by the House Judiciary Committee, and also has the unanimous support of the House Intelligence Committee, and is ripe for action on the House floor. I would like to thank you for your help in fashioning this legislation, and just mention to my colleagues that this is a way to share information now, stripping out sources and methods so that those without security clearances can receive it. It would cover the FBI, the CIA, and all those agencies not in this new department, and would get their information down to first responders who desperately need to understand better what our threats are. So, thank you for your help with this. That's one thing we can do now.

The other thing we can do now, I think, relates to interoperability. When Saxby Chambliss and I visited your excellent emergency facilities some weeks back, at your invitation, we saw state-of-the-art technology that you have been putting together. There is still an enormous amount of work down the road, and we all agree about hooking in private sector, cutting-edge technology into this new department. But meanwhile, there exists now integrating devices that can bring together the different frequencies and different handheld communication devices in an emergency. This would create interoperability, which we absolutely need for first responders from different police and fire and EMT agencies to come together at the scene of a terrorist attack in somebody's hometown. As you point out, all terrorist attacks are local.

There is a device called the ACU-1000, which is built in North Carolina, and which many communities are using. Its problem is that it is too small to handle the requirements of large metropolitan areas like Los Angeles County. Yesterday, in front of this building I saw in a van a technology developed by a large aerospace company that wraps this ACU-1000, a technical term meaning adds to it, and can connect five or more vans to cover the frequencies that an entire metropolitan area might need to use in an emergency.

Example: L.A. County has 88 cities, 55 police departments, 33 fire departments. It could, they allege, cover L.A. County.

My question to you is, how do we get to these bridging technologies—they may not be the perfect answer, but they sure are better than where we are—now? How do we make things like this happen right now, even before this department is up and running? Because, as our Chairman points out, these terrorists are among us and could attack us in 20 minutes from now.

Mr. RIDGE. First of all, Congresswoman, I think your point about bridging technologies and systems integration now, as we develop even more robust technologies and better systems down the road, is very appropriate, because I think it will take us—once we determine what our mission is and how we are going to achieve our goals, I think we can have the technology overlay, but we still have to work out some of these—some of these matters before we take advantage of the entrepreneurial nature of this country and our extraordinary technology sector.

I would suggest that there are a couple of things that we have done and we can do. One, our Office of Homeland Security has been working with the President's Office of Science and Technology Policy. And my recommendation would be that we take a look at the technology application that you have just discussed, make it available to this—to these groups, and have them give us an assessment as to the impact on particularly urban communication systems where there remains a huge gap. Obviously, we need interoperable communications, we need a bridging system now. Down the road, we hope to have a unified system not only within urban America, but within the country.

The second thing I would recommend, and I say this with enormous respect, the $3.5 billion first responder money is sitting in the 2003 budget. So, as Congress sets its priorities in dealing with the budget proposal in 2003, if we could make the homeland security portion, or many of those portions, available to local communities as quickly as possible, once there is a stamp of approval, once there is an imprimatur on pieces of equipment like this that it does the job it claims it can do, then we'll be in a position to buy these technologies immediately.

Ms. HARMAN. Thank you. My time is up.

Mr. Chairman, I just want to note—it is going to 10 seconds—that at our Conference on Technology and Terrorism last week, Dr. Marburger was there from the Office of Science and Technology Policy. He was talking in terms of this whole effort coming on line in 2004.

I think this effort is on line this minute, and bridging technologies, as you have just said, are the answer; and I would hope you would encourage him to be thinking with a little more urgency of the need to tap these various technologies in our country to confront the various terrorists in our country now. Thank you, Mr. Chairman.

Thank you, Governor Ridge.

Mr. BURR [presiding]. The gentlelady's time has expired. The Chair will recognize himself at this time. Let me welcome you and apologize—I was not here for opening statements—but also say that I am supportive of the President's proposal. There are a number of areas of the bioterrorism bill that we took a tremendous amount of time in trying to integrate. Where we knew there were strengths in agencies, we tried to beef up those strengths; where there were weaknesses, we tried to compensate, through the legislation, to make sure that the tools and resources were there for that in fact to be a success.

And I think that many of those areas, as we anticipated, would be encompassed in the new homeland security agency; and I think, in most cases, we are very supportive of that. My questions are going to deal more with the areas where not 100 percent of the responsibility of that area that we saw, where it might have been weak to start with, is shifting over and whether we thought through exactly the consequences of stealing half the responsibility and leaving the other half.

The new department is a security entity first and foremost. Tasking it with the disaster mitigation and response and to a certain extent research and development might distract from the security responsibilities that homeland security has.

Do you have any reservations about the pieces that you pick up that deal with research and development and mitigation of disaster response?

Mr. RIDGE. I believe the President's proposal tries to encompass the broadest range of homeland security matters under one agency, and that is from prevention and detection through preparation and response. And it is for that reason that you see the—this is a multitasked agency, and it covers the full spectrum of activities that would be appropriately associated with securing our homeland.

And I think, in time, the integration of these different responsibilities—the establishment of a strategic plan dealing with research and development clearly has implications for the new analytical unit potentially, for the border aggregation clearly, and for the preparedness and response. So I think you can see that if you take a look at the different units, they are not really stovepiped. At the end of time, there is really a relationship among all of them.

Mr. BURR. We looked very closely at things like that, the national medical response teams that we had. We tried to explore why they weren't more effective, that they are very crucial to our entity today; and I think through our efforts on bioterrorism, we felt there was a need to create an assistant secretary at HHS to sort of shepherd those areas. Now we sort of shift those responsibilities.

I guess my question is, do you still think there are enough areas at HHS that we need that assistant secretary there, or can you envision the need, whether it is HHS or other agencies, where you have pulled in jurisdiction and responsibilities, do you need an assistant secretary there as a liaison for homeland security?

Mr. RIDGE. I know the committee was very concerned about creating that capacity within Health and Human Services, and I would leave it to your good judgment to determine whether or not you would want to create another one to work as a liaison. Clearly, given the dual nature of the infrastructure that both a Department of Homeland Security and HHS would be using; clearly, given the benefit of many of the research dollars and the need for communication and coordination, I am going to leave that to your best judgment as to whether or not you think it would enhance that collaborative effort to create a similar position now in HHS as we bring this position over to the Department of Homeland Security.

Mr. BURR. Clearly, there are areas—I think section 905 of the President's proposal, and 906, deal with pharmaceutical stockpiles and select agent registration. Select agent registration was something that in the last administration was by default handed over to CDC because we found we didn't have a successful means to keep up with it.

I am a little bit concerned. We all believe there needs to be a list that is kept, one that the appropriate people have access to, one that we don't question its accuracy.

The difficulty that exists is that CDC seems to still be responsible for allowing these agents out for the purposes of research, but there is the problem of making sure that, in fact, that information gets from CDC to Homeland Security where, in fact, the registration of where that product has gone would have to be.

Do you have any concerns about that?

Mr. RIDGE. I think, for security reasons, the select agent list must be—should be part of the Homeland Security function and any regulations attendant to the preservation and maintenance of that list. But CDC continues to have that public health responsibility and would continue to do the research on these pathogens and continue to oversee the work done, whether it is done at CDC or elsewhere in conjunction with the Department of Homeland Security.

Mr. BURR. I truly do not raise it as a criticism, but there is a link where we are almost relying on the system we had 5 years ago of somebody making a notification to another agency when the decision is made to let one of the pathogens go out for research purposes. And I know we were all faced with a shocking reality when the anthrax scare came, and we tried to track down how many places might have had anthrax under research.

Mr. RIDGE. And we weren't sure.

Mr. BURR. Title VII of the bill deals with the coordination with non-Federal entities, the IG and the Secret Service. My only concern in section 701, which requires the secretary to direct and supervise grant programs of the Federal Government for State and local emergency response providers. And it is not a lack of confidence in Homeland Security to make those grants.

I guess the question that I would have, how much input will the agencies that currently have that responsibility have, since a lot of the grant, a lot of the research, a lot of the programs that the grant money will be for might still be the responsibility of the other agency.

Mr. RIDGE. If I might, Congressman, give you a good example, the folks at the local level generally would like to go to one Federal agency to get emergency preparedness and response grants. They also recognize that they take many forms. There is a bioterrorism response initiative that HHS has. There is an Office of Domestic Preparedness that actually has even more dimensions, but that is in the Department of Justice. And then, obviously, FEMA.

What I think is proposed under this legislation is, one, that we have by statute continued the collaboration with Health and Human Services so when these dollars go out they do go out in collaboration with Health and Human Services as it relates to the bioterrorism prevention and public health prevention.

Two, the Office of Domestic Preparedness and the Department of Justice where it is envisioned that that entire operation would become a more robust and more muscular agency that FEMA becomes when they have responsibility for in excess of $3 billion under the President's 2003 budget. And then clearly FEMA has been reaching out over the past several months working with States and local communities trying to work with them to set up a framework through which these multiple grants can be issued. So FEMA has also undertaken as part of its longer term goal the establishment of the kind of relationship they need with the States and the local communities to help frame the issuance of these grants.

The goal here is to buildup a national capacity of some sort around the country. Obviously, it will not be done in a year. Congresswoman Harman pointed out the need for interoperability of communications. My sense, in talking to FEMA and a lot of other people is, that may be the No. 1 priority. If you're going to save lives, it is predicated on time. The best way you minimize time is better communication; and unfortunately, we don't have integrated communications systems in too many places in this country.

Having said that, FEMA is working with State and local governments to develop these plans. And what we are, what the President is hopeful of as it relates to the 2003 budget—and I know I am going off just a bit, but I say this to members who will be appropriating the dollars—is that the moneys that would be issued, not just in 2003, but in future years as we buildup a capacity to respond to terrorist activity, that we build it up consistent with plans that begin at the local level and then take it to the regional level and move up to the State, that we begin to develop a capacity around mutual aid packs, a capacity built on standards that are designed after consultation within the departments and agencies that are also designed based on threat assessments and vulnerabilities.

So we still have a lot of work to do. And the purpose of the President's integration of all these agencies is to give some strategic focus not only to the efforts of the men and women that have been providing homeland security services for this country for a long time, but also give strategic focus to the dollars and technology and the kinds of equipment that we provide to this country to prepare for a potential response to a terrorist act.

Mr. BURR. Governor, thank you. My time has expired.

One more time I want to commend you personally for the job that you have done. You were asked to step in at a—I can't think of a more difficult time to take on a task that was then undefined and not understood. You were asked to do it with a limited group of people, and I think that you have done an extraordinary job. My hope is that as we take up this legislation and, hopefully, pass it in an expedited way that you, like we, remember that we can do things of this magnitude without growing bureaucracies that are bigger than the last one.

And I know that the President's legislation chooses a secretary and a deputy and five under secretaries and no more than six assistant secretaries, but there is room for an additional 10 assistant secretaries. My hope is you will always think smaller from the standpoint of the internal structure up here and, in fact, remember what I think you learned very early on, that most of the intelligent folks and the best ideas happen in the localities around the country that are ultimately the ones that we need to communicate with in real time, so less emphasis is spent up here and more around the country.

The Chair would recognize the gentleman from Massachusetts, Mr. Markey, for questions.

Mr. MARKEY. Thank you, Mr. Chairman, very much.

Governor Ridge, the Nuclear Regulatory Commission and the Departments of Energy and Defense have historically had jurisdiction over nuclear facilities whether they be civilian or government. And they have had the responsibility for constructing the design basis threat against which each of these facilities has to be protected, and they also have responsibility for conducting the force-on-force test against those facilities.

Now, in the overriding—in the legislation you have sent up it says that this new department will have primary responsibility for infrastructure protection. And so the question is, what does that mean in terms of the agency, yours or the NRC or the Department of Energy or Defense that will have primary responsibility over the security around nuclear facilities once the legislation is passed?

Mr. RIDGE. Congressman, I believe that your question highlights a characteristic of homeland security that can't be underscored enough, and that is the continuing need for intergovernmental and interdepartmental communication and coordination. It is a point you make very effectively. DOD and DOE and the Nuclear Regulatory Commission have multiple responsibilities with regard to the security of our nuclear facilities whether they be power plants or storage systems for nuclear weapons. That will continue to be the case.

However, this new department, working particularly with the Nuclear Regulatory Commission on the design threat assessment as it relates to the potential vulnerabilities that exist, will play a very important role as we go about matching threats against vulnerabilities and taking prescriptive actions.

Mr. MARKEY. So, for example, the Nuclear Regulatory Commission 9 months after September 11 have refused to begin a new design basis threat rulemaking, even though we know it moved from nonsuicidal, nontechnically sophisticated handfuls of terrorists that had to be protected against before September 11 to something which is suicidal, technically sophisticated, heavily armed and large numbers.

Would, under the new system, the Office of Homeland Security have responsibility for ordering the design basis threat regulation to be upgraded, or would that still remain with the Nuclear Regulatory Commission? Who would have the ultimate authority, the NRC or the Office of Homeland Security?

Mr. RIDGE. Ultimately, Congressman, if the Department of Homeland Security felt that the Nuclear Regulatory Commission hadn't moved either quickly enough or effectively enough vis-a-vis the threat you are talking about, one would hope that the new Cabinet Secretary, in conjunction with the chairman of the Nuclear Regulatory Commission, can resolve that.

Clearly, the President has said that he seeks to retain as part of the White House apparatus the Assistant to the President for Homeland Security that has been tasked with coordinating that activity and resolving differences of opinion. But if there is a difference of opinion finally, you get one tie breaker, and that is the President of the United States.

Mr. MARKEY. The tie breaker is the President. The tie breaker is not whoever heads up the Office of Homeland Security?

Mr. RIDGE. I think the new Secretary of Homeland Security is going to be empowered with enormous authority and responsibility to deal with vulnerability assessments.

Mr. MARKEY. I guess all I am saying is, if you identify a flaw in the security at Livermore or at Diablo Canyon and you go to the NRC or the Department of Energy and you say, upgrade, they say, no, we are not going to upgrade, we are not going to go to a new system, you are saying that the head of the Office of Homeland Security can't say, upgrade.

Then it goes to the President to resolve the dispute between the two offices?

Mr. RIDGE. Well, first of all, I think it is important that we always play out the worst case scenario. And my judgment, Congressman, is that if the vulnerability assessment is significant, we won't have any difficulty getting the cooperation.

But if you want to go to the worst case scenario——

Mr. MARKEY. Yes.

Mr. RIDGE. [continuing] the matter would—since the assets themselves—none of the national labs are part of the infrastructure of the Department of Homeland Security.

You talked about having problems at—the national lab at Livermore or Los Alamos does not have direct command and control over those entities. The first responsibility is to identify the vulnerability, convince them of the vulnerability and get them to do something about the vulnerability. If there remains a conflict, it would be resolved presumably within the—by the Assistant to the President for Homeland Security. There is a coordinating function, and that function remains within the White House.

Mr. MARKEY. That would be someone on the President's staff that would resolve it?

Mr. RIDGE. Assistant to the President.

Mr. MARKEY. That is the job to get then.

Mr. RIDGE. It's a pretty good job. It is the one I have right now. You are addressed with a great deal of authority.

Mr. MARKEY. When you——

Mr. GREENWOOD. The time of the gentleman has expired.

Mr. MARKEY. Could I have 1 more minute?

Mr. GREENWOOD. Unanimous consent, the gentleman is granted an additional minute.

Mr. MARKEY. When you say, presumably the person on the President staff will then break the tie between the Office of Homeland Security and the NRC or the DOE, is that going to be written into the statute?

Mr. RIDGE. It is a function of the executive order signed by the President of the United States creating the office on October 8.

I am going to say the other leverage that you have on any department or agency changing its direction or focus is also, the Congress of the United States would have to be—could be a potential partner in that enterprise as well. But if we are—as we've said before, this is an enterprise within which we are all engaged, and I guess I can imagine a worst case scenario, and I guess we have to plan for it, but I think it is very unlikely.

Mr. MARKEY. Thank you very much. We appreciate your being here.

Mr. GREENWOOD. The Chair thanks the gentleman and recognizes the gentleman from New Hampshire, Mr. Bass, for 5 minutes.

Mr. BASS. Thank you very much, Mr. Chairman. And thank you, Governor, for coming here. This must be a very interesting time in your life and certainly one of the most important issues that this Congress will deal with.

I have a question having to deal with DOE's nuclear emergency support teams, the NEST teams. I served on the Intelligence Committee, and we had some involvement with this issue in prior years.

Now, it is my understanding that the President's proposal transfers the control of DOE's nuclear response teams to the new Secretary in the event of an attack or emergency, and also gives the new Secretary the authority to set standards for DOE's group, as well as conduct training and exercises for these teams. But as I understand it, these DOE teams also always—almost always work in concert with DOD, and usually conduct joint exercises with DOD, FBI, State and other agencies, and that is because of their responsibility to deal with more than just a nuclear issue.

Will the new Secretary coordinate the exercises and training of all of these interagency components or just the DOE, Department of Energy, portion?

Mr. RIDGE. I believe it is envisioned from time to time that we would want to deploy all of these agencies in a realistic drill or exercise. So depending on the circumstances and the nature of the drill, Congressman, it could very well oversee an exercise involving all those agencies and serving in a coordinating function.

Mr. BASS. Okay. That is good.

I also understand that DOE's radiological assistance teams, which are spread out regionally throughout the country, are currently authorized to respond to requests from State and local officials for assistance and need not wait until the Secretary of Energy formally calls them into action.

Will the President's proposal change that requiring action by the new Secretary before these teams can be deployed for any reason?

Mr. RIDGE. Congressman, in that change in the—I cannot give you a specific answer to the change in the historical relationship. I will get back to you on that. That is the way they used to be deployed. I think there is a lot to be said for maintaining that kind of a relationship, but I will have to get back to you for a specific answer.

Mr. BASS. I appreciate that and I yield back to the chairman.

Mr. GREENWOOD. The Chair recognizes for 5 minutes the gentleman from California, Mr. Waxman.

Mr. WAXMAN. Thank you very much, Mr. Chairman.

Mr. Ridge, in your own home State of Pennsylvania, a newspaper reporter for the Pittsburgh Tribune-Review conducted an investigation to determine how vulnerable chemical facilities were to terrorists after September 11; and I don't know if this article came to your attention, but it is pretty shocking. According to that article, which was published on April 7, the security was so lax at 30 sites that in broad daylight a Trib reporter wearing a press pass and carrying a camera could walk or drive right up to tanks, pipes and control rooms considered key targets for terrorists. And I want to read to you specifically what they found.

"Absent dilapidated or unfinished fence lines or carelessly opened gates allowed access to 18 sites. Inside the sites no one stopped the reporter from going wherever he wanted, even into control rooms and up to tanks and train switching and derailing levers. No security at the potentially deadliest plants of the 123 plants nationwide that individually could endanger more than a million people; two are in western Pennsylvania. The reporter spent more than an hour walking through each without encountering a guard or an employee."

Now, I wrote to the President on this issue on September 26, 2001, asking him to use just $7 million out of the $40 billion of the Emergency Supplemental Appropriations Act for recovery and response to terrorist attacks to examine the vulnerability of these facilities to attack. Congress required these vulnerability assessments to be completed by this August, yet apparently the administration has not even begun them.

I am also concerned the administration has failed to make any proposal to address these significant risks. Does the administration support Congress, requiring decisive action to address these risks, and if so, why isn't it in your proposal?

Mr. RIDGE. Congressman, your reference to that—the critical infrastructure and the potentially devastating consequences associated with the terrorist attack on chemical facilities is something that the Office of Homeland Security has been focused on and clearly will become a priority of the new Department of Homeland Security. And I think, clearly, that not only this President, but previous Presidents have called on, and I believe the Congress of the United States has called on, the private sector and others to do a—perform critical infrastructure assessments and then take action to deal with the vulnerabilities.

Obviously, the pace of the change within some sectors of the economy and within some companies hasn't been what you or I or most Americans would like.

At the end of the day, when you have a Department of Homeland Security, Congressman, whose responsibility is to match threats with vulnerabilities and to work with other agencies within the Federal Government to harden these targets that are owned by the private sector, I think that will certainly accelerate the changes that are needed. And until such time, we continue to—the administration continues to work with all industry sectors to identify vulnerabilities and get them committed to taking action.

I refer to a conversation that I had with some folks with regard to these vulnerabilities across the board in various sectors. And I think one of the ways, Congressman, that we can make sure that those chemical facilities or some of these other facilities in your neighborhood and my neighborhood, your State or mine, everybody else's, is up to the standard that we seek is to have our first responders in those communities visit and work with those companies to make sure that the standards are met, because these are the men and women who are going to have to show up if these facilities are attacked.

Mr. WAXMAN. With all due respect, you just said we want this new department to be sure to do this job, we want the cooperation in the private sector to run these plants to be sure they're doing the job, and then we want the first responders to be doing the job. But you have been head of the Office of Homeland Security, and one of the mandates from Congress was to look at these vulnerabilities and do something about them.

So does it strike you that maybe I am hearing you just point your finger at everybody else, but not taking responsibility for getting this done?

Mr. RIDGE. Oh, no. I wouldn't want you to interpret it that way. I suspect that there has been sufficient follow-up by Congress, and I would assure you there has been sufficient follow-up within the Office of Homeland Security.

As part of the President's directive to our office, we were to—in the designing of a national strategy, we were to work with both the public and the private sector to do a critical infrastructure vulnerability assessment. That process is an ongoing process. It is something that needed to be done for a long, long time, and we are in the process of doing that, and that will be part of the national strategy that we will present to the President and to the Congress and to the public in the next several weeks.

Mr. WAXMAN. Just one last short question. Was I incorrect when I said this was required to have been completed by August, but the administration has not even begun the assessment of the risk at these facilities?

Mr. RIDGE. The administration began that some time ago. It has been a work in progress within the Office of Homeland Security; and my recollection of the executive order creating our office, there was no specific timetable. We created our internal timetable and are trying to get most of it done before we submit the strategy to the President, to the Congress and the people sometime in July. But you can——

Mr. WAXMAN. What is your own internal deadline?

Mr. RIDGE. We have said we are going to get the strategy to the President for his eyes by the 1st of July, mid-July. We are working on it.

Mr. WAXMAN. That is a strategy, but there is a vulnerability.

Mr. RIDGE. Congressman, the enormity of that task, we don't shy away from it in any manner, shape or form. But this is a process that I believe Congress has been and probably will be working on years and years as well. We have taken advantage of some of the work that Congress has done, but our own internal work started several months ago. It will need a few more months to be completed to give you the kind of specificity that I think you are looking for.

But we are doing our job, and when Congress completes its work and when the other agencies complete that work, I think we are going to have a pretty good system of determining where the vulnerabilities are and working together to come up with the means to harden those targets and reduce the vulnerability.

Mr. GREENWOOD. The time of the gentleman has expired.

Mr. WAXMAN. But assessments required by Congress are to be completed by August 2002?

Mr. GREENWOOD. The time of the gentleman has expired. The Chair would note that the mandate from Congress to do the vulnerability assessment of the chemical facilities was passed in 1999, and it was the Clinton Administration that did nothing subsequent to that.

The Chair thanks the Governor for your presence with us and for your testimony and for your guidance.

Mr. WAXMAN. That is a little cheap, Mr. Chairman.

Mr. GREENWOOD. The Chair has the floor and the gentleman may or may not be recognized in the future.

The Chair notes, Governor, that you are thanked for your service many times a day for good reason because you have given us such a sense of confidence.

But I would like to take the opportunity, as your friend, to thank your wife, Michelle, to thank your daughter, Leslie, and your son, Tommy. I know that after 10 years or so in the Congress, 8 years as Governor of Pennsylvania, they were probably and you were probably expecting to take off the mantle of responsibility and hang it up in the home cabinet for awhile. And I know it is only because of the dire circumstances that we faced and your sense of duty to your country that you put that mantle—and a large mantle it is—back on your broad shoulders, and we thank you for that. And we want to thank your family for the sacrifices they make every day in letting you do this job. Thank you. Thank you very much.

The Chair then calls forward the second panel consisting of the Honorable Claude Allen, Deputy Secretary of the Department of Health and Human Services, as well as General John Gordon, Administrator of the National Nuclear Security Administration. Gentlemen, welcome. We thank you for being with us this morning. Thank you for your forbearance. Let me begin by saying that I believe you are aware that the committee is holding an investigative hearing and, when doing so, has had the practice of taking testimony under oath.

Do either of you have any objection of giving testimony under oath?

Chair then advises, under the Rules of the House and the rules of the committee, you are entitled to be advised by counsel. Do either of you care to be advised by counsel?

Seeing negative responses, the Chair would ask that you rise and raise your right hand, and I will swear you in.

[Witnesses sworn.]

Mr. GREENWOOD. Thank you; you are under oath. And, Mr. Allen, I believe we will begin with your testimony.

TESTIMONY OF HON. CLAUDE A. ALLEN, DEPUTY SECRETARY, U.S. DEPARTMENT OF HEALTH AND HUMAN SERVICES; AND JOHN A. GORDON, ADMINISTRATOR, NATIONAL NUCLEAR SECURITY ADMINISTRATION

Mr. ALLEN. Thank you, Mr. Chairman, and members of the committee for the opportunity to appear before you to discuss the proposed Department of Homeland Security and how it will interface with the Department of Health and Human Services. Secretary Thompson and I support strongly the initiative that the President announced earlier this month and feel that this is the best direction for the Nation to move in order to ensure our homeland security.

The threat of terrorism has become a part of our daily lives since September 11, and this new Department of Homeland Security will enable us to make significant advances in protecting the American public from terrorism. We are pleased that the Congress is giving the President's proposal such prompt and thorough review and attention. And Secretary Thompson and I look forward to working with you to ensure the passage of this important legislation.

The President's proposal will transfer several terrorism-related activities that are housed currently within HHS to the new Department of Homeland Security. Homeland security will assume responsibility also for setting goals and providing strategic direction for other relevant public health and medical activities, but will rely upon HHS to implement and operate them on a day-to-day basis. First, I want to talk with you about the activities that will go to homeland security. Those areas include the Select Agent registration enforcement program, the Office of the Assistant Secretary for Public Health Emergency Preparedness and the Strategic National Stockpile.

Right now, the Centers for Disease Control and Prevention regulates the transfer of certain dangerous pathogens and toxins commonly referred to as "Select Agents" from one registered facility to another. These agents, such as the bacterium that caused anthrax, the bacterium that causes Plague, and the viruses that causes Ebola are used widely in the research laboratories across America. These Select Agents are prime examples and candidates for use by would-be bioterrorists, so when they are used in research, they must be kept under constantly safe and secure conditions.

The Public Health Security and Bioterrorism Preparedness and Response Act of 2002 authorized HHS to promulgate and to enforce regulations concerning the possession and use of Select Agents as well as their transfer. While CDC has done its best to manage the Select Agent program, CDC is a public health agency and not a regulatory body. Therefore, we believe that the new department is better suited to prevent Select Agents from falling into the wrong hands.

HHS will be prepared to provide homeland security with whatever scientific expertise and other technical expertise they may need to manage the program. In fact, under the administration bill, the Secretary of Homeland Security would administer the Select Agents program in consultation with the HHS Secretary, and HHS would continue to make key medical and scientific decisions, such as which biological agents should be included in the Select Agent list.

Let me talk about the Office of the Assistant Secretary for Public Health and Emergency Preparedness. The Public Health Security and Bioterrorism Preparedness and Response Act of 2002 also created the HHS Office of the Assistant Secretary for Public Health Emergency Preparedness. The responsibilities of this new office include the supervision of the Office of Emergency Preparedness, the

National Disaster Medical System and the Metropolitan Medical Response Systems, as well as related HHS emergency management functions. By having this office within the Department of Homeland Security, we will have a seamless integration of our national public health and medical emergency management assets with the Nation's new preparedness and response infrastructure.

Third, the National Pharmaceutical Stockpile, which currently CDC manages: The stockpile consists of 12 "push packages" of pharmaceuticals and medical supplies and equipment which are located strategically across the United States, and additional lots of pharmaceuticals and caches of medical materiel are maintained also by manufacturers under special contractual arrangements.

The Secretary and I are proud of the job that CDC has done in managing our Strategic National Stockpile, which was evidenced in our ability to get a push package into New York City on September 11. This fine work has set the stage for smooth integration of the stockpile with our other national emergency preparedness and response assets within Homeland Security.

The Secretary of Homeland Security will assume responsibility for continued development, maintenance and deployment of the National Stockpile, while the HHS Secretary will continue to determine its contents. This arrangement will ensure effective blending of our public health expertise with the logistical and emergency management expertise of Homeland Security.

With the strong integration and cooperation that exists between HHS and Homeland Security, two functions of the new department will be carried out by HHS unless otherwise directed by the President. The first is Homeland Security's civilian human health-related biological, biomedical and infectious disease defense research and development work.

We recognize the expertise, successful track record and unique capabilities of the National Institutes of Health and the Centers for Disease Control and Prevention. The Secretary of Homeland Security, in consultation with the HHS Secretary, shall have the authority to establish the research and development program that will be implemented through HHS. This means that Homeland Security will provide strategic direction regarding the Nation's biological and biomedical countermeasure research priorities.

Certain public health-related activities will also be directed by Homeland Security and carried out through HHS. This would include activities like enhancing the bioterrorism preparedness of State and local governments and non-Federal public and private health care facilities and providers. The object of this provision is to continue the important role that CDC plays, that the Health Resources and Service Administration plays and other elements of HHS play in assisting States and local governments and the hospitals and public health community in preparing for and responding to large-scale public health emergencies.

As with the research program, the Secretary of Homeland Security, in consultation with HHS Secretary, will have the authority to establish the Nation's antiterrorism preparedness and response program. But the implementation of the public health components of that program will be carried out largely through HHS.

Mr. Chairman, members of the committee, our Nation needs a Department of Homeland Security. The Secretary and I strongly support the President's proposal and look forward to doing whatever is necessary to effect a smooth and swift transition of responsibilities and operations. We believe that the President's proposal strikes the right balance by playing to the strength of HHS and recognizing this agency's core mission that is the protection of the Nation's public health, while capitalizing on the strategic and logistical strength of the new Homeland Security. We will ensure that HHS fulfills its obligation to the new department and provides that whatever public health, medical and scientific expertise it may require.

At this time, I would be happy to answer any questions that the committee may have.

[The prepared statement of Hon. Claude A. Allen follows:]

PREPARED STATEMENT OF HON. CLAUDE A. ALLEN, DEPUTY SECRETARY, DEPARTMENT OF HEALTH AND HUMAN SERVICES

Thank you, Mr Chairman and members of the Committee for giving me the opportunity to appear before you today to discuss the proposed Department of Homeland Security. Secretary Thompson and I strongly support the reorganization initiative that the President announced earlier this month.

The threat of terrorism in its myriad forms has become an ever-present part of our daily lives. The new Department will enable us to make further significant advances in protecting the American people from those who are bent upon inflicting death, destruction, and social disorder to achieve their ideological ends. We are pleased that the Congress is giving the President's proposal prompt and thorough attention. Secretary Thompson and I look forward to working with this and other Committees to ensure passage of the legislation for the new Department.

The President's proposal deals with certain terrorism-related activities that currently are the responsibility of the Department of Health and Human Services (HHS). Some of these HHS activities would be transferred to the Department of Homeland Security (DHS). For other relevant public health and medical activities, DHS would assume responsibility for setting goals and providing strategic direction but would rely upon HHS to implement and operate the activities on a day-to-day basis.

I will discuss examples from each group of activities in turn.

EXAMPLES OF ACTIVITIES PROPOSED FOR TRANSFER FROM HHS TO DHS

HHS functions conveyed to the new Department in the President's proposal include:

- The Select Agent registration enforcement program;
- The Office of the Assistant Secretary for Public Health Emergency Preparedness; and
- The Strategic National Stockpile.

Select Agent Registration Program

Within HHS, the Centers for Disease Control and Prevention (CDC) currently regulates the transfer of certain dangerous pathogens and toxins—commonly referred to as "Select Agents"—from one registered facility to another. These agents are widely used in research laboratories across America. Examples are the bacterium that causes anthrax, the bacterium that causes Plague, and the virus that causes Ebola, a lethal hemorrhagic fever. Select Agents are prime candidates for use by would-be bioterrorists and thus, when used in research, must be kept constantly under safe and secure conditions.

The recently enacted Public Health Security and Bioterrorism Preparedness and Response Act of 2002 authorized HHS to promulgate and enforce regulations concerning the possession and use of Select Agents, as well as their transfer. While CDC has done its best to manage the Select Agent program, CDC is a public health agency and not a regulatory body. We believe that the new department, with its strong multi-purpose security and regulatory infrastructure, will be well-suited to prevent nefarious or other irresponsible uses of Select Agents. HHS will be prepared to provide DHS with whatever scientific expertise and other technical assistance it may seek to help it manage the program. Under the Administration bill, the Secretary of Homeland Security would administer the select agents program in consultation with the HHS Secretary, and HHS would continue to make key medical and scientific decisions, such as which biological agents should be included in the select agents list.

Office of the Assistant Secretary for Public Health Emergency Preparedness

The Public Health Security and Bioterrorism Preparedness and Response Act of 2002 created the HHS Office of the Assistant Secretary for Public Health Emergency Preparedness. The responsibilities of this new office include the supervision of the Office of Emergency Preparedness, the National Disaster Medical System, the Metropolitan Medical Response Systems, and related HHS emergency management functions. This cluster of activities is a logical and proper candidate for transfer to DHS—thereby enabling seamless integration of national public health and medical emergency management assets with the Nation's new preparedness and response infrastructure at DHS. The Public Health Service Officers and other HHS employees who have faithfully performed disaster relief work over the years have done a wonderful service for our Nation. They are a credit to HHS as they surely will be to the new Department.

National Pharmaceutical Stockpile

CDC currently manages 12 "push packages" of pharmaceutical and medical supplies and equipment strategically located around the United States; additional lots of pharmaceuticals and caches of medical materiel are maintained by manufacturers under special contractual arrangements with CDC. You may recall that one of the push packages was dispatched to New York City on September 11th and that elements of the stockpile were used to respond to the anthrax attacks. The Secretary and I strongly believe that CDC has done an exemplary job managing what is now called the Strategic National Stockpile and this fine work has set the stage for integration of the Stockpile with other national emergency preparedness and response assets at DHS.

The President's proposal is designed to achieve this integration by tapping the strengths of DHS and HHS in a precisely coordinated way. Thus, the Secretary of Homeland Security will assume responsibility for continued development, maintenance, and deployment of the Stockpile—making it an integral part of the larger suite of federal response assets managed by FEMA and other future DHS components—while the Secretary of Health and Human Services will continue to determine its contents. The arrangement will ensure effective blending of the public health expertise of HHS with the logistical and emergency management expertise of DHS.

DHS FUNCTIONS TO BE CARRIED OUT THROUGH HHS

Certain specific program level details and administrative choices are still being studied in order to ensure the most seamless transition, and to give the greatest possible levels of efficiency and effectiveness to our fight against the threat of biological warfare and to protect the public health. However, the President's proposal clearly designates the following two activity areas that the Secretary of Homeland Security will carry out through the Department of Health and Human Services:

1. Civilian Human Health-Related Biological, Biomedical and Infectious Disease Defense Research and Development

The President's proposal provides that the new Department's civilian human health-related biological, biomedical, and infectious disease defense research and development work shall—unless the President otherwise directs—be carried out through HHS. Under the President's proposal, the Secretary of Homeland Security, in consultation with the Secretary of Health and Human Services, shall have the authority to establish the research and development program that will be implemented through HHS. Thus, as the agency responsible for assessing threats to the homeland, DHS, in consultation with the HHS Secretary, will provide strategic direction regarding the Nation's biological and biomedical countermeasure research priorities.

2. Certain Public Health-Related Activities

The President's proposal provides that the new Department shall—unless otherwise directed by the President—carry out through HHS certain public health related activities (such as programs to enhance the bioterrorism preparedness of state and local governments and non-federal public and private health care facilities and providers). The object of this provision is to continue the important role that HHS plays in assisting state and local governments and the hospital and public health community in preparing for and responding to large scale public health emergencies. As with the research program, the Secretary of Homeland Security, in consultation with the Secretary of Health and Human Services, will establish the Nation's anti-terrorism preparedness and response program and priorities, but the implementation of the public health components of that program will be carried out largely through HHS.

CONCLUSION

Mr. Chairman and members of the Committee, our Nation needs a Department of Homeland Security. The Secretary and I strongly support the President's proposal and look forward to doing whatever is necessary to effect a smooth and swift transition of responsibilities and operations. The Secretary and I believe that the President's proposal strikes the right balance: it plays to the strengths of HHS and recognizes this agency's core mission—the protection of our Nation's public health—while capitalizing on the strategic and logistical strengths of the new Department of Homeland Security. We will ensure that HHS fulfills its obligations to the new Department and provides it with whatever public health, medical, and scientific expertise it may require.

At this time, I would be happy to answer your questions.

Mr. GREENWOOD. Thank you very much, Mr. Secretary.

General Gordon you are recognized for your opening statement

TESTIMONY OF JOHN A. GORDON

Mr. GORDON. Thank you, Mr. Chairman. Again, on behalf of Secretary Abraham, we offer full support for the Homeland Security Act. My remarks this morning will focus primarily on what is Title V. We can go beyond that in the questions if you like.

The President's proposal to organize the Department of Homeland Security is really quite visionary and enjoys the full support of the Secretary and I. It will significantly improve the way the government responds to threats.

And the President's plan makes good sense. Centralizing the responsibility for our response to weapons of mass destruction can leverage resources currently spread across the government and allow us to operate more effectively and more efficiently. At the same time, leaving the nuclear response assets home-based in DOE and the National Nuclear Security Administration will allow us to maintain their considerable expertise and make them available for other potential responses.

We at NNSA are proud of the role we have had so far in the fight against terrorism, especially WMD terrorism, and look forward to working with the Congress and the administration to make a smooth transition to this new department. NNSA has really attracted over the years the world's premier nuclear scientists, technicians, engineers and designers, and they manage the national nuclear weapons program. These capabilities and these assets and the training have been applied toward Homeland Security and counterterrorism before 9/11, as well.

In short, we have the responsibility to operate and maintain a strong technical capability to respond quickly to discrete, specific nuclear and radiological emergencies. People and equipment are trained and they're standing alert, along with unique transportation assets, ready to respond now.

These capabilities were designed for short-term events, not 24-7-365 operations. With that said, they responded remarkably well to 9/11 and to specific taskings following that, such as the Salt Lake

Olympics. And, importantly, we are seeking to make them more responsive than they have been in the past by moving assets forward and realigning them to coincide better with the Federal districts.

There are seven organizations that make up this capability. The first and most widely known is, in fact, the Nuclear Emergency Support Team, NEST. They do the search, the identification of nuclear materials, diagnostics, suspect devices, technical operations to render them safe and packaging for transport. We have an aerial measurement system with helicopters and fixed-wing aircraft to provide a rapid response to detect and measure radioactive material.

There's an Accident Response Group that provides scientific and technical expertise to a U.S. nuclear accident or an incident. The real-time assessments of the consequences of potential radiation releases made by the Atmospherical Release Advisory Capability. The Radiological Assistance Program was established in the late 1950's and it comprises some 26 teams across the United States that are DOE and NNSA first responders to provide for the search, detection, and identification and advice to State, local, tribal, industry and even private citizens. They're actually called out about 24 times a year.

The Radiation Emergency Assistance Center really works with the medical diagnostics and provides the basis for understanding the radiological and physiological response to radiation. And, finally, the Federal Government maintains an extensive response capability for radiological response, assessment and monitoring. This organization assures the hand-off from crisis response to longer-term consequence management and monitoring and that that hand-off is accomplished smoothly and effectively.

Through these tailored and responsive teams, NNSA is able to marshal highly trained, unique scientific and technical expertise drawn across the NNSA nuclear weapons complex and the DOE as a whole. More than 900 individuals are on call to respond in the event of a nuclear or radiological emergency. Only about 70 of these are full-time.

The ability to call upon professionals from across the complex brings the depth of the nuclear/radiological response into this program and the full depth and breadth of the weapon's complex expertise and staffing can be brought to bear.

Response teams are staffed with nuclear professionals who undertake this work as additional duty. Day-to-day, these individuals ensure the safety and reliability of our nuclear weapons stockpile, and with few exceptions, these individuals work other full-time jobs at DOE and NNSA, but they are on call as a response team when one is needed anywhere in the country. In that sense, nuclear incident response teams are analogous perhaps to the National Guard.

The capabilities of the program are maintained and improved because of their cutting edge knowledge and because of their intimate relationship. These are the people who design and work on the weapons and the systems every day, and they are the ones we also bring into the fight, to the problem, in an incident. They have unique capabilities, but they are quite limited. Many years of hands-on work in some cases, going back to the Manhattan Project provides the knowledge and the insight and the background to draw upon.

How will these teams work with the Department of Homeland Security? We believe that they will work very much as they do now. The team members will work at their regular jobs at DOE and NNSA unless they're activated. Under the Atomic Energy Act, the FBI is responsible to the United States for investigating illegal activities involving nuclear materials, including terrorist threats involving special nuclear materials. Executive Order 12656 provides the authority for DOE to assist in conducting, directing, and coordinating search and recovery operations for materials, weapons or devices in assisting and identifying and deactivating what we would call an improvised nuclear device or Radiological Dispersal Device. The State Department, Mr. Chairman, plays a similar role for overseas international events and has the authority to reach back to our teams for assistance. So when requested, NNSA-DOE response teams are activated and deployed in support or resolution of the crisis.

Under the bill to establish Homeland Security, the new Secretary would coordinate responses to WMD incidents, including nuclear or radiological functions. We do not anticipate that the NNSA capabilities as a response to a nuclear or radiological accident or incident will be compromised in any way by this transfer of responsibility. What Homeland Security can add in addition to a centralized response to terrorism is a new and focused effort to set stronger standards for the capabilities of our teams, to strengthen training standards to ensure their inoperability, and to conduct joint exercises. There would be a single agency responsible for ensuring that we have the right assets available by setting nationally understood requirements and priorities.

In summary, DOE and NNSA nuclear radiological response capabilities are critical in any domestic response to a nuclear radiological incident. But they are also vital to the DOE and to NNSA's capability to respond to an accident or incident within the weapons complex or the nuclear energy sector. With the teams organized essentially as they are now, subject to the call of the Secretary of Homeland Security, they can continue to function to support DOE and NNSA, the State Department and Homeland Security professionally, effectively and in a cost-efficient manner.

Mr. Chairman, I will be pleased to turn to your questions.

[The prepared statement of John A. Gordon follows:]

PREPARED STATEMENT OF JOHN A. GORDON, UNDER SECRETARY OF ENERGY AND ADMINISTRATOR FOR NUCLEAR SECURITY, NATIONAL NUCLEAR SECURITY ADMINISTRATION, U.S. DEPARTMENT OF ENERGY

Thank you, Mr. Chairman. It's a pleasure to be here today to discuss Title V of the Homeland Security Act as it applies to the National Nuclear Security Administration (NNSA) at the Department of Energy (DOE).

The President's proposal to organize the Department of Homeland Security (DHS) is at once visionary and down-to-earth. It will significantly improve the way the government responds to threats against the United States. Centralizing responsibility for our response to weapons of mass destruction will leverage resources currently spread across the government. The President's plan simply makes good sense. We at NNSA are proud of our role in the fight against terrorism, and we look forward to working with Congress and the Administration to make a smooth transition to a new department.

The Department of Energy (DOE)/National Nuclear Security Administration (NNSA) develops and attracts the world's premiere nuclear scientists, technicians, and nuclear weapon designers as a result of over 50 years of managing the nation's nuclear weapons program. Many of these capabilities and assets have been applied toward homeland security and counter terrorism challenges long before 9/11, as well as since then.

Under the Atomic Energy Act of 1954, as amended, the Federal Bureau of Investigation (FBI) is responsible, within the United States, for investigating illegal activities involving the use of nuclear materials, including terrorist threats involving the use of special nuclear materials. Executive Order 12656 provides authority for DOE to assist the FBI in conducting, directing, and coordinating search and recovery operations for nuclear materials, weapons, or devices, and assisting in identifying and deactivating an Improvised Nuclear Device (IND) or a Radiological Dispersal Device (RDD). Today's operations have been updated to address the threat of terrorists using weapons of mass destruction (WMD). When requested DOE/NNSA response teams are activated and deploy to support resolution of the WMD crisis.

Under the Bill to establish the Department of Homeland Security, the new Secretary would coordinate responses to WMD incidents, including nuclear and/or radiological support function. We do not anticipate that the DOE/NNSA capabilities or response to a nuclear/radiological accident or incident will be compromised in any way by this transfer of responsibility.

Through tailored and responsive teams, DOE/NNSA is able to marshal highly trained and unique scientific and technical expertise in support of the Lead Federal Agency (LFA). This expertise is made up of 70 full time and 870 part time personal that draws from across the nuclear weapons complex and is composed of 29 full time and 118 part time Federal officials; 29 full time and 320 part time National Laboratory staff; and, 11 full time and 450 part time contractor staff.

Although nearly 900 individuals are involved with the nuclear/radiological incident response teams, through extensive matrixing and leveraging of resources, the cost to the government is only equivalent to 212 full time employees. This matrixing makes the response programs stronger and keeps the costs very low. The response teams are staffed with volunteers who, for the most part, work on ensuring the safety and reliability of the Nation's nuclear stockpile day in and day out. These professionals respond to staff a response team when called, much like a volunteer firefighter, or a National Guard member.

Individuals from fifteen various DOE/NNSA sites/facilities or National Laboratories across the nation are on call to respond in the event of a nuclear/radiological incident or emergency. The ability to call upon professionals from across the weapons complex brings depth to the nuclear/radiological response programs. The full depth and breadth of the weapons complex experience and staffing are brought to bear in the event of a significant incident or an emergency.

The capabilities of the response programs are improved because of the cutting edge knowledge of the stockpile stewardship program that these scientists bring with them when they respond to a call. This knowledge is gained over years of working with the stockpile stewardship program on a daily basis and cannot be duplicated—neither to replace the scientists on the response teams nor on the stockpile stewardship program. These very unique scientific/technical resources are extremely limited. Only the fundamental concepts of the stockpile stewardship programs are taught in a university. Many years of hands on work, in some cases going back to the Manhattan Project, provides knowledge, insights and background to draw upon that are invaluable.

THE NUCLEAR/RADIOLOGICAL INCIDENT RESPONSE PROGRAMS

As the steward of the nation's nuclear weapons program, DOE/NNSA brings the knowledge and expertise of the world's leading nuclear scientists, technicians, and nuclear weapon designers in response to a significant nuclear/radiological incident or emergency. When the need arises, DOE/NNSA is prepared to respond immediately anywhere in the world with seven unique response capabilities.

The response capability most widely known of is the Nuclear Emergency Support Team (NEST). The NEST program was initiated in 1974 as a means to provide technical assistance to the Lead Federal Agency (LFA). NEST is our program for preparing and equipping specialized response teams to deal with the technical aspects of nuclear or radiological terrorism. NEST capabilities include search for and identification of nuclear materials, diagnostics and assessment of suspected nuclear devices, technical operations in support of render safe procedures, and packaging for transport to final disposition. NEST response team members are drawn from throughout the nation's nuclear weapons complex. Response teams vary in size from a five person technical advisory team to a tailored deployment of dozens of searchers and scientists who can locate and then conduct or support technical operations on a suspected nuclear device. NEST personnel and equipment are ready to deploy worldwide at all times.

A Nuclear/Radiological Advisory Team deploys as part of an FBI-led Domestic Emergency Support Team (DEST) or as part of a State Department-led Foreign Emergency Support Team (FEST) is an incident occurs overseas to provide nuclear scientific and technical advice to the LFA.

If the location of a suspected nuclear or radiological device is not known, search operations may be required. NEST search teams are routinely configured to detect and locate a radiological source using a variety of methods ranging from hand-carried to vehicle-mounted search equipment. The basic building block for NEST search operations is the Search Response Team (SRT). The Search Response Team is prepared to deploy on either civilian or military aircraft. Upon arrival on-scene, the Search Response Team can begin searching immediately or can equip and train local responders, who are already familiar with the search area.

When a device is located, the specific resolution is dependent upon the political, technical, and tactical situation. The ultimate goal in resolving a nuclear terrorism crisis is to keep the terrorist device from producing a nuclear yield. This involves special explosive ordnance disposal (EOD) procedures conducted by highly-trained technical personnel. DOE/NNSA Joint Technical Operations Teams have been designated to work with military EOD teams during all phases of the crisis response. This approach also draws upon the personnel and equipment resources of the Accident Response Group (ARG).

The Accident Response Group (ARG) mission is to manage the resolution of accidents or significant incidents involving nuclear weapons that are in DOE's custody at the time of the accident or incident. ARG will also provide timely, worldwide support to the Department of Defense in resolving accidents or significant incidents involving nuclear weapons in DoD's custody. Scientists, engineers, technicians, health physics and safety professionals from the National Laboratories and production facilities make up the ARG team. These skilled professionals from 30 different areas of technical expertise are ready to respond immediately. ARG members deploy with highly specialized, state-of-the-art equipment is used for monitoring, assessing or removing nuclear weapons, components or debris. Once the weapon leaves the site, the ARG mission is complete. Monitoring and assessment activities would most likely continue using other DOE/NNSA assets such as the Aerial Measuring System (AMS), the Atmospherical Release Advisory Capability (ARAC), the Federal Radiological Monitoring and Assessment Center (FRMAC), the Radiological Assistance Program (RAP), and the Radiation Emergency Assistance Center/Training Site (REAC/TS).

The Aerial Measuring System (AMS) aircraft carry radiation detection systems, which provide real-time measurements of ground and airborne contamination—even very low radiation levels. AMS can also provide detailed aerial photographs and multi-spectral imagery and analysis of an accident site. AMS provides a rapid response to radiological emergencies with helicopters and fixed-wing aircraft equipped to detect and measure radioactive material deposited on the ground and to sample and track airborne radiation. The AMS uses a team of DOE/NNSA scientists, technicians, pilots and ground support personnel. Maps of the airborne and ground hazards are developed very rapidly which enables the scientists to determine ground deposition of radiological materials and project the radiation doses to which people and the environment are exposed. This information gives the decision-making officials, e.g., the Federal Emergency Management Agency (FEMA), the Environmental Protection Agency (EPA), and state, local, or Tribal emergency management officials, information they need to effectively respond to the emergency. The AMS capability can also be used to locate lost or stolen radiological materials.

The Atmospheric Release Advisory Capability (ARAC) role in an emergency begins when a nuclear, chemical, or hazardous material is released into the atmosphere. ARAC's main function is to provide near real-time assessments of the consequences of actual or potential radiation releases by modeling the movement of hazardous plumes to provide emergency response officials with the vital immediate information they need to rapidly evaluate airborne and ground contamination projections and thus effectively protect people and the environment. ARAC staff have vast databases available for a variety of data, including: a worldwide library of potential accident sites such as nuclear power plants and fuel-cycle facilities and a terrain database covering most of the world at a resolution of one-half kilometer.

Upon receiving a request for support, ARAC's specialists begin downloading the most recent regional and site weather data for input into the model calculations.

On-scene emergency response officials provide critical information such as the time and exact location of the release and the type of accident or incident causing the emergency. After ARAC team members have downloaded the regional weather information and received site input, computer codes simulate the release from the explosion, fire, vent or spill with dispersion models, which show the spread of the material. These dispersion models take into consideration the effects from the local terrain or topography and complex meteorology. ARAC staff scientists prepare graphic contour plots of the contamination overlaid on the local maps. These plots are distributed to emergency response officials and also provided to DOE/NNSA response teams such as: AMS, ARG, FRMAC, RAP, REAC/TS, and NEST.

In addition to accidental radiological releases, ARAC has assessed natural disasters such as volcanic ash cloud and earthquake-induced hazardous spills, manmade disasters such as the Kuwaiti oil fires, and toxic chemical releases from a wide spectrum of accidents.

The Federal government maintains an extensive response capability for radiological monitoring and assessment. In the unlikely event of a major radiological incident, the full resources of the U.S. government can support state, local and Tribal governments. The FBI, as the Lead Federal Agency for domestic incidents, is responsible for leading and coordinating all aspects of the Federal response. DOE/NNSA may respond to a state or LFA request for assistance by deploying a RAP team. If the situation requires more assistance than RAP can provide, DOE/NNSA will alert or activate a Federal Radiological Monitoring and Assessment Center (FRMAC). FRMAC activities include: coordinating Federal offsite radiological environmental monitoring and assessment activities; maintaining technical liaison with state, local and Tribal governments; maintaining a common set of all offsite radiological monitoring data; and providing monitoring data and interpretations to the LFA, state, local and Tribal governments. The main DOE/NNSA emergency response assets that supplement and are integrated into FRMAC capabilities are: RAP, ARAC, AMS, and REAC/TS. These assets are employed to detect and monitor radiation, measure the concentration of radiation in the air and on the ground, and to evaluate current weather conditions and forecasts, which may affect the radiation impacts. Other Federal agencies provide key professionals specializing in technical areas of importance to the Federal monitoring assessment activities.

The Radiological Assistance Program (RAP), established in the late 1950's, is composed of 26 teams spread across the United States, RAP is often the first-responding DOE/NNSA resource in assessing an emergency situation and advising decision-making officials. A RAP response is tailored based on the scale of the event. Specific areas of expertise include: assessment, area monitoring, and air sampling, exposure and contamination control. RAP team members are trained in the hazards of radiation and radioactive materials to provide initial assistance to minimize immediate radiation risks to people, property, and the environment. Their equipment includes the most advanced radiation detection and protection equipment available.

Since 1980, the Radiation Emergency Assistance Center/Training Site (REAC/TS) has been a World Health Organization Collaboration Center for Radiation Emergency Assistance. REAC/TS focuses on providing rapid medical attention to people involved in radiation accidents and is a resource to doctors around the world. DOE/NNSA's REAC/TS radiation experts are on call 24 hours a day for consultation to give direct medical and radiological advice to health care professionals at the REAC/TS treatment facility or an accident site. If needed, additional REAC/TS physicians and other team members can be deployed to the accident scene. This highly trained and qualified team can provide advice regarding assessment and treatment of contamination, conduct radiation dose estimates, diagnose and provide prognosis of radiation-induced injuries, conduct medical and radiological triage, perform decontamination procedures and therapies for external and internal contamination, and calculate internal radiation doses from medially induced procedures.

REAC/TS is also the recognized center for training national and foreign medical, nursing, paramedical, and health physics professionals for the treatment of radiation exposure. As a World Heath Organization Collaborating Center, REAC/TS is prepared to serve as a central point for advice and possible medical care in cases of radiation injuries; set up a network of available equipment and staff specializing in radiopathology; develop medical emergency plans in the event of a large-scale radiation accident; develop and carry out coordinated studies on radiopathology; prepare radiation documents and guidelines; and provide consultation or direct medical assistance to foreign governments if an actual radiation accident occurs.

In summary, the DOE/NNSA nuclear/radiological response capabilities are critical in any domestic response to a nuclear/radiological incident, but they are also vital to the DOE and NNSA's ability to respond to an accident or incident within the weapons complex or nuclear energy sector. With the teams organized as they are now, subject to the call of the Secretary of Homeland Security, they can continue to function to support DOE and NNSA and Homeland Security in an efficient, cost-effective manner.

The DOE/NNSA has more than 50 years of nuclear weapons experience that continue to provide the nation with an extensive base for science & technology, systems engineering, and manufacturing that has application across a broad set of national security missions, including homeland security and counter terrorism. Creation of a cabinet level Homeland Security agency holds promise for dramatic acceleration of improved capabilities against domestic threats. We in the DOE/NNSA are committed to the success of this new Department, and will work to facilitate it.

I would be pleased to answer any questions.

Mr. GREENWOOD. Thank you for your testimony, General.

The Chair recognizes himself for 5 minutes for inquiry. Let me start with you, Secretary Allen.

In order to speed the development of priority countermeasures, such as new vaccines and drugs, the Secretary of HHS is going to have to expedite approvals under the Federal Food, Drug, and Cosmetic Act. Moreover, some research efforts will be important both to counterterrorism and to advance public health research generally. We need to make sure that general research priorities are not diminished.

How will HHS assure proper priority and coordination on the regulatory front with the new department?

Mr. ALLEN. Mr. Chairman, the question is a very important one. It really goes to the heart of the mission of HHS, and in terms of time in dealing with bioterrorism.

We don't believe the mission will change significantly at all in that regard for the very mere fact that HHS right now prioritizes the research, prioritizes how we are going to be addressing the need for getting new products to market. So we don't anticipate there will be much change at all, if any, in regards to how the FDA will move in terms of getting products approved for their use whether that be for a bioterrorism response or whether it is for a general civilian response in terms of the use.

And to give you a good example, Congress just passed and the President has signed—as part of the bioterrorism legislation was included the passage of legislation that included the user fees for pharmaceutical products that would go to market. We believe that that will continue to be a part of that. But recognize that those products, those pharmaceutical products, while they serve a general purpose, using Cipro as an example—in terms of just an infection, they were used specifically in response to the anthrax outbreak which was a bioterrorism agent. So we don't anticipate there will be a significant change in how we were.

The question we will have is that the department will need to coordinate with the Department of Homeland Security as we are looking at products that will be coming to market, that FDA will need to approve and review for approval; and that is going to be a function that will have to be conducted again at a very senior level within the department. But FDA will continue to be involved in that process, and we will just need to create a liaison to work with Homeland Security to ensure the speed and accuracy of getting that information between the departments and getting the products to market.

Mr. GREENWOOD. The MDMS is being transferred from HHS to the new department, but as I understand it, these teams often have to be coordinated with other HHS elements, such as the Public Health Service.

Will the separation of the MDMS from the Public Health Service present problems in your opinion, and if not, how will continued coordination be assured?

Mr. ALLEN. We don't anticipate it will create problems in terms of the ultimate function of the MDMS system. While indeed the legislation under section 502 transfers that function to the new department, we do believe that as it currently exists in HHS, it was transferred from under the Assistant Secretary of Health to the Office of Public Health Preparedness, what would be the Assistant Secretary for Public Health Emergency Preparedness, and there had to be coordination even within the department of those assets and resources.

So we would anticipate that there would be an ongoing coordination with, now, the Department of Homeland Security that had already existed between HHS, VA, the Veterans' Administration, FEMA, DOD and other agencies that were involved in the MDMS system.

So we don't anticipate much change, but we would work through agreements, working with the Department of Homeland Security to ensure a smooth transition to ensure that those responses continue.

Mr. GREENWOOD. Just a question or two to you, General Gordon.

With respect to the NEST, the President's proposal leaves these teams under DOE authority generally, except for emergency situations when they would be under the new Secretary's authority. In our discussions with those who make up these teams at the labs, there is some sense of confusion as to the exact dividing line.

Can you shed some additional light on that question, based on your understanding of the administration's views?

Mr. GORDON. Mr. Chairman, now, today, if a team were to deploy to a situation under Federal control—a nuclear incident, a suspected weapon—that team would "chop," in the military term, would "deploy" under the control and command of the lead Federal agency, which in most circumstances would be the FBI.

Under this act, I think there is still a bit of a sorting out to be done on exactly how that relationship between the Secretary and the FBI works out. But the NEST teams will chop to the lead Federal agency.

Mr. GREENWOOD. Sorting out requires some fine tuning of the legislative language.

Mr. GORDON. I think it is just a decision. Whether it is legislation or within the administration, I think it is a decision.

My sense is it's not going to have any measurable effect on the operation or the effectiveness of the teams. They are going to work for someone who is in charge of the overall action.

Mr. GREENWOOD. Will this new bill require that the new Secretary authorize any deployments of these teams, or components of these teams, which I understand is not all that uncommon? Or will the DOE Secretary or the regional commanders of these teams remain authorized to deploy assets when deemed necessary or upon request of State or local officials?

Mr. GORDON. We view these very much as dual-use assets in that regard. If there's a national incident that requires the team, the teams will provide it then. However, these are individuals with qualities and capabilities that we need to be able to deploy to an energy or national lab incident that we can deploy ourselves. There are not a huge number of teams, but certainly enough to handle more than one incident at a time.

Mr. GREENWOOD. My time has expired, but before I yield to the ranking member, I would just ask that both Secretary Allen and General Gordon commit to us that your staffs will work diligently with us in the short, truncated period that we have to get this legislation prepared for the House floor.

Mr. GORDON. Absolutely.

Mr. ALLEN. Absolutely.

Mr. DEUTSCH. Thank you, Mr. Chairman.

Secretary Allen, I believe you were here throughout the entire comments by the Governor and the questions. And I really wanted to follow up a little bit about that. Besides myself, a number of other members, I think, are just really trying to inquire and really enter into a dialog into the changes of some of these responsibilities from HHS to this new department.

Under this proposed governmental structure, what public health responsibilities are left in HHS?

Mr. ALLEN. Actually, Congressman Deutsch, the vast majority of the public health responses are left in HHS. It does not dramatically impact the Public Health Service Act that exists right now to focus on HHS' public health responsibilities. What it does do is set some particular areas that will be dual use.

What is transferred from HHS under the proposal are, one, the national pharmaceutical stockpile, which includes the procurement, the maintenance, and deployment of the stockpile; second, the transfer of the Office of the Assistant Secretary for Public Health Emergency Preparedness, which includes the National Disaster Medical System, includes the Metropolitan Medical Response teams, includes our Disaster Medical Assistance teams. Those assets which would also be part of our Office of Emergency Preparedness will transfer. And then last, the select agent regulations will transfer.

So the vast majority of the functions of HHS will continue and will not be fully transferred over to the new department.

Mr. DEUTSCH. Could you specifically respond to, I guess one of the questions I also asked Governor Ridge, regarding the grant program, the billion dollar grant program for public health preparedness established by Secretary Thompson and authorized by the 2002 Public Health Security and Bioterrorism Preparedness Act? How will that change in terms of the proposals?

Mr. ALLEN. Under the proposals, the Department of Homeland Security will have the responsibility for those State and local programs; however, they will do that through contracting with HHS to run those programs. Certainly, the Administration did not want to disrupt what was accomplished in the public health, the act, the bioterrorism act, to disrupt what has already been taking place, and that is, is getting resources to State and local communities. We will still be in essence the grant managers in that sense actually working with State and local governments. It's simply that the strategic decisionmaking will be primarily the responsibility of the

Department of Homeland Security, and they will consult with and contract with through—and through memorandums of understanding with the Department in carrying out of those functions.

Mr. DEUTSCH. Now, our understanding is—my understanding as well is that for budgeting purposes, these two infrastructures that we are just describing cannot cost more than the single one. How is that possible? And is that correct?

Mr. ALLEN. It's possible, because, for example, in terms of what we are already doing, the functions will—the functions, the personnel will remain at the Health Resources Services Administration, which is working on the possible preparedness issues, and will remain at the Center for Disease Control which is working with State and local health departments in terms of the functions there.

So, in essence, the money is going to be funded through the Department of—the Department of Homeland Security, and they will contract with HHS to carry out those functions.

Mr. DEUTSCH. In your testimony, you stated that HHS would continue to decide what agents would be on the select agent list. Could you cite the legislation or the provision for that?

Mr. ALLEN. Actually, under the legislation, the scientific work that is being done, the medical expertise that is necessary right now to determine what the select agents are would be accomplished by working with the scientists who currently exist at H HS. Under the section 502, it transfers—subsection 5, it transfers the work of the Office of Assistant Secretary for Public Health Emergency Preparedness, but also transfers—and all their functions in the strategic and national stockpile is also transferred.

With regards to the select agent rule, I have to find the specific records.

Mr. DEUTSCH. You can provide that to us, if you can.

Mr. ALLEN. Sure. I will be glad to do that.

Mr. DEUTSCH. Again, I just see my time is running out, so let me go through two other questions very quickly.

What percentage of public health service officers are actually supposed to go over to the new agency?

Mr. ALLEN. We don't have a number of actual individuals. I can give you the number of individuals who are supposed to transfer over.

With regards to—if you will hold on for a second. Under the select agent rule, for example, we will be transferring seven FTEs. Those are the individuals who actually worked at CDC who worked on the select agent transfer program. We also—under the Office of the Assistant Secretary for Public Health Preparedness, that would include approximately 116 staff and detailees who are currently on board, including 87 individuals who are at the Office of Emergency Preparedness. And for the functions in terms of the national, the national pharmaceutical stockpile would include currently about 28 individuals.

Mr. DEUTSCH. Thank you.

Mr. ALLEN. And your cite for the select agent rule, I do have that for you. It's under section 502. 302, I'm sorry. Section 302, subsection 1. It says that the select agent registration enforcement programs and activities of the Department of Health and Human

Services, including the functions of the Secretary of HHS relating thereto, will transfer over.

Mr. WHITFIELD. General Gordon, one of the laboratories in their written testimony asked a very good question about how NEST's effectiveness depends in large part on the continued R&D and technology improvement efforts under way at DOE. If you divide—if the NEST teams are divorced in some way from the R&D component, whether by transfer of NEST or transfer of those R&D components to Homeland Security, in your opinion, what would the impact of that be? And does that concern you?

Mr. GORDON. Mr. Chairman, it's not our intent to break that link at all. The labs have a huge capacity to do this R&D. It's very important to us. And they are, of course, the ones who provide the experts for NEST.

As we discussed in the statement, the NEST will continue to operate and live as an organic unit within the National Nuclear Security Administration and DOE, and be available as a national asset, as the demand requires. We intend to keep them linked tightly together.

Mr. WHITFIELD. Okay. On these NEST teams, is it—many people devote time voluntarily to this. Is that correct? Or——

Mr. GORDON. Of the 900 or so people that are identifiable on the full range of nuclear incident response teams, which goes beyond NEST, there is probably only about 70 full-time employees. The others, I'm not sure I would call them volunteers so much as additional duty. They accept this duty, they accept this responsibility. They train to it and exercise to it.

But the point being, from my perspective, Mr. Chairman, the point being that's one of the reasons you just can't pick this thing up lock, stock, and barrel, and move it elsewhere. Their expertise, their currency is actually from the jobs they do day to day.

Mr. WHITFIELD. You know, some people have described this situation as following the National Guard model in which equipment and supplies are centrally managed—in this case, by the new Secretary—while the personnel remain under the general authority of the respective departments—in this case DOE—except when called to duty. Is that your understanding of the approach embodied in this bill?

Mr. GORDON. I might use a different analogy but toward the same end. Military service today, their responsibility is to organize, train, and equip.

Mr. WHITFIELD. Right.

Mr. GORDON. And then they are then fought by a commander in chief. I think that there is an analogy here pretty strong to that point, that we would organize, train, and equip to standards that I would hope that the new department would help sharpen, help strengthen, and work the interoperability perhaps better than we do today.

Mr. WHITFIELD. I was wondering if you would elaborate just a little bit on these joint tactical operations teams. Actually, what is their mission?

Mr. GORDON. What they would be doing is we would be augmenting the individuals who were hands-on attempting to deal with or dismantle a weapon. So, basically, in those instances, Mr.

Chairman, what we do is we bring in the technical expertise that sits behind the bomb squad.

Mr. WHITFIELD. Okay.

Now, Secretary Allen, if we move some of the key functions of the new Assistant HHS Secretary for Public Health Preparedness—and maybe you all touched on this earlier. But if we moved that to the new department, does that eliminate the need for that assistant secretary entirely, or would there be remaining functions, such as coordination, that would need to be done?

Mr. ALLEN. Clearly, the need for coordination within the Department of HHS of these activities will not be eliminated. Whether that is the requirement of having an assistant secretary level function, that is something that remains to be addressed. Clearly, the department under Secretary Thompson following 9/11, he created the Office of Public Health Preparedness, and had a director of that office to coordinate those functions. But it was certainly the wisdom of Congress to create an office of an assistant secretary. So we would be flexible to work with it, but there will need to have very senior leadership coordinating the activities of the department to work with Homeland Security to ensure the continuity of those programs.

Mr. WHITFIELD. Thank you very much. I see my time has expired. We will recognize the gentleman from Michigan for 5 minutes.

Mr. STUPAK. Thank you, Mr. Chairman.

General Gordon, I think in your opening statement you commented, or maybe it was in response to a question, about Salt Lake City Olympics. Did you—or, not you. But were there radiation detection devices at the Salt Lake Olympics?

Mr. GORDON. We didn't set up specifically. The emphasis on the Salt Lake Olympics was more in the area of some biological response, which I would prefer to discuss in a different session.

Mr. STUPAK. Sure. But in answer to my question, so there wasn't any radiation detection devices at Salt Lake that you know of?

Mr. GORDON. We did not set up specific portals.

Mr. STUPAK. Right. My question is, do you know if there were any radiation detection devices? I know you didn't set them up, but were there?

Mr. GORDON. I just don't know the answer to your question.

Mr. STUPAK. Okay.

Mr. GORDON. I will provide you a response.

[The following was received for the record:]

At the request of the U.S. Secret Service and the Federal Bureau of Investigation, and in support of the Utah Olympic Public Safety Command, the Department of Energy deployed the Nuclear/Radiological Advisory Team (NRAT) and members of the Radiological Assistance Program (RAP) team from Region 6 (Idaho) with portable radiation detection equipment to the Salt Lake City 2002 Winter Olympic Games. The equipment deployed included small pager-sized radiation detectors, detectors carried in briefcases and backpacks, and vehicle-mounted detectors. Identification units, which are used to identify the specific type of radiological material, were also sent. No radiation portal monitoring was conducted at any time.

Prior to the arrival of the athletes, NRAT and RAP conducted radiological surveys around Salt Lake City and the high security areas. Surveys of this type are useful in cataloging the radiological signature of the surrounding areas, saving critical response time in the event of an actual incident. During the survey process several locations revealed an elevated radiation signature. In each instance, the NRAT scientists deployed with identification units and determined that the readings were due to natural background radiation, a normal occurrence. Once the Olympics began, the radiological surveying stopped and the teams assumed a response posture. There were no incidents requiring the use of NRAT or RAP personnel or equipment during the Olympics.

Mr. STUPAK. Okay. But the only point I was driving at—it wasn't a trick question—is my impression is that there were radiation detection devices we used at Salt Lake City. In the earlier panel with Governor Ridge here, we were talking a lot about radiation detection devices. If they were set up and used in Salt Lake City and if there is concerns we should have them elsewhere in this country, why aren't we using them? That's all I'm trying to get at.

Mr. GORDON. Again, I would really like to discuss this in a different session.

Mr. STUPAK. Sure. Let me put it this way. When I was asking—we were talking about it before, myself and Governor Ridge, we talked about how Customs wanted these devices, and then contractors gave them to DOE, and DOE has now gone to one of the labs to try to get some standards and get some development going, and we are already down the three levels. And in response to the question, it was like, "Well, Congressman, that's sort of the way the Federal bureaucracy works." I didn't get a warm, fuzzy feeling when I got that answer.

I guess if we are going to do this new Homeland Security, Department of Homeland Security, how are things going to be different?

Mr. GORDON. I want to sign up to exactly what I think you are getting at, sir. We had proposed and suggested at the beginning that there be developed in effect a lead technical agency that could bring together the disparate variety of activities that are under way in this with some national standards, with some national priorities that are set up for where we are going. That is, in my understanding, what is to be incorporated into this new department. Because what we have now, even in our own areas for the Department of Energy and NNSA, is some very specific capabilities that were put together for some very specific and somewhat narrow uses. We have now expanded those, I think, with considerable expertise and a little bit of alacrity in response to 9/11. The pagers, the sort of small radiation detection pagers that are used at airports have been made available to the extent that we could get them fast enough or cause them to be produced fast enough, deployed in a number of locations with a number of different forces.

I think there is a good effort across the board in where we are using and deploying some systems, which I would be glad to talk with you in a smaller group, but it is time to pull it together in an aggressive program.

Mr. STUPAK. Okay. Again, maybe it would be appropriate in a closed session, and, again, just a little bit. But I'm still trying to get at if we create this new department how is it going to be different? How are we going to have accountability, responsibility, and make sure the job is getting done, and we don't have finger-pointing after an incident? That's what I'm driving at.

Mr. GORDON. We bring it together in one place with individuals who are charged to look at it nationally——

Mr. STUPAK. Okay.

Mr. GORDON. [continuing] who are designed to set up what are the priorities that you want us to spend our research dollars and our production dollars on, and take that in an aggressive step and just work right down a strategic plan.

Mr. STUPAK. I'm sure, Mr. Chairman, when we get into the radiation detection, I would suggest that might be a place we want to go in closed session. I know I have some more questions, but I am going to leave that issue right now and go to another spot.

Well, let's take the NEST teams. I don't know of any significant problems that have been evident by the way these teams have been presently structured or how their command and control has worked in the past. So if you move NEST teams over to the new department, how is that going to improve them or improve their functionality?

Mr. GORDON. I think the point, sir, is that they don't move over; that they become part of the coordinated units that are available to respond to a crisis upon the direction of the Secretary.

Mr. STUPAK. So the teams wouldn't move over to Homeland Security?

Mr. GORDON. The teams do not move as a unit. They stay where they are because—they need, in fact, to stay inside the organization because they are not full-time personnel that deploy. These are actually the experts that are working on our stockpile stewardship program, working on our weapons, working the intelligence side. So we bring them together, as the Chairman had suggested, in a National Guard way or in a military service way to respond to individual crisis.

Mr. STUPAK. Okay. All right. I was under the impression, and maybe wrongly so, that NEST teams are going to be moved to Homeland Security.

Mr. GORDON. No, sir. They would be available under the command of the Secretary of Energy upon call for national issues. They also would be available to the Secretary or myself for an DOE-NNSA incident where they had to respond. And we need them to stay tied in to their current work, because they are not full-time NEST employees, on the whole.

Mr. STUPAK. Okay. They stay where they are, but additional people can employ them, if need be.

Mr. GORDON. And that's effectively the way it is today. If there were an incident this moment that involved a nuclear weapons or terrorist attack, the FBI would be responsible for commanding that incident, and we would deploy our forces to the FBI for their use.

Mr. GREENWOOD. The time of the gentleman has expired.

Thank you, Secretary Allen, thank you, General Gordon, for your testimony, for responding to our questions, to your pledges of co-operation as we work through this legislation. Thank you again, and you are excused.

Mr. GORDON. Thank you, Mr. Chairman.

Mr. GREENWOOD. The Chair then calls forward our third panel for this hearing. We have Ms. Jan Heinrich, who is the Director of Health Care and Public Health Issues at the U.S. General Accounting Office; Dr. Harry C. Vantine, Program Leader, Counterterrorism and Incident Response at the Lawrence Livermore National Laboratory; Dr.—or Mr. David Nokes, Director, Systems Assessment and Research Center, Sandia National Laboratories; Dr. Donald D. Cobb, Associate Director for Threat Reduction, Los Alamos National Laboratory; Dr. Lew Stringer, Medical Director, Division of Emergency Management, the North Carolina Department of Crime Control and Public Safety; and Mr. Edward P. Plaugher, Chief of the Arlington County Fire Department, and also Executive Agent, Washington Area, National Medical Response Team.

Lady and gentlemen, we welcome you, and thank you for joining us this morning. And I would—you are aware that this committee is holding an investigative hearing, and when doing so, it is our practice to take testimony under oath. Do any of you have any objections to giving your testimony under oath? No? You are also, under the rules of this committee and the House, entitled to be represented by counsel. Do any of you wish to be represented by counsel this morning? Okay.

Is Dr. Stringer not here? Doctor, take your time and hurry on up to the table.

Welcome, Dr. Stringer. As I indicated to the other witnesses, sir, you are aware that this committee is holding an investigative hearing, and you are aware that, pursuant to our practices, we take testimony under oath. And I should ask you, do you have any objection to giving your testimony under oath?

Mr. STRINGER. No, sir.

Mr. GREENWOOD. Then for all of you, you are entitled under the rules of the House and the committee to be represented by counsel. Do any of you wish to be represented by counsel? Okay. In that case, if you would each stand, and all stand and raise your right hands, I will swear you in.

[Witnesses sworn.]

Mr. GREENWOOD. Okay. You are all the under oath. And Ms. Heinrich, you are recognized for 5 minutes for your opening statement. Thank you for being with us.

TESTIMONY OF JANET HEINRICH, DIRECTOR, HEALTH CARE AND PUBLIC HEALTH ISSUES, U.S. GENERAL ACCOUNTING OFFICE; HARRY C. VANTINE, PROGRAM LEADER, COUNTERTERRORISM AND INCIDENT RESPONSE, LAWRENCE LIVERMORE NATIONAL LABORATORY; K. DAVID NOKES, DIRECTOR, SYSTEMS ASSESSMENT AND RESEARCH CENTER, SANDIA NATIONAL LABORATORIES; DONALD D. COBB, ASSOCIATE DIRECTOR FOR THREAT REDUCTION, LOS ALAMOS NATIONAL LABORATORY; LLEWELLYN W. STRINGER, JR., MEDICAL DIRECTOR, DIVISION OF EMERGENCY MANAGEMENT, NORTH CAROLINA DEPARTMENT OF CRIME CONTROL AND PUBLIC SAFETY; AND EDWARD P. PLAUGHER, CHIEF, ARLINGTON COUNTY FIRE DEPARTMENT, EXECUTIVE AGENT, WASHINGTON AREA NATIONAL MEDICAL RESPONSE TEAM

Ms. HEINRICH. Mr. Chairman and members of the subcommittee, I appreciate the opportunity to be here today to discuss the proposed creation of the Department of Homeland Security. Since the terrorist attacks of September 11 and the subsequent anthrax incidents, there has been concern about the ability of the Federal Government to prepare and coordinate an effective public health response to such events. Our earlier work found that more than 20 Federal departments and agencies carry some responsibility for bioterrorism preparedness and response, and that their efforts are fragmented.

Emergency response is further complicated by the need to coordinate actions with agencies at the State and local level where much of the response activity would occur. My remarks will focus on the aspects of the proposal concerned with public health preparedness and response, and the two primary changes to the current system found in title 5 of the proposed bill.

First, the proposal would transfer certain emergency preparedness and response programs, as we have already heard.

Second, it would transfer the control over but not the operation of other public health preparedness assistance programs, such as providing emergency preparedness planning assistance to State and local governments from HHS to the new department.

The consolidation of Federal agencies and resources for medical response to an emergency as outlined in the proposed legislation has the potential to improve efficiency and accountability for these activities at the Federal level, as well as the State and local levels. The programs to be consolidated have already been identified for you. As Governor Ridge has stated, issues of coordination will remain, however.

The proposed transfer of the Metropolitan Medical Response System does not address the need for enhanced regional communication and coordination, for example. The National Disaster Medical System functions as a partnership among HHS, the Department of Defense, the Department of Veterans Affairs, FEMA, State and local governments, and the private sector. Thus, coordination across departments will still be required.

Similarly, the Strategic National Stockpile will involve the VA for purchase and storage, and HHS, in regards to the medical contents.

Although the proposed department has the potential to improve emergency response functions, its success is contingent on merging the perspectives of the various programs that would be integrated under the proposal. We are concerned that the lines of authority of the different parties in the event of emergency still need to be clarified.

As an example, in the recent anthrax events, local officials complained about differing priorities between the FBI and public health officials handling suspicious specimens. The FBI viewed the specimens as evidence in a criminal case, while public health officials' first priority was contacting physicians to ensure effective treatment was begun promptly.

The President's proposal to shift the authority, funding, and priority-setting for all programs assisting State and local agencies and public health emprgency Preparedness from HHS to the new department raises concerns because of the dual purpose nature of these activities. These programs include, as we have heard, the CDC's bioterrorism and preparedness programs and the HRSA Bioterrorism Hospital Preparedness Program. Functions funded through these programs are central to investigations of naturally occurring infectious disease outbreaks and to regular public health communications, as well as to identifying and responding to a bioterrorism event. Just as with the West Nile virus outbreak in New York City, which initially was feared to be the result of bioterrorism, when an unusual case of disease occurs, public health officials must investigate. Although the origin of the disease may not be clear at the outset, the same public health resources are needed, regardless of the source.

The recently enacted Public Health Security and Bioterrorism Preparedness and Response Act of 2002 recognized that these dual purpose programs are needed in State and local communities. Now States are beginning to plan to expand laboratory capacity, enhance their ability to conduct infectious disease surveillance and epidemiological investigations, and develop plans for communicating with the public. While under the proposal, the Secretary of Homeland Security would be given control over these assistance programs, their implementation would continue to be carried out by H HS.

The proposal also authorizes the President to direct that these programs no longer be carried out in that manner without addressing the circumstances under which such authority would be exercised.

We are concerned that this approach may disrupt the synergy that exists in these dual purpose programs. We are also concerned that the separation of control over the programs from their operations would lead to difficulty in balancing priorities. Although the HHS programs are important for homeland security, they are just as important to the day-to-day needs of public health agencies and hospitals, such as reporting on meningitis outbreaks and providing alerts to the medical community on influenza. The current proposal does not clearly provide a structure that ensures that both the goals of homeland security and public health will be met.

In summary, many aspects of the proposal are in line with our previous recommendations to consolidate programs, coordinate functions, and provide a statutory basis for leadership of homeland security. However, we do have concerns, as we have noted.

Mr. Chairman, this completes my prepared statement. I am happy to respond to any questions you or other members may have.

[The prepared statement of Janet Heinrich follows:]

PREPARED STATEMENT OF JANET HEINRICH, DIRECTOR, HEALTH CARE AND PUBLIC HEALTH ISSUES, U.S. GENERAL ACCOUNTING OFFICE

Mr. Chairman and Members of the Committee: I appreciate the opportunity to be here today to discuss the proposed creation of the Department of Homeland Security. Since the terrorist attacks of September 11, 2001, and the subsequent anthrax incidents, there has been concern about the ability of the federal government to prepare for and coordinate an effective public health response to such events, given the broad distribution of responsibility for that task at the federal level. Our earlier work found, for example, that more than 20 federal departments and agencies carry some responsibility for bioterrorism preparedness and response and that these efforts are fragmented.[1] Emergency response is further complicated by the need to coordinate actions with agencies at the state and local level, where much of the response activity would occur.

---

[1] U.S. General Accounting Office, *Bioterrorism: Federal Research and Preparedness Activities*, GAO-01-915, (Washington, D.C.: Sept. 28, 2001).

The President's proposed Homeland Security Act of 2002 would bring many of these federal entities with homeland security responsibilities—including public health preparedness and response—into one department, in an effort to mobilize and focus assets and resources at all levels of government. The aspects of the proposal concerned with public health preparedness and response would involve two primary changes to the current system, which are found in Title V of the proposed bill. First, the proposal would transfer certain emergency preparedness and response programs from multiple agencies to the new department. Second, it would transfer the control over, but not the operation of, other public health preparedness assistance programs, such as providing emergency preparedness planning assistance to state and local governments, from the Department of Health and Human Services (HHS) to the new department.[2]

In order to assist the committee in its consideration of this extensive reorganization of our government, my remarks today will focus on Title V of the President's proposal and the implications of (1) the proposed transfer of specific public health preparedness and response programs currently housed in HHS into the new department and (2) the proposed transfer of control over certain other public health preparedness assistance programs from HHS to the new department. My testimony today is based largely on our previous and ongoing work on federal, state, and local preparedness in responding to bioterrorist threats,[3] as well as a review of the proposed legislation.

In summary, we believe that the proposed reorganization has the potential to repair the fragmentation we have noted in the coordination of public health preparedness and response programs at the federal, state, and local levels. As we have recommended, the proposal would institutionalize the responsibility for homeland security in federal statute. We expect that, in addition to improving overall coordination, the transfer of programs from multiple agencies to the new department could reduce overlap among programs and facilitate response in times of disaster. However, we have concerns about the proposed transfer of control from HHS to the new department for public health assistance programs that have both basic public health and homeland security functions. These dual-purpose programs have important synergies that we believe should be maintained. We are concerned that transferring control over these programs, including priority setting, to the new department has the potential to disrupt some programs that are critical to basic public health responsibilities. We do not believe that the President's proposal is sufficiently clear on how both the homeland security and the public health objectives would be accomplished.

BACKGROUND

Federal, state, and local government agencies have differing roles with regard to public health emergency preparedness and response. The federal government conducts a variety of activities, including developing interagency response plans, increasing state and local response capabilities, developing and deploying federal response teams, increasing the availability of medical treatments, participating in and sponsoring exercises, planning for victim aid, and providing support in times of disaster and during special events such as the Olympic games. One of its main functions is to provide support for the primary responders at the state and local level, including emergency medical service personnel, public health officials, doctors, and nurses. This support is critical because the burden of response falls initially on state and local emergency response agencies.

The President's proposal transfers control over many of the programs that provide preparedness and response support for the state and local governments to a new Department of Homeland Security. Among other changes, the proposed bill transfers HHS's Office of the Assistant Secretary for Public Health Emergency Preparedness to the new department. Included in this transfer is the Office of Emergency Preparedness (OEP), which currently leads the National Disaster Medical System (NDMS)[4] in conjunction with several other agencies and the Metropolitan Medical

---

[2] These changes are primarily covered by Sections 502 and 505, respectively, in Title V of the President's proposed legislation.

[3] See "Related GAO Products" at the end of this testimony.

[4] In the event of an emergency, the National Disaster Medical System has response teams that can provide support at the site of a disaster. These include specialized teams for burn victims, mental health teams, teams for incidents involving weapons of mass destruction, and mortuary teams that can be deployed as needed. About 2,000 civilian hospitals have pledged resources that could be marshaled in any domestic emergency under the system.

Response System (MMRS).[5] The Strategic National Stockpile,[6] currently administered by the Centers for Disease Control and Prevention (CDC), would also be transferred, although the Secretary of Health and Human Services would still manage the stockpile and continue to determine its contents.

Under the President's proposal, the new department would also be responsible for all current HHS public health emergency preparedness activities carried out to assist state and local governments or private organizations to plan, prepare for, prevent, identify, and respond to biological, chemical, radiological, and nuclear events and public health emergencies. Although not specifically named in the proposal, this would include CDC's Bioterrorism Preparedness and Response program and the Health Resources and Services Administration's (HRSA) Bioterrorism Hospital Preparedness Program. These programs provide grants to states and cities to develop plans and build capacity for communication, disease surveillance, epidemiology, hospital planning, laboratory analysis, and other basic public health functions. Except as directed by the President, the Secretary of Homeland Security would carry out these activities through HHS under agreements to be negotiated with the Secretary of HHS. Further, the Secretary of Homeland Security would be authorized to set the priorities for these preparedness and response activities.

REORGANIZATION HAS POTENTIAL TO IMPROVE COORDINATION

The consolidation of federal assets and resources in the President's proposed legislation has the potential to improve coordination of public health preparedness and response activities at the federal, state, and local levels. Our past work has detailed a lack of coordination in the programs that house these activities, which are currently dispersed across numerous federal agencies. In addition, we have discussed the need for an institutionalized responsibility for homeland security in federal statute.[7] The proposal provides the potential to consolidate programs, thereby reducing the number of points of contact with which state and local officials have to contend, but coordination would still be required with multiple agencies across departments. Many of the agencies involved in these programs have differing perspectives and priorities, and the proposal does not sufficiently clarify the lines of authority of different parties in the event of an emergency, such as between the Federal Bureau of Investigation (FBI) and public health officials investigating a suspected bioterrorist incident. Let me provide you more details.

We have reported that many state and local officials have expressed concerns about the coordination of federal public health preparedness and response efforts.[8] Officials from state public health agencies and state emergency management agencies have told us that federal programs for improving state and local preparedness are not carefully coordinated or well organized. For example, federal programs managed by the Federal Emergency Management Agency (FEMA), Department of Justice (DOJ), and OEP and CDC all currently provide funds to assist state and local governments. Each program conditions the receipt of funds on the completion of a plan, but officials have told us that the preparation of multiple, generally overlapping plans can be an inefficient process.[9] In addition, state and local officials told us that having so many federal entities involved in preparedness and response has led to confusion, making it difficult for them to identify available federal preparedness resources and effectively partner with the federal government.

The proposed transfer of numerous federal response teams and assets to the new department would enhance efficiency and accountability for these activities. This would involve a number of separate federal programs for emergency preparedness and response, including FEMA; certain units of DOJ; and HHS's Office of the Assistant Secretary for Public Health Emergency Preparedness, including OEP and its NDMS and MMRS programs, along with the Strategic National Stockpile. In our

---

[5] The Metropolitan Medical Response System is a program that provides support for local community planning and response capabilities for mass casualty and terrorist incidents in metropolitan areas.

[6] The stockpile, previously called the National Pharmaceutical Stockpile, consists of two major components. The first component is the 12-Hour Push Packages, which contain pharmaceuticals, antidotes, and medical supplies and can be delivered to any site in the United States within 12 hours of a federal decision to deploy assets. The second component is the Vendor Managed Inventory.

[7] U.S. General Accounting Office, *Homeland Security: Responsibility and Accountability for Achieving National Goals*, GAO-02-627T (Washington, D.C.: Apr. 11, 2002).

[8] U.S. General Accounting Office, *Bioterrorism: Federal Research and Preparedness Activities*, GAO-01-915, (Washington, D.C.: Sept. 28, 2001).

[9] U.S. General Accounting Office, *Combating Terrorism: Intergovernmental Partnership in a National Strategy to Enhance State and Local Preparedness*, GAO-02-547T (Washington, D.C.: Mar. 22, 2002).

previous work, we found that in spite of numerous efforts to improve coordination of the separate federal programs, problems remained, and we recommended consolidating the FEMA and DOJ programs to improve the coordination.[10] The proposal places these programs under the control of one person, the Under Secretary for Emergency Preparedness and Response, who could potentially reduce overlap and improve coordination. This change would make one individual accountable for these programs and would provide a central source for federal assistance.

The proposed transfer of MMRS, a collection of local response systems funded by HHS in metropolitan areas, has the potential to enhance its communication and coordination. Officials from one state told us that their state has MMRSs in multiple cities but there is no mechanism in place to allow communication and coordination among them. Although the proposed department has the potential to facilitate the coordination of this program, this example highlights the need for greater regional coordination, an issue on which the proposal is silent.

Because the new department would not include all agencies having public health responsibilities related to homeland security, coordination across departments would still be required for some programs. For example, NDMS functions as a partnership among HHS, the Department of Defense (DOD), the Department of Veterans Affairs (VA), FEMA, state and local governments, and the private sector. However, as the DOD and VA programs are not included in the proposal, only some of these federal organizations would be brought under the umbrella of the Department of Homeland Security. Similarly, the Strategic National Stockpile currently involves multiple agencies. It is administered by CDC, which contracts with VA to purchase and store pharmaceutical and medical supplies that could be used in the event of a terrorist incident. Recently expanded and reorganized, the program will now include management of the nation's inventory of smallpox vaccine. Under the President's proposal, CDC's responsibilities for the stockpile would be transferred to the new department, but VA and HHS involvement would be retained, including continuing review by experts of the contents of the stockpile to ensure that emerging threats, advanced technologies, and new countermeasures are adequately considered.

Although the proposed department has the potential to improve emergency response functions, its success is contingent on several factors. In addition to facilitating coordination and maintaining key relationships with other departments, these include merging the perspectives of the various programs that would be integrated under the proposal, and clarifying the lines of authority of different parties in the event of an emergency. As an example, in the recent anthrax events, local officials complained about differing priorities between the FBI and the public health officials in handling suspicious specimens. According to the public health officials, FBI officials insisted on first informing FBI managers of any test results, which delayed getting test results to treating physicians. The public health officials viewed contacting physicians as the first priority in order to ensure that effective treatment could begin as quickly as possible.

NEW DEPARTMENT'S CONTROL OF ESSENTIAL PUBLIC HEALTH CAPACITIES RAISES CONCERN

The President's proposal to shift the responsibility for all programs assisting state and local agencies in public health emergency preparedness and response from HHS to the new department raises concern because of the dual-purpose nature of these activities. These programs include essential public health functions that, while important for homeland security, are critical to basic public health core capacities.[11] Therefore, we are concerned about the transfer of control over the programs, including priority setting, that the proposal would give to the new department. We recognize the need for coordination of these activities with other homeland security functions, but the President's proposal is not clear on how the public health and homeland security objectives would be balanced.

Under the President's proposal, responsibility for programs with dual homeland security and public health purposes would be transferred to the new department. These include such current HHS assistance programs as CDC's Bioterrorism Preparedness and Response program and HRSA's Bioterrorism Hospital Preparedness

---

[10] U.S. General Accounting Office, *Combating Terrorism: Selected Challenges and Related Recommendations,* GAO-01-822 (Washington, D.C., Sept. 20, 2001).

[11] The recently enacted Public Health Security and Bioterrorism Preparedness and Response Act of 2002 (P.L.107-188) cited core public health capacities that state and local governments need, including effective public health surveillance and reporting mechanisms, appropriate laboratory capacity, properly trained and equipped public health and medical personnel, and communications networks that can effectively disseminate relevant information in a timely and secure manner.

Program. Functions funded through these programs are central to investigations of naturally occurring infectious disease outbreaks and to regular public health communications, as well as to identifying and responding to a bioterrorist event. For example, CDC has used funds from these programs to help state and local health agencies build an electronic infrastructure for public health communications to improve the collection and transmission of information related to both bioterrorist incidents and other public health events.[12] Just as with the West Nile virus outbreak in New York City, which initially was feared to be the result of bioterrorism,[13] when an unusual case of disease occurs public health officials must investigate to determine whether it is naturally occurring or intentionally caused. Although the origin of the disease may not be clear at the outset, the same public health resources are needed to investigate, regardless of the source.

States are planning to use funds from these assistance programs to build the dual-purpose public health infrastructure and core capacities that the recently enacted Public Health Security and Bioterrorism Preparedness and Response Act of 2002[14] stated are needed. States plan to expand laboratory capacity, enhance their ability to conduct infectious disease surveillance and epidemiological investigations, improve communication among public health agencies, and develop plans for communicating with the public. States also plan to use these funds to hire and train additional staff in many of these areas, including epidemiology.

Our concern regarding these dual-purpose programs relates to the structure provided for in the President's proposal. The Secretary of Homeland Security would be given control over programs to be carried out by another department. The proposal also authorizes the President to direct that these programs no longer be carried out in this manner, without addressing the circumstances under which such authority would be exercised. We are concerned that this approach may disrupt the synergy that exists in these dual-purpose programs. We are also concerned that the separation of control over the programs from their operations could lead to difficulty in balancing priorities. Although the HHS programs are important for homeland security, they are just as important to the day-to-day needs of public health agencies and hospitals, such as reporting on disease outbreaks and providing alerts to the medical community. The current proposal does not clearly provide a structure that ensures that both the goals of homeland security and public health will be met.

CONCLUDING OBSERVATIONS

Many aspects of the proposed consolidation of response activities are in line with our previous recommendations to consolidate programs, coordinate functions, and provide a statutory basis for leadership of homeland security. The transfer of the HHS medical response programs has the potential to reduce overlap among programs and facilitate response in times of disaster. However, we are concerned that the proposal does not provide the clear delineation of roles and responsibilities that we have stated is needed. We are also concerned about the broad control the proposal grants to the new department for public health preparedness programs. Although there is a need to coordinate these activities with the other homeland security preparedness and response programs that would be brought into the new department, there is also a need to maintain the priorities for basic public health capacities that are currently funded through these dual-purpose programs. We do not believe that the President's proposal adequately addresses how to accomplish both objectives.

Mr. Chairman, this completes my prepared statement. I would be happy to respond to any questions you or other Members of the Committee may have at this time.

For further information about this testimony, please contact me at (202) 512-7118. Marcia Crosse, Greg Ferrante, Deborah Miller, and Roseanne Price also made key contributions to this statement.

RELATED GAO PRODUCTS

Homeland Security

---

[12] These include the Health Alert Network (HAN), a nationwide system that facilitates the distribution of health alerts, dissemination of prevention guidelines and other information, distance learning, national disease surveillance, and electronic laboratory reporting, and Epi-X, a secure Web-based disease surveillance network for federal, state, and local epidemiologists that provides tools for searching, tracking, discussing, and reporting on diseases and is therefore a key element in any disease investigation.

[13] U.S. General Accounting Office, *West Nile Virus Outbreak: Lessons for Public Health Preparedness*, GAO/HEHS-00-180 (Washington, D.C.: Sept. 11, 2000).

[14] P.L. 107-188.

Homeland Security: Key Elements to Unify Efforts Are Underway but Uncertainty Remains. GAO-02-610. Washington, D.C.: June 7, 2002.

Homeland Security: Responsibility and Accountability for Achieving National Goals. GAO-02-627T. Washington, D.C.: April 11, 2002.

Homeland Security: Progress Made; More Direction and Partnership Sought. GAO-02-490T. Washington, D.C.: March 12, 2002.

Homeland Security: Challenges and Strategies in Addressing Short- and Long-Term National Needs. GAO-02-160T. Washington, D.C.: November 7, 2001.

Homeland Security: A Risk Management Approach Can Guide Preparedness Efforts. GAO-02-208T. Washington, D.C.: October 31, 2001.

Homeland Security: Need to Consider VA's Role in Strengthening Federal Preparedness. GAO-02-145T. Washington, D.C.: October 15, 2001.

Homeland Security: Key Elements of a Risk Management Approach. GAO-02-150T. Washington, D.C.: October 12, 2001.

Homeland Security: A Framework for Addressing the Nation's Efforts. GAO-01-1158T. Washington, D.C.: September 21, 2001.

Public Health

Bioterrorism: The Centers for Disease Control and Prevention's Role in Public Health Protection. GAO-02-235T. Washington, D.C.: November 15, 2001.

Bioterrorism: Review of Public Health Preparedness Programs. GAO-02-149T. Washington, D.C.: October 10, 2001.

Bioterrorism: Public Health and Medical Preparedness. GAO-02-141T. Washington, D.C.: October 9, 2001.

Bioterrorism: Coordination and Preparedness. GAO-02-129T. Washington, D.C.: October 5, 2001.

Bioterrorism: Federal Research and Preparedness Activities. GAO-01-915. Washington, D.C.: September 28, 2001.

Chemical and Biological Defense: Improved Risk Assessment and Inventory Management Are Needed. GAO-01-667. Washington, D.C.: September 28, 2001.

Combating Terrorism: Need for Comprehensive Threat and Risk Assessments of Chemical and Biological Attacks. GAO/NSIAD-99-163. Washington, D.C.: September 14, 1999.

West Nile Virus Outbreak: Lessons for Public Health Preparedness. GAO/HEHS-00-180. Washington, D.C.: September 11, 2000.

Chemical and Biological Defense: Program Planning and Evaluation Should Follow Results Act Framework. GAO/NSIAD-99-159. Washington, D.C.: August 16, 1999.

Combating Terrorism: Observations on Biological Terrorism and Public Health Initiatives. GAO/T-NSIAD-99-112. Washington, D.C.: March 16, 1999.

Combating Terrorism

National Preparedness: Technologies to Secure Federal Buildings. GAO-02-687T. Washington, D.C.: April 25, 2002.

National Preparedness: Integration of Federal, State, Local, and Private Sector Efforts Is Critical to an Effective National Strategy for Homeland Security. GAO-02-621T. Washington, D.C.: April 11, 2002.

Combating Terrorism: Intergovernmental Cooperation in the Development of a National Strategy to Enhance State and Local Preparedness. GAO-02-550T. Washington, D.C.: April 2, 2002.

Combating Terrorism: Enhancing Partnerships Through a National Preparedness Strategy. GAO-02-549T. Washington, D.C.: March 28, 2002.

Combating Terrorism: Critical Components of a National Strategy to Enhance State and Local Preparedness. GAO-02-548T. Washington, D.C.: March 25, 2002.

Combating Terrorism: Intergovernmental Partnership in a National Strategy to Enhance State and Local Preparedness. GAO-02-547T. Washington, D.C.: March 22, 2002.

Combating Terrorism: Key Aspects of a National Strategy to Enhance State and Local Preparedness. GAO-02-473T. Washington, D.C.: March 1, 2002.

Chemical and Biological Defense: DOD Should Clarify Expectations for Medical Readiness. GAO-02-219T. Washington, D.C.: November 7, 2001.

Anthrax Vaccine: Changes to the Manufacturing Process. GAO-02-181T. Washington, D.C.: October 23, 2001.

Chemical and Biological Defense: DOD Needs to Clarify Expectations for Medical Readiness. GAO-02-38. Washington, D.C.: October 19, 2001.

Combating Terrorism: Considerations for Investing Resources in Chemical and Biological Preparedness. GAO-02-162T. Washington, D.C.: October 17, 2001.

Combating Terrorism: Selected Challenges and Related Recommendations. GAO-01-822. Washington, D.C.: September 20, 2001.

Combating Terrorism: Actions Needed to Improve DOD Antiterrorism Program Implementation and Management. GAO-01-909. Washington, D.C.: September 19, 2001.

Combating Terrorism: Comments on H.R. 525 to Create a President's Council on Domestic Terrorism Preparedness. GAO-01-555T. Washington, D.C.: May 9, 2001.

Combating Terrorism: Accountability Over Medical Supplies Needs Further Improvement. GAO-01-666T. Washington, D.C.: May 1, 2001.

Combating Terrorism: Observations on Options to Improve the Federal Response. GAO-01-660T. Washington, DC: April 24, 2001.

Combating Terrorism: Accountability Over Medical Supplies Needs Further Improvement. GAO-01-463. Washington, D.C.: March 30, 2001.

Combating Terrorism: Comments on Counterterrorism Leadership and National Strategy. GAO-01-556T. Washington, D.C.: March 27, 2001.

Combating Terrorism: FEMA Continues to Make Progress in Coordinating Preparedness and Response. GAO-01-15. Washington, D.C.: March 20, 2001.
Combating Terrorism: Federal Response Teams Provide Varied Capabilities; Opportunities Remain to Improve Coordination. GAO-01-14. Washington, D.C.: November 30, 2000.
Combating Terrorism: Need to Eliminate Duplicate Federal Weapons of Mass Destruction Training. GAO/NSIAD-00-64. Washington, D.C.: March 21, 2000.
Combating Terrorism: Chemical and Biological Medical Supplies Are Poorly Managed. GAO/T-HEHS/AIMD-00-59. Washington, D.C.: March 8, 2000.
Combating Terrorism: Chemical and Biological Medical Supplies Are Poorly Managed. GAO/HEHS/AIMD-00-36. Washington, D.C.: October 29, 1999.
Combating Terrorism: Observations on the Threat of Chemical and Biological Terrorism. GAO/T-NSIAD-00-50. Washington, D.C.: October 20, 1999.
Combating Terrorism: Need for Comprehensive Threat and Risk Assessments of Chemical and Biological Attacks. GAO/NSIAD-99-163. Washington, D.C.: September 14, 1999
Combating Terrorism: Use of National Guard Response Teams Is Unclear. GAO/T-NSIAD-99-184. Washington, D.C.: June 23, 1999.
Combating Terrorism: Observations on Growth in Federal Programs. GAO/T-NSIAD-99-181. Washington, D.C.: June 9, 1999.
Combating Terrorism: Analysis of Potential Emergency Response Equipment and Sustainment Costs. GAO/NSIAD-99-151. Washington, D.C.: June 9, 1999.
Combating Terrorism: Use of National Guard Response Teams Is Unclear. GAO/NSIAD-99-110. Washington, D.C.: May 21, 1999.
Combating Terrorism: Observations on Federal Spending to Combat Terrorism. GAO/T-NSIAD/GGD-99-107. Washington, D.C.: March 11, 1999.
Combating Terrorism: Opportunities to Improve Domestic Preparedness Program Focus and Efficiency. GAO/NSIAD-99-3. Washington, D.C.: November 12, 1998.
Combating Terrorism: Observations on the Nunn-Lugar-Domenici Domestic Preparedness Program. GAO/T-NSIAD-99-16. Washington, D.C.: October 2, 1998.
Combating Terrorism: Observations on Crosscutting Issues. GAO/T-NSIAD-98-164. Washington, D.C.: April 23, 1998.
Combating Terrorism: Threat and Risk Assessments Can Help Prioritize and Target Program Investments. GAO/NSIAD-98-74. Washington, D.C.: April 9, 1998.
Combating Terrorism: Spending on Governmentwide Programs Requires Better Management and Coordination. GAO/NSIAD-98-39. Washington, D.C.: December 1, 1997.

Disaster Assistance

Disaster Assistance: Improvement Needed in Disaster Declaration Criteria and Eligibility Assurance Procedures. GAO-01-837. Washington, D.C.: August 31, 2001.
Chemical Weapons: FEMA and Army Must Be Proactive in Preparing States for Emergencies. GAO-01-850. Washington, D.C.: August 13, 2001.
Federal Emergency Management Agency: Status of Achieving Key Outcomes and Addressing Major Management Challenges. GAO-01-832. Washington, D.C.: July 9, 2001.

Budget and Management

Budget Issues: Long-Term Fiscal Challenges. GAO-02-467T. Washington, D.C.: February 27, 2002.
Results-Oriented Budget Practices in Federal Agencies. GAO-01-1084SP. Washington, D.C.: August 2001.
Managing for Results: Federal Managers' Views on Key Management Issues Vary Widely Across Agencies. GAO-01-592. Washington, D.C.: May 25, 2001.
Determining Performance and Accountability Challenges and High Risks. GAO-01-159SP. Washington, D.C.: November 2000.
Managing for Results: Using the Results Act to Address Mission Fragmentation and Program Overlap. GAO-AIMD-97-146. Washington, D.C.: August 29, 1997.
Government Restructuring: Identifying Potential Duplication in Federal Missions and Approaches. GAO/T-AIMD-95-161. Washington, D.C.: June 7, 1995.
Government Reorganization: Issues and Principles. GAO/T-GGD/AIMD-95-166. Washington, D.C.: May 17, 1995.
Grant DesignGrant Programs: Design Features Shape Flexibility, Accountability, and Performance Information. GAO/GGD-98-137. Washington, D.C.: June 22, 1998.
Federal Grants: Design Improvements Could Help Federal Resources Go Further. GAO/AIMD-97-7. Washington, D.C.: December 18, 1996.
Block Grants: Issues in Designing Accountability Provisions. GAO/AIMD-95-226. Washington, D.C.: September 1, 1995.

Mr. GREENWOOD. Thank you very much.

Dr. Vantine, you are recognized for 5 minutes.

TESTIMONY OF HARRY C. VANTINE

Mr. VANTINE. Well, thank you, Mr. Chairman, and members of the committee, for asking me to speak before you today. It's a pleasure to be here. My name is Harry Vantine. I head the Counterterrorism and Incident Response Program at Lawrence Livermore National Laboratory. Our program at Livermore covers the waterfront, chemical, biological, nuclear, radiological. Today, my remarks are going to concentrate on nuclear and radiological, but I think similar remarks could be made for the chem-bio program.

Let me start by saying that my overall reaction to this legislation was that it is very broad, it's very inclusive. I think that's a very good thing. It's clear to me that as we go into the establishment of this Homeland Security Department, we are going to learn by doing, we are going to have to be able to change and adapt, and I think the legislation allows us to do that.

What I would like to do is stress this morning some of the elements that I think are really important in countering terrorism. There are several elements that I see. One is that we need a layered approach to counterterrorism. There is no one silver bullet that is going to solve this problem. So, a layered approach. I mean, we've got to look at beginning—we've got to look at indications and warnings. We have got to try and see the threats. We have got to protect the materials, the nuclear materials that—or the weapons that might be diverted for terrorist use. We need to have response teams that search, that disable. We need to have consequence management teams. We need to do the whole spectrum, and that's what I call a layered approach. Any one of them won't work. It's a big problem. It's a huge problem.

And so, you know, the second point I want to get to is because it's such a large problem, how do we solve that? We are going to need new and innovative approaches. And the way that—coming from a technology laboratory like Livermore, the way I see new technologies, new approaches being developed is through R&D technology. I think we are going to have to rely very heavily on R&D to find those new solutions.

Next, I would like to come to the issue of funding. When I look at R&D funding in industries, if I look at pharmaceuticals, biotechnologies, those type of industries, it's not unusual in some of the pharmaceutical industries to invest 20 percent of your revenues in R&D. We are going to have to have a very aggressive investment strategy and new approaches. Other companies invest 10 to 20 percent—10 to 15 percent. DOD is in that category. DOD invests in RDT&E something like 10 percent. So I think that's another approach going forward.

The fourth point I want to make is that I think we need clear lines of authority in this department. One of the drawbacks in the current system is that the current response system is somewhat a response that's clues together from different agencies. I think with this new department we have the ability to have people really dedicated to this mission, they know it's their job, and they're going to do it, and they're going to know what they have to do. They have clear authority.

The final general—the general attribute I think this homeland security strategy needs is strategic planning. We really have to do planning on big systems. We have to take a big systems approach to how we do this. The planning has got to be based on risk assessment to protect entire infrastructures. At the laboratories, we've put together these big ideas in the past, we've put together ideas such as model city protection, the basis program for biological detection, protection system for protecting metros, detection and tracking system for looking at nuclear materials, a national test bid for cargo inspection. These are the kind of ideas that we need, overreaching ideas that really cover the waterfront.

Information synthesis, I think, is also an important area. We are going to have to pull together the different intelligence functions from the different agencies. I think the new Office of Homeland— the Department of Homeland Security is going to have to have access to the intelligence data, the raw intelligence data it needs to process that information, to put it together, and understand the threats.

And another program that's been brought over from the existing programs is the nuclear assessment program. It's an NNSA program that has actually run—operated all three of the national weapons laboratories, headed by Livermore, though, that—and these people have been real heroes since September 11, working hard to look and assess nuclear threats.

Let me say in summary that I really think we are going to have to make a sustained investment in science and technology to win the war on terrorism. It's an enormous task. It's a task that the laboratories are eager to do, and with your help and with your planning, we think we can do it.

[The prepared statement of Harry C. Vantine follows:]

PREPARED STATEMENT OF HARRY C. VANTINE, PROGRAM LEADER FOR COUNTERTERRORISM AND INCIDENT RESPONSE, LAWRENCE LIVERMORE NATIONAL LABORATORY

Mr. Chairman and members of the committee, thank you for the opportunity to appear before you today. I lead the program in Counterterrorism and Incident Response at the Lawrence Livermore National Laboratory (LLNL). However, the opinions that I present today represent my views and not necessarily those of the Laboratory or the National Nuclear Security Administration. Today I would like to focus on nuclear and radiological response activities proposed for transfer to the Department of Homeland Security. There are analogies for chemical and biological response.

IMPORTANCE OF THE CBRNI (CHEM/BIO/RADIOLOGICAL/NUCLEAR/INFORMATION) MISSION

The threat of covert/terrorist delivery of weapons of mass destruction (WMDs) is a concern of the utmost gravity. There are many important government missions, but there is none more important than the Homeland Security mission. Witnessing the changes in the past 20 years, the bio-technology revolution, the breakup of the Soviet Union, the information explosion on the web, my conviction has only gotten stronger that Homeland Security is an enduring national security mission.

ESSENTIAL ELEMENTS OF A RESPONSE STRATEGY

What can we do to protect the U.S. against terrorist acquisition and use of WMDs? As with every other aspect of the terrorism problem, there is no silver bullet.

We see the following as essential elements.

- A layered strategy is required, addressing the various stages on this threat.
- This strategy will rely heavily on R&D. Only new solutions will offer adequate level of protection and be affordable.
- Adequate funding is needed. Industries, such as information technologies, biotechnologies, and pharmaceuticals, invest heavily in R&D: 10 to 15% of their budget. DOD has a similar profile of RDT&E investment.
- Clear lines of authority. This will shorten the time to get new capabilities to the field. Multi-group, multi-level approvals and negotiations will be curtailed.
- Strategic planning. Planning, based on risk assessment, is needed to protect entire infrastructures. Included in this planning are ideas such as Model City Pro-

tection, Detection and Tracking Systems, and the National Testbed for cargo inspection.

NUCLEAR INCIDENT RESPONSE

The Nuclear Incident Response Program has a broad charter to train for and respond to nuclear threats at the local, regional, and national level. The program is multi-agency. In the DHS legislation, it appears that there are three Under Secretaries who deal with various aspects of nuclear counterterrorism: Sec.301 is Nuclear Countermeasures, Sec. 401 is Border and Transportation Security, and Sec 501 is Emergency Preparedness and Response. The activities of there three need to be closely tied together so that there is one coordinated operational mission.

The advent of monitoring systems, first responder reach back ("Triage"), expanded regional response (RAP or Radiation Assistance Program) capability will require more robust communication systems and a robust fusion cell manned by technical experts. We will need to respond rapidly to assess the level of threat while waiting for the arrival of advanced technical assets. To maximize this capability it is critical that the proper equipment be with the first responders, who need to be practiced in their interactions with the fusion cell. The Nuclear Laboratories have the capability of making rapid and detailed analyses if sufficient information is transmitted to them. Thus it is critical that the equipment for the first response assets be carefully screened to maximize its capability. At the same time the capability and technical personnel at LLNL and LANL need to be expanded to provide the proper coverage and response capability to any scenario which occurs.

RECOMMENDATIONS FOR NUCLEAR INCIDENT RESPONSE

1. Training should be realistic, with preparation and training aids that challenge the responder. Results of training exercises should be used to improve system response.
2. Training should mimic actual response operations. "Practice like you play."
3. The operational architecture should include all levels of response from the first responder, to the regional and national responders.
4. A strategy to transition new technology into capable, prototype operational systems is essential. Technology developers must be included in the operational planning process.
5. Technical aspects of Nuclear Counter Terrorism should be managed by the laboratories with technical capabilities in this area, i.e. LLNL, LANL, SNL, and RSL. One laboratory should be in charge of coordinating and managing these technical activities among all the labs.

NUCLEAR ASSESSMENT PROGRAM

The Department of Homeland Security will have responsibilities for receiving and analyzing all source information in order to understand the nature and scope of the terrorist threat to the American homeland. This must involve access to both law enforcement and intelligence information at the most sensitive levels if the Department is to be successful in developing a strategic national plan for securing key resources and critical infrastructures, as well as responding to pending threats and attacks as they are detected. The terrorist threat is dynamic and global in nature. Understanding it and anticipating its countermoves will be an ongoing process that would benefit from interaction with other existing government programs analyzing and tracking a number of "classic" nuclear, chemical and biological threats and proliferation concerns. Essential intelligence information needed to support the Department's roles and missions must be quickly obtained, distributed, and analyzed so that protective priorities can be adjusted and/or warnings issued.

The Department faces major information analysis challenges. The number and diversity of these suggest that it would be appropriate to generously size and support the Department's strategic law enforcement and intelligence analysis programs including the nuclear assessment program. It will certainly require some "fully cleared" people, direct intelligence oversight and specific infrastructure to comply with DCID policies and guidance. New protocols for sharing and integrating law enforcement information with intelligence data may have to be developed. Furthermore, it seems highly likely that, sooner or later, it will require some additional supporting communication infrastructure.

INFORMATION ANALYSIS

The rapid advances in computer and information technology have enabled our society to generate massive amounts of data and information, but frequently we end up drowning in this sea of data because we lack the ability to select out the information or the relationships between information that is relevant. It is possible to develop computing tools and architectures that will enable us to progress beyond information overload to credible insights that can be used by decision-makers. The need for this "Information-to-Insight (I2I)" capability spans many national security areas and most of the Laboratory's programs. I2I will create a fundamental shift in the way that we relate critical information. The impact will be especially great for combating threats to our national security where anticipating and characterizing specific threats based upon detailed data from many varied sources are prerequisites for taking preventative action before it is too late.

We envision addressing questions and problems that require the ability to rapidly access massive amounts of data from disparate sources in such a way that one can uncover the critical linkages and insights hidden therein. Effectively linking the vast number of disparate and complex data sources that government decision makers and analysts must use to address U. S. national security issues is a major R&D challenge. Because our goal is to provide timely insights, the knowledge management system also needs to be able to constantly update itself.

OTHER SPECIFIC RECOMMENDATIONS

The new agency needs to have access to Restricted Data as defined in the Atomic Energy Act of 1954. This category of information has its own unique requirements compared to National Security Information and Law Enforcement Sensitive information. It would be reasonable to include within Sec. 203 (Access to Information) that any Restricted Data shared under that section is transmitted, retained, and disseminated consistent with the authority of the Secretary of Energy to protect Restricted Data. (This is similar to the approach taken for both intelligence information and law enforcement sensitive information.)

The new agency needs to have access to radioactive materials for purposes of testing and evaluating equipment. This includes Special Nuclear Materials (SNM) in various forms (e.g., oxides and metals) and test objects that are in nuclear explosive-like configurations containing SNM. The new department should be given the authority to specify and order such sources from DOE, own the sources (transfer them from DOE), and determine where the sources will be used. The new agency should be required to conform to security requirements comparable to those of the Department of Energy for these types and quantities of material.

The new agency needs to have the authority to work with the Director of Central Intelligence in setting priorities for intelligence gathering activities that may be critical to the security of the United States' homeland. In this way the new agency will not only be able to assess gathered information, but influence the type and priorities of information gathered by other agencies to make it more useful to the homeland security mission.

SCOPE OF THE PROBLEM

We must make a sustained investment in the science and technology to win the war on terrorism. It is an enormous task.

In closing, let me assure you that we at Lawrence Livermore National Laboratory have long been concerned about the terrorist WMD threat. We have built on our historical nuclear weapons mission and developed unique expertise, capabilities, and technologies to meet these emerging threats. LLNL is already providing critical elements of the nation's defense against nuclear, chemical, and biological terrorism, many of which were called into action post-September 11. We are committed to using our world-class scientific and technological resources—people, equipment, and facilities—to meet the nation's national security needs today and in the future.

Mr. GREENWOOD. Thank you, Dr. Vantine.

Mr. Nokes for 5 minutes.

TESTIMONY OF K. DAVID NOKES

Mr. NOKES. Mr. Chairman, distinguished members of the committee, thank you for allowing me to——

Mr. GREENWOOD. I think your microphone is not on, sir. There we go.

Mr. NOKES. Mr. Chairman and distinguished members of the committee, thank you for the opportunity to testify today. I am David Nokes. I am Sandia's director for our Systems Assessment and Research Center, and coordinator for our Homeland Security and Combating Terrorism Activities. I would like to briefly highlight some of the points I have made in my written testimony today.

Sandia, as well as the other NNSA labs, were able to respond to the events of September 11 very quickly, with good technology. And the reason they did that is because of the investments that have been made by the NNSA nuclear weapons program, the Armed Control and Treaty Verification Programs, and the sponsorship of many other government agencies to our work or other's program. And that is the technology that has been harvested by the Nation from the laboratories to address the problems of homeland security.

Perhaps you were aware that the decontamination foam that Sandia developed and licensed was used here on Capitol Hill to decontaminate or help decontaminate the anthrax. That was work that was done under our laboratory directed research and development program several years ago. And there are many other examples of work that was applicable directly to the events immediately post 9/11.

Let me turn now to the challenging problems of chemical and nuclear and biological detection and the weapons of mass destruction. One of the specialties that we have are nuclear sensors that rely on spectral analysis. That's important because those sensors reduce the nuisance alarms, the false alarms, and have an excellent record of detecting malevolent nuclear devices. We believe that there are sensor technologies that we have that are ready now for commercialization that could be transferred to industry and could be produced in quantities at this time.

We have also developed portable chemical and biological sensors, sensors that detect biotoxins, chemical agents, and recently we prototyped a system that would detect anthrax and identify anthrax in about a 5-minute timeframe. These are also in prototype stage, but they could join the suite of sensors that's available to first responders.

An area that we have developed almost unique technology is in the system of tools that are used to dismantle and disable explosive devices, and these are devices that could be used as the foundation for a weapon of mass destruction. Sandia's tools have been deployed widely. We run schools and we have trained over 750 first responders in the use of these high-tech tools that are useful in dismantling explosive devices. We are a full participant in the emergency response, the NEST teams of the Department of Energy. At Sandia, we have about 90 folks who are members of the response teams, in addition to the normal job. These are additional duties that they have elected to take on. They have been the very core of our design activities, and that's why they are useful as they go out and try to assess and render safe the various nuclear incidents.

We think it's going to be important for the Office of Homeland Security—the Department of Homeland Security to have a full portfolio of research activities, and it has to serve two parts. One is, we must provide the technology that's in hand to solve the current and emergent problems. And that's a transfer into industry so they can make these technologies available to the folks who need them.

Second, an equally important part is a longer range vision of what we can do in research and development to make great security affordable and sustainable, because otherwise you will end up with a system that is unsustainable and unaffordable, and that's a challenge for the new department to establish that research agenda.

I think that there is some bureaucratic problems that might harm the way the laboratories can be constructively engaged in the problems of the Office of—or the Department of Homeland Security. One that would be useful, if the NNSA were explicitly given the mission of developing technologies around homeland security, that would allow them to bring the force of the laboratories together, and it would be very useful if the Department of Homeland Security were able to task the laboratories directly as the agencies within the Department of Energy do. That would eliminate much of the bureaucratic problems that we have working with the government agencies.

On behalf of the folks at Sandia, I applaud your efforts. I think this is going to be a very important step in actual national and homeland security. I thank you, and I would be happy to respond to your questions.

[The prepared statement of K. David Nokes follows:]

PREPARED STATEMENT OF K. DAVID NOKES, SANDIA NATIONAL LABORATORIES

INTRODUCTION

Mr. Chairman and distinguished members of the committee, thank you for the opportunity to testify on the Administration's proposal to create a Department of Homeland Security, and specifically, the radiological, chemical, and biological response activities that may be of value to the new department. I am David Nokes, Director of Sandia National Laboratorie' Systems Assessment and Research Center. I have more than forty years experience in the nuclear weapons program, and currently head Sandia's activities that support our nation's intelligence community as well as the laboratory's activities in homeland security and the war against terrorism. I will shortly assume responsibility for all of Sandia's arms control, threat assessment, security technology, nonproliferation, and international cooperative programs as Vice President of Sandia's National Security and Arms Control Division.

Sandia National Laboratories is managed and operated for the National Nuclear Security Administration (NNSA) of the U.S. Department of Energy (DOE) by Sandia Corporation, a subsidiary of the Lockheed Martin Corporation. Sandia's unique role in the nation's nuclear weapons program is the design, development, qualification, and certification of nearly all of the nonnuclear subsystems of nuclear warheads. We perform substantial work in programs closely related to nuclear weapons, including intelligence, nonproliferation, and treaty verification technologies. As a multiprogram national laboratory, Sandia also conducts research and development for other national security agencies when our special capabilities can make significant contributions.

At Sandia National Laboratories, we perform scientific and engineering work with a mission in mind—never solely for its own sake. Even the fundamental scientific work that we do (and we do a great deal of it) is strategic for the mission needs of our sponsors. Sandia's management philosophy has always stressed the ultimate linkage of research to application. When someone refers to Sandia as "the nation's premier engineering laboratory," that statement does not tell the whole story: We are a science and engineering laboratory with a focus on developing technical solutions to the most challenging problems that threaten peace and freedom.

My statement will describe Sandia National Laboratories' contributions and capabilities in homeland security and discuss our technologies for radiological, chemical, and biological sensing. I will also describe our role in nuclear incident response and comment on the proposed relationship of that function to the Department of Homeland Security. Finally, I will offer suggestions for how the new department can efficiently access and manage the scientific and technology development resources it will require to support its mission.

SANDIA'S CONTRIBUTIONS TO HOMELAND SECURITY AND THE WAR AGAINST TERRORISM

Like most Americans, the people of Sandia National Laboratories responded to the atrocities of September 11, 2001, with newfound resolve on both a personal and professional level. As a result of our own strategic planning and the foresight of sponsors to invest resources toward emerging threats, Sandia was in a position to immediately address some urgent needs.

For example, by September 15, a small Sandia team had instrumented the K9 rescue units at the World Trade Center site to allow the dogs to enter spaces inaccessible to humans while transmitting live video and audio to their handlers. This relatively low-tech but timely adaptation was possible because of previous work we had done for the National Institute of Justice on instrumenting K9 units for SWAT situations.

You may perhaps be aware that a formulation developed by Sandia chemists was one of the processes used to help eliminate anthrax in the Hart, Dirksen, and Ford buildings on Capitol Hill and at contaminated sites in New York and in the Postal Service. Sandia had developed the non-toxic formulation as a foam several years ago and licensed it to two firms for industrial production in 2000. The formulation neutralizes both chemical and biological agents in minutes.

An array of devices invented by explosives experts at Sandia have proved to be effective for safely disarming several types of terrorist bombs. For the past several years, our experts have conducted training for police bomb squads around the country in the techniques for using these devices for safe bomb disablement. The shoe bombs that Richard Reid allegedly tried to detonate onboard a trans-Atlantic flight from Paris to Miami were surgically disabled with an advanced bomb-squad tool originally developed at Sandia. That device, which we licensed to industry, has become the primary tool used by bomb squads nationwide to remotely disable handmade terrorist bombs while preserving them for forensic analysis.

Sandia is a partner with Argonne National Laboratory in the PROTECT program (Program for Response Options and Technology Enhancements for Chemical/Biological Terrorism), jointly funded by DOE and the Department of Justice. PROTECT's goal is to demonstrate systems to protect against chemical attacks in public facilities, such as subways and airports. For more than a year, a Sandia-designed chemical detector test bed has been operating in the Washington D.C. Metro. The system can rapidly detect the presence of a chemical agent and transmit readings to an emergency management information system. We successfully completed a demonstration of the PROTECT system at a single station on the Washington Metro. The program has since been funded to accelerate deployment in multiple Metro stations. DOE has also been requested to implement a PROTECT system for the Metropolitan Boston Transit Authority.

Another major worry for homeland security is the potential for acts of sabotage against municipal water supplies. In cooperation with the American Water Works Association Research Foundation and the Environmental Protection Agency, Sandia developed a security risk assessment methodology for city water utilities. This tool has been employed to evaluate security and mitigate risks at several large water utilities. We have used similar methodologies to evaluate risks for other critical infrastructures such as nuclear power-generation plants, chemical storage sites, and dams.

These and other contributions to homeland security and the war against terror are possible because of strategic planning we had conducted years ago and early investment in the capabilities that were needed to respond to emerging threats. The outstanding technology base supported by NNSA for its core missions is the primary source of this capability. We also made strategic decisions to invest laboratory-directed research and development funds (LDRD) in the very things that we knew were urgent needs: items to the Afghanistan theater, the decontamination foam, the sensors we have deployed, and special-purpose robotics that we have developed. In recent months, requests for Sandia's services from federal agencies other than DOE for work in emerging areas of need have increased. Approximately twenty-eight percent of our total laboratory operating budget is now provided by federal agencies other than DOE.

SANDIA CAPABILITIES FOR HOMELAND SECURITY

Sandia National Laboratories and the other nuclear weapon laboratories constitute a broad, multidisciplinary technology base in nearly all the physical sciences and engineering disciplines. We seek to leverage those capabilities to support other national security needs germane to our missions, including homeland security, when our capabilities can make significant contributions.

Nuclear Sensing

A terrorist with a nuclear weapon and the knowledge and skill to use it, will use it if he is not stopped. The Department of Homeland Security will be responsible for preventing an attack on the United States by a terrorist with a nuclear weapon of mass destruction (WMD). The Department must prepare for this type of attack by reducing the vulnerability of the United States to nuclear terrorism through detection, identification, and interdiction of the nuclear materials that could be used in such an attack.

Nuclear weapons that could be used by a terrorist organization can be divided into three categories:

- A stolen or purchased *functional* nuclear warhead. Such a device has a high level of sophistication and the probability that it would detonate is high. The damage it would cause would be great, with large-scale loss of life, environmental devastation, and economic ruin.
- A weapon indigenously crafted, by a terrorist organization, from *stolen or purchased* plutonium or uranium. This device would have a moderate level of sophistication and a lower probability that it would detonate. However, if it did detonate, the damage could be great, perhaps similar to that caused by a stolen or purchased weapon.
- A radiation dispersal device (RDD) often referred to as a "dirty bomb." This is not a nuclear weapon, but consists of radioactive material (of any type) packaged with conventional explosives. It is designed simply to disperse radioactive material over a target area. The level of sophistication may be very low, but the probability that it would work is high, although the results desired by the perpetrator may be difficult to achieve. The actual damage a weapon of this type would cause is relatively small, compared to a nuclear detonation; however, it would result in radioactive contamination and could cause public panic and fear.

A nuclear bomb is a product of science and technology, and it is this same technology that must be used to protect against its use by terrorists. Scientists and engineers at the nation's nuclear weapon laboratories understand nuclear weapons—how they work, how to build them, what they can do. More importantly for homeland security, they know how to detect them, what characteristics to look for, how to sense their emissions, how to interpret what the sensors detect, and how to disable them.

Sandia National Laboratories has more than fifty years of experience in the nuclear weapons arena and an extensive knowledge of nuclear weapon science and technology. In addition to our mission of nuclear weapons stewardship, we have long been committed to safeguarding the nuclear weapons stockpile and actively supporting nonproliferation. The terrorist attack at the 1972 Munich Olympics focused our awareness on our nation's vulnerability to terrorist attacks abroad and, in particular, on the need to protect our stored nuclear weapons. This led to our work in access delay and denial at weapons storage sites and improving the security of weapon storage vaults. More recently, we have turned our physical protection expertise to protection and control of nuclear materials in Russia and the former Soviet Union.

If a terrorist intends to detonate a nuclear or radiological device in the United States, then he must deliver that device to his target. The device will emit radiation that can be detected with a radiation sensor. If his nuclear device was acquired or built outside the United States and smuggled into the country, we must find it before it enters or as it crosses into the country. If it originates in the United States, then we must detect it when it is being transported to the target site.

There are many different types of radiation detectors. The one that usually comes to mind is the Geiger counter, a simple device that can detect the presence or absence of some types of radiation. But it can't tell you very much about what type of material is emitting the radiation. Because there are many naturally occurring, medical, and industrial radioactive materials, knowing what type of material is emitting the radiation is crucial in order to avoid false and nuisance alarms and to zero-in on only those objects that pose a threat. For this purpose you need a spectral sensor.

Sandia National Laboratories produces radiation sensors for a variety of government customers. One of our specialties is spectral sensor systems that provide automatic radioactive material identification using special algorithms developed by Sandia. These systems detect and analyze nuclear materials quickly, in real time, in indoor or outdoor environments, and with a high degree of precision that provides a high level of confidence. We have produced a wide variety of sensor systems, from very large, fixed installations to small, rugged, portable battery-powered units.

Sandia's Radiation Assessment Identification and Detection (RAID) System was originally conceived, built, and tested before the tragic events of September 11, 2001. However, it meets the post-9/11 need to help safeguard our nation from nuclear terrorism. This system is designed to detect and identify radioactive materials transported through portals at passenger and package terminals at international ports of entry. RAID uses a commercial sodium iodide scintillation spectrometer and associated electronics, along with Sandia-developed analysis algorithms, to detect and identify radioactive materials passing within several meters of the sensor. A video image of the detection event scene is displayed on a base-station computer. The system automatically and continuously updates and recalibrates for background phenomena and can identify a radioactive source even if the source is shielded.

Based on our experience with RAID and other more advanced nuclear sensing systems, we believe the state of development of our nuclear sensors is such that the technology could be quickly transferred to commercial producers and widely and rapidly deployed at a cost of less than $50,000 per unit. These deployed systems would have a very high probability of detecting a smuggled nuclear weapon or an RDD if properly deployed. Nuclear sensing systems could be placed at ports of entry, around likely targets, or even scattered throughout a city to scan people, packages, and vehicles. Since these sensors are passive devices, they don't emit a signal and, consequently, are very difficult to detect. In other words, a terrorist can't use a radar detector to determine if one of these sensors is present. Unbeknownst to a terrorist, an alarm from one of these sensors could alert law enforcement personnel to the presence or movement of a weapon that employs radioactive material.

Of course, challenges exist in transitioning any technology from the laboratory to mass-produced industrial products. However, as we have demonstrated many times with technologies that we have transferred to industry in the past, Sandia works closely with industrial partners to work through the design challenges associated with manufacturing engineering and commercialization.

Another important tool in the war against nuclear terrorism is the Department of Energy's Second Line of Defense (SLD) program. Its purpose is to minimize the risk of nuclear proliferation and terrorism through cooperative efforts with foreign governments to strengthen their overall capability to detect and deter illicit trafficking of nuclear material across their borders. Here too, the nation's nuclear weapons laboratories have brought to bear their technical expertise in nuclear physics and engineering. Short-term, the Second Line of Defense program has adapted commercially available radiation detection equipment, security systems, and communications equipment to work comprehensively with Russian Customs and other foreign agencies to stop nuclear smuggling now. It is effective in detecting both weapons material and radiological dispersal devices.

Long-term, the Second Line of Defense program will deploy radiation detection equipment optimized for border use, integrate it with local, regional, and national-level communication systems geared for quick response, and cooperatively train foreign officials in use of the systems. Long-term sustainability is planned into every level of the program to ensure continued training and equipment maintenance.

Chemical and Biological Agent Sensing

Sandia is developing a variety of technical solutions to counter the threat posed by chemical and biological agents. This activity is supported by the DOE Chemical/Biological Nonproliferation Program (CBNP) and includes threat and response analysis, environmental sensing and monitoring, facility protection and biosecurity, advance chem/bio-terror warning systems, reagent design, and decontamination technology.

Sandia has developed a portable bio-sensor to put into the hands of first responders. Configured to detect toxins such as ricin and botulinum, the device uses microfabricated "chips" as a miniature chemical analysis lab to isolate and identify biological agents. This system has been demonstrated to also reliably and rapidly detect a variety of chemical weapon agents in realistic situations where obscurants to mask the signature are present. The system is being modified to analyze viruses and bacteria. We have identified commercial partners to produce and market the unit.

A prototype handheld detector under development at Sandia can identify anthrax in less than five minutes. The instrument analyzes fatty acid esters vaporized from the cell walls of bacteria and compares them with cataloged signatures indicative of anthrax or other pathogens. This technique has been used to identify pathogens at the genus level and often at the species level. Identifying the bacillus in minutes, rather than the hours currently necessary, is a crucial step toward developing bio-attack warning systems and defenses such as foam dispersal systems in public facilities similar to the PROTECT system that is being deployed in the Washington

Metro and other locations. We have applied for a patent on this detector and expect to license the technology to industry for commercial development and manufacture. Sandia's Laboratory-Directed Research and Development program supported this work.

Sandia is engaged in an accelerated development effort for a standoff biological weapons detection system to provide advance warning of a biological weapon threat. The system will employ ultraviolet laser-induced fluorescence to scan for and to discriminate clouds of biological agents over a broad field of view. Prototypes of this system have been demonstrated on various mobile and fixed platforms and have demonstrated excellent standoff range and sensitivity. Under NNSA sponsorship, we are moving toward the demonstration phase of the system development in the next several months.

Explosives Detection

Today, a commercially produced, walk-through portal for detecting trace amounts of explosive compounds on a person is available for purchase and installation at airports and other public facilities. The technology for this device was developed, prototyped, and demonstrated by Sandia National Laboratories over a period of several years and licensed to Barringer Instruments of Warren, New Jersey, for commercialization and manufacture. The instrument is so sensitive that microscopic quantities of explosive compounds are detected in a few seconds.

Using similar technology, we have developed and successfully tested a prototype vehicle portal that detects minute amounts of common explosives in cars and trucks. Detecting explosives in vehicles is a major concern at airports, military bases, government facilities, and border crossings. The system uses Sandia's patented sample collection and preconcentrator technology that has previously been licensed to Barringer for use in screening airline passengers. The same technology has been incorporated into Sandia's line of "Hound™" portable and hand-held sensors, capable of detecting parts-per-trillion explosives and other compounds. These devices can be of great value to customs and border agents at ports of entry.

Bomb Disablement Technology and Training

As first responders, American firefighters, police, and emergency personnel will be called upon to be America's first line of defense against terrorist attacks. These men and women must be prepared for the full range of terrorist threats, from improvised explosive devices to chemical, biological, radiological, and nuclear weapons of mass destruction. It will be the responsibility of the Department of Homeland Security to ensure their preparedness by providing them with the training and tools they need to do their jobs.

Sandia National Laboratories began holding advanced bomb-disablement technology workshops for bomb squad technicians in 1994. Since then, Sandia has transferred advanced bomb-disablement technology to more than 750 workshop participants through Operation America and its predecessors, Operation Riverside and Operation Albuquerque. Operation America is a series of ongoing regional workshops hosted by a local police department in the state where the event is held and supported by regional FBI offices. Participants come from bomb squads, police and fire departments, and emergency response organizations throughout the United States, including most of our major metropolitan cities and the U.S. Capitol Police. They also come from other government agencies, all branches of the U.S. military, and, internationally, from our allies in some of the world's terrorism hotspots. Participants come to learn applied explosives technology and advanced bomb-disablement logic, tools, and techniques. Technical classroom presentations, live-range demonstrations, hands-on training, and special high-risk scenarios give them the knowledge and technology they need to respond to terrorist threats involving explosives.

Most of the bomb-disablement technologies demonstrated in Operation America were developed by Sandia National Laboratories as part of the DOE Laboratory-Directed Research and Development program and our work for other federal agencies. These tools include the Percussion-Actuated Nonelectric (PAN) Disrupter used to dismantle suspected explosive devices and preserve forensic evidence. The device was used at the Unabomber's cabin in Montana and was available at the 1996 Summer and 2002 Winter Olympic Games. More recently, Massachusetts State Police, with the assistance of the FBI, used the Sandia-developed PAN Disrupter to disable the alleged shoe bombs removed from an American Airlines flight from Paris to Miami.

The PAN disrupter, as well as other advanced disablement tools developed by Sandia, are currently in use by local bomb squads and could be used against terrorist threats such as radiological dispersal devices (RDDs) and other weapons of mass destruction. Most of these bomb-disablement tools are relatively simple to assemble in the field, can be used safely from a distance, and are affordable, and they are currently in use throughout the bomb-disablement community. These tools disrupt and "render-safe" explosive packages without initiating the explosives or destroying forensic evidence.

Once Sandia has researched, developed, and tested a bomb-disablement tool, it begins the process of transferring the technology to the first-responders community, putting the technology in the hands of the men and women who need it. Operation America sponsors include Sandia National Laboratories, the National Institute of Justice, and DOE.

Critical Infrastructure Protection

National security and the quality of life in the United States rely on the continuous, reliable operation of a complex set of interdependent infrastructures consisting of electric power, oil and gas, transportation, water, communications, banking and finance, emergency services, law enforcement, government continuity, agriculture, health services, and others. Today, they are heavily dependent on one another and becoming more so. Disruptions in any one of them could jeopardize the continued operation of the entire infrastructure system. Many of these systems are known to be vulnerable to physical and cyber threats and to failures induced by system complexity.

In the past, the nation's critical infrastructures operated fairly independently. Today, however, they are increasingly linked, automated, and interdependent. What previously would have been an isolated failure, today could cascade into a widespread, crippling, multi-infrastructure disruption. As the documented cases of attacks on vital portions of the nation's infrastructure grow, there is a sense of urgency within industry and government to understand the vulnerabilities.

The National Infrastructure Simulation and Analysis Center (NISAC)—which would be transferred to the Department of Homeland Security under the Administration's bill—is a comprehensive capability to assess the nation's system of infrastructures and their interdependencies. NISAC's partners are Sandia National Laboratories and Los Alamos National Laboratory, both of which possess extensive supercomputer resources and software expertise. NISAC will provide reliable decision support analysis for policy makers, government leaders, and infrastructure operators. It will perform modeling, simulation, and analysis of the nation's infrastructures, with emphasis on the interdependencies.

Sandia pioneered probabilistic risk assessment (PRA) as a tool for evaluating the risks associated with high-consequence systems such as nuclear weapons and nuclear power generation plants. We apply this tool to risk assessments for critical infrastructures such as dams, water utilities, chemical plants, and power plants. Combined with our expertise in security systems for nuclear facilities, we have helped utilities and industrial associations create security assessment methodologies that can guide owners and operators through the assessment process to determine vulnerabilities and identify mitigation options. Methodologies have been developed for water utilities, chemical storage facilities, dams, power plants, and electrical power transmission systems.

Cyber Sciences

Computer systems and networks are attractive targets of attack by terrorists, foreign governments, or high-tech criminals. Government functions, commerce, and the military increasingly rely on cyber networks in their operations. Computerized supervisory control and data acquisition (SCADA) systems often control the operations of critical infrastructures such as power utilities and distribution networks and municipal water supplies.

Sandia has significant activities in the technologies intended to protect cyber and network resources and the information that resides on such systems. Programs that assess the vulnerabilities associated with these systems are in place for our own resources as well as for those at other federal government agencies. Sandia operates a SCADA laboratory to study such cyber control systems and to determine effective protection strategies. We conduct red-teaming to challenge cyber and information systems and identify and remove vulnerabilities. Our objectives are to enhance the robustness of cyber systems and critical information systems and develop solutions for survivability and response options for systems under attack. Our understanding of the issues associated with computer and network vulnerabilities is enhanced by the microelectronic design and fabrication capability resident at Sandia as well as the state-of-the-art work performed as part of NNSA's Advanced Simulation and Computing (ASC) campaign.

NUCLEAR INCIDENT RESPONSE

The President's bill to establish a Department of Homeland Security defines a Nuclear Incident Response Team that includes entities of the Department of Energy and the Environmental Protection Agency that perform nuclear and/or radiological emergency support functions (Section 504).

NNSA plays a vital support role in combating acts of nuclear terrorism through its Nuclear Emergency Support Team (NEST). NEST provides the FBI and other federal and state agencies with technical assistance in response to terrorist use or threat of use of a nuclear or radiological device in the United States. NEST also supports the Department of State in a similar role for incidents overseas. Another NNSA team, the Accident Response Group (ARG), has the different mission of providing technical support in response to accidents involving U.S. nuclear weapons while they are either in the custody of DOE or the military services. The ARG and NEST teams draw from the same pool of experts at the NNSA laboratories, all of whom are volunteers.

NEST maintains a fast-response capability for a radiological emergency involving dispersal of radioactive debris—for example, from the detonation of a so-called "dirty bomb" or radiological dispersal device (RDD). The NNSA's Radiological Assistance Program (RAP) provides initial responders who can be on the scene in a matter of hours. Their support role is to characterize the radiological environment, provide technical advice to the FBI, FEMA, and other emergency response agencies, and to assist with decontamination and material recovery. NNSA is in the process of enhancing the Radiological Assistance Program to perform radiological weapons detection and device characterization missions on a regional basis consistent with the FEMA response regions.

The Joint Technical Operations Teams (JTOTs) are major operational elements of NEST that directly assist military units and crisis response operations. These teams are trained and equipped to support render-safe operations and advise on stabilization, packaging, and disposition procedures.

In addition to the NEST and ARG capabilities, NNSA maintains Consequence Management Teams that are available to provide assistance to federal and state agencies that require radiological emergency assistance after an event has occurred. The teams are trained and equipped to support incident assessment, monitoring and sampling activities, laboratory analysis, and health and safety support to incident responders.

Sandia National Laboratories contributes approximately ninety team members to the various elements of NEST, ARG, RAP, and Consequence Management. Sandia's role focuses largely on RAP incident response, device characterization, render-safe techniques, assessment and prediction of consequences from radiological incidents and accidents, and methods for containment of radiological materials. Sandia is the only NNSA laboratory that maintains the capability for containment of particulates that would be released in an RDD explosion.

The President's bill would place the Nuclear Incident Response Team under the author-ity and control of the Secretary of Homeland Security during an actual or threatened terrorist attack or other emergency. During such a time, it would operate as an organizational unit of the Department of Homeland Security. At all other times, DOE/NNSA would be responsible for organizing, training, equipping, and exercising authority and control over NEST, ARG, and the Consequence Management Teams. This arrangement is not ideal, but it makes sense in this case because the volunteer NEST and ARG experts are integrated with the nuclear design activities of the DOE/NNSA laboratories. It would not be possible, for example, to transfer the NEST/ARG functions to the Homeland Security Department on a permanent basis because the personnel who constitute those teams are full-time weapon scientists, engineers, and technicians.

Consequently, it will be important to establish and exercise a clearly understood process for deploying the Nuclear Incident Response Team elements to avoid interagency conflicts over roles and authorities. The process should be designed to minimize the layers of federal offices involved in both management and deployment.

SCIENCE AND TECHNOLOGY DEVELOPMENT FOR HOMELAND SECURITY MISSIONS

The national laboratories of the NNSA are widely regarded as the premier science and technology laboratories in the federal government. These institutions have a long history of excellence in research and development in nuclear weapons and other national security applications. They are uniquely able to deploy multidisciplinary teams on complex problems in a way that integrates science, engineering, and design with product.

In a world where threats are increasingly insidious—with worrisome developments in chemical and biological weapons, cyber warfare, and proliferation—it is important that the NNSA laboratories be major contributors in the national effort to address these threats. These national laboratories can provide enormous value to homeland security challenges. They are also the logical entities to perform technology evaluation on the many products and proposals that will inevitably be advocated to the Department of Homeland Security from countless vendors.

Unfortunately, established bureaucratic structures and regulations that insulate agencies from one another will stand in the way of effective utilization of the NNSA laboratories for homeland security unless legislative action is taken to remove the barriers. As a first step, it would be helpful to explicitly authorize NNSA to carry out research and development for homeland security by adding that activity to the NNSA's authorized missions listed at Title 42, Section 2121 of the United States Code. Next, the Homeland Security Act should give the Department of Homeland Security the power to task the NNSA laboratories directly, just as the Science, Energy, Environmental, and other non-NNSA offices of DOE are able to do. That authority would eliminate the bureaucratic red tape and additional costs associated with the Work-for-Others (WFO) process that inhibits access and utilization of the laboratories by non-DOE sponsors.

It will be important for the Homeland Security Department to have the authority to determine for itself how and where to make its research and development investments to support its mission goals. There will be some laboratories and institutions that will seek to be designated as homeland security laboratories or as centers of excellence for this or that homeland security mission area. The Department will need to look beyond labels to demonstrated capabilities and a track record of deliverables. Its research and development program should encourage a competition of ideas among many performers, including industrial firms, universities, and federal laboratories, and then fund the development of the best ideas based on considerations of technical merit and not on who the performer is. The Defense Advanced Research Projects Agency (DARPA) uses such an approach, and it may be an effective model for the Homeland Security Department to emulate.

Under the President's bill, the research and development program for the entire Department would be directed by the Under Secretary for Chemical, Biological, Radiological, and Nuclear Countermeasures. Certainly that official will have formidable R&D challenges, but he or she must also be cognizant of the science and technology needs for the other mission areas of homeland security, including information analysis and infrastructure protection, borders and transportation security, and emergency preparedness and response. As an alternative, it may be useful to consider a chief scientist position reporting to the Secretary with authority for coordinating and directing the Department's overall research and development program. Each Under Secretary may benefit from a dedicated R&D element focused on the challenges peculiar to his mission.

SUMMARY AND CONCLUSION

Sandia National Laboratories and the other NNSA laboratories constitute a broad, multidisciplinary technology base in nearly all the physical sciences and engineering disciplines. We are eager to leverage those capabilities to support the science and technology needs of the Department of Homeland Security when our capabilities can make significant contributions.

Sandia possesses strong competencies in nuclear, chemical, and biological sensors and engineered systems suitable for transfer to industry and deployment in homeland security applications. We have been proactive in supporting our nation's first responders and addressing the challenges of infrastructure protection. We have a track record of anticipating emerging homeland security threats and investing in technology development to counter them through our Laboratory-Directed Research and Development program and sponsor-directed programs. We are one of the premier laboratories for working with industry to transition laboratory technologies into deployable commercial applications.

Bureaucratic and regulatory roadblocks exist that limit access to the DOE/NNSA national laboratories by other federal agencies, and those obstacles should be removed by the homeland security legislation in order to facilitate direct access to those resources. The Homeland Security Department needs the authority to manage a research and development program that encourages competition of ideas among many performers—including industrial firms, universities, and federal laboratories—and then fund the development of the best ideas based on technical merit and applicability to mission needs.

On behalf of the dedicated and talented people who constitute Sandia National Laboratories, I want to emphasize our commitment to strengthening United States security and combating the threat to our homeland from terrorism and weapons of mass destruction. It is our highest goal to be a national laboratory that delivers technology solutions to the most challenging problems that threaten peace and freedom.

Thank you, Mr. Chairman. I would be pleased to respond to any questions you may have.

Mr. GREENWOOD. Thank you, Mr. Nokes.

Dr. Cobb for 5 minutes.

TESTIMONY OF DON COBB

Mr. COBB. Thank you, Mr. Chairman and members of the committee. It is a pleasure for me to be here and talk about a very important part of the establishment of the new Department of Homeland Security, namely, the part that's associated with the ability to respond to threats of weapons of mass destruction, terrorism against our own country.

My name is Don Cobb. I'm the Associate Director for Threat Reduction at Los Alamos. I have about 30 years experience in dealing with various kinds of threats, working in arms control, nonproliferation, and counterterrorism. Over that period of time I have had experience in developing technologies, from radiation technologies to satellite-based technologies.

Los Alamos, about one quarter of the laboratory, something over 20 percent of the laboratory, is involved in these kind of threat reduction activities across the board. As you know, Los Alamos is operated by the University of California for the Department of Energy; has been for the last 60 years. So we are uniquely, along with our brethren at the other labs, operated for the country to do major missions that are broad S&T-based, like the homeland security issue is today.

What I want to do is confine my remarks to the Nuclear Emergency Research Team and try to elucidate some of the issues that I think are most important in the setting up of this new department to preserve the capability and hopefully enhance our nuclear response capabilities.

First, let me say Los Alamos is involved in virtually every aspect of nuclear emergency response, from threat analysis, analyzing all source information to understand what the threat is, to fielding detection diagnostics, radiation sensors, and so forth, to neutralizing the threat, to making recommendations how to—how to safe the device, whatever it may be. This is a shared responsibility that I have primarily with the other two NNSA laboratories.

The main point that I want to make, and I think General Gordon made it earlier, is—made it for me, is that the NEST tech base is not something that you can isolate as a piece and transfer it to the new department. It does not stand alone. It's the synergy of that tech base with the nuclear weapons and threat reduction program at the laboratories.

For example, to give you the idea, there are over 100 people at Los Alamos that work at the Nuclear Emergency Support Team. Only about seven of these are full-time people. The rest of them are nuclear weapon designers, they're nuclear weapon engineers, they're people who do radiological detection development for sensors and systems. And it's those skills, and also the specialized facilities that we have where you can actually make measurements and utilize nuclear materials, that make this a unique support capability. We need to keep that synergy in the transition.

Let me talk to three specific issues that I think are important to us that will matter but that can be resolved, I think—or, perhaps not through legislation, but through just negotiating the right roles and responsibilities between the existing DOE, the laboratories, and the new Department of Homeland Security.

First, about command and control relating to NEST. It has to be clear, when NEST is under the authority of the new Department of Homeland Security, under what conditions it remains under the authority of the DOE. For example, under a heightened threat condition, we may be deploying people or looking at threats as part of our NEST responsibilities; we will call people in to work on that. Under that condition, we need to understand whether we are reporting to the DOE or whether we are reporting to the Department of Homeland Security.

Similarly, the RAP program, the Radiation Assistance Program, has similar kinds of response to maybe State and local responders. We need to understand whether they continue to do that under the DOE.

So the roles and responsibilities, and to clarify under what conditions these various responsibilities will occur between the departments has to be worked out. And then we need to jointly do exercises and drills and practices so we can understand how this actually plays together in case and when these assets are needed and they are called upon. So that's one. The command and control structure needs to be clarified.

The second one has also been previously mentioned, but I want to raise it again because it is very important. The R&D that generates the technology that goes into the NEST programs quite often comes from other programs, not necessarily directly through the NEST program. It relies on and leverages other investments that are being made in parallel that develop related technology. Heretofore the DOE has accepted that responsibility and understands that kind of relationship.

If the NEST R&D is rolled over to the Department of Homeland Security as part of a total R&D package, it will sever some of that leveraging, and it would have to be done very carefully. I would argue in favor of keeping the R&D and the technology integration as part of the NEST package and keep that as part of the current DOE structure.

Then the third one I want to mention is legal issues. We currently, working for the University of California, have clear indemnification and liability protection for our people and our institution in participating and supporting NEST activities. That's because of our M&O contractual relationship that's spelled out very clearly. If we move that over to the Department of Homeland Security, again, we would have to examine all those legal issues again, and at least they would have to be redone, preserved in another way.

So my final comment is we currently work—when we are called out, we have a DOE lead person in the field who leads our NEST teams. That lead person for the DOE interacts with the lead Federal agency. It might be the FBI, depending on what kind of incident it is. So there is a clear mechanism for doing this. The Department of Homeland Security could easily be the—could be the lead Federal agency in certain emergency situations, and we'd still have our DOE NEST team responding in similar fashion. If we do that, if that's the nature of the relationship that's set up, then I think all of the issues that I've raised here are pretty straightforward in terms of being able to handle them. If we don't, it's going to be much more complicated.

So thank you. And I would be happy to answer questions.

[The prepared statement of Don Cobb follows:]

PREPARED STATEMENT OF DON COBB, ASSOCIATE DIRECTOR, THREAT REDUCTION, LOS ALAMOS NATIONAL LABORATORY

INTRODUCTION

Thank you Mr. Chairman and distinguished members of the House Energy and Commerce Subcommittee on Oversight and Investigations, for inviting me here today to discuss the important issue of the creation of the Department of Homeland Security and what its proposed role will be in terms of dealing with chemical, biological, radiological and nuclear emergency response activities.

I am Don Cobb, Associate Director for Threat Reduction at the Department of Energy National Nuclear Security Administration's Los Alamos National Laboratory. Los Alamos is one of the three NNSA laboratories responsible for maintaining the nation's nuclear stockpile. At Los Alamos, I am responsible for all programs directed at reducing threats associated with weapons of mass destruction. I personally have more than 30 years experience working to reduce these threats.

Today, I would like to discuss with you the emergency response activities at Los Alamos National Laboratory, focusing on our involvement and work with nuclear emergency response efforts, primarily the Department of Energy's Nuclear Emergency Support Team (NEST). In addition to NEST, I also will discuss Los Alamos' efforts in responding to biological threats and incidents, in particular the Biological Aerosol Sentry and Information System (BASIS). Responding to the biological threat is an area in which our national capability is not as mature as the capabilities that we have in dealing with the nuclear threat.

NUCLEAR EMERGENCY SUPPORT TEAM (NEST)

Los Alamos plays an important role within the area of nuclear emergency response. The largest and the most well-known team in this area is the DOE-managed NEST team. NEST was created in 1975 in response to concerns over nuclear terrorism activity. Its effectiveness is due to well-established interagency relationships including significant Department of Defense and FBI collaboration. NEST is focused on responding to a threatened act involving radiological or nuclear materials or devices. Among the range of potential terrorist threats involving weapons of mass destruction, the nuclear response infrastructure and capabilities are the most mature and capable of addressing the threat. NEST includes the capabilities to search for, diagnose, and disable an improvised nuclear device.

NEST depends on a team of highly dedicated individuals at the national laboratories and facilities throughout the DOE-complex who volunteer their expertise to this program. Los Alamos' NEST and related activities are funded at approximately $10 million in fiscal year 2002. More than 100 Los Alamos scientists and engineers are involved in various aspects of the NEST program. Nearly all are involved in other parts of the Laboratory's research in nuclear weapons or threat reduction. Many of the employees who work part-time on NEST are involved with more than one team within the NEST program.

It is important to note that NEST is more than a group of scientists who stand at the ready with pagers on their belts, waiting to be contacted to respond to a crisis. NEST team members at the DOE and NNSA laboratories, including Los Alamos, are involved in a wide range of related activities including research and development into diagnostic tools, disablement techniques, and computer simulations and modeling; working with the intelligence and law enforcement communities on the analysis of threats and the development of analytical tools; training of employees from other government agencies in environments that allow hands-on work with the actual nuclear materials that they might encounter in the field; and providing subject-matter experts when required. Los Alamos has the lead within NEST for development of nuclear diagnostic tools to help determine the nature of the suspected threat device and for maintenance of what is called the "home team," a group of experts parallel to those that would be deployed in the field who can provide analysis, advice and technical support.

Los Alamos is involved to varying degrees in all aspects of the national NEST program. The activities of the national team, and Los Alamos' role, are as follows:

- Search activities—Los Alamos is primarily involved in research and evaluation of detectors used for search.
- Joint Tactical Operations Team (JTOT)—JTOT is a partnering of DOE and DoD expertise that provides advice or direct assistance to render safe a suspect malevolent employment of a nuclear device by terrorists or others and to perform a nuclear safety assessment for the eventual safe disposition of the device. Los Alamos plays a major role in the JTOT mission and is involved in maintaining management oversight, render-safe capability, diagnostics capability, emergency response home team capability, a watchbill (a group of experts who are on call 24 hours a day, seven days a week, year-round), communications support and deployable equipment, and contingency planning.
- Accident Response Group (ARG)—ARG is responsible for dealing with incidents involving a U.S. weapon, commonly referred to as a "Broken Arrow." Los Alamos has experts on the ARG roster who may be called upon if their particular set of knowledge is necessary to deal with the given situation.
- Disposition—These assets support both the JTOT and the ARG team, making decisions about the ultimate disassembly and disposition of a device after it has been made safe to move and ship to a remote location.
- Consequence Management—Following an incident, this team is involved in the immediate monitoring of any potential radiological dispersal and in monitoring and forecasting that can advise responders on issues of evacuation and treatment.
- Attribution—This area involves drawing upon capabilities from the U.S. weapons testing program to analyze samples and draw forensic inferences about a threat device.
- Radiological Assistance Program (RAP)—Related to but separate from NEST, DOE and Los Alamos maintain response plans and resources to provide radiological assistance to other federal agencies; state local, and tribal governments; and private groups requesting such assistance in the event of a real or potential radiological emergency. The Los Alamos RAP organization provides trained personnel and equipment to evaluate, assess, advise, and assist in the mitigation of actual or perceived radiological hazards or risks to workers, the public, and the environment. This Los Alamos capability supports associated activities throughout RAP Region Four: Kansas, Oklahoma, Texas, Arizona, and New Mexico.

BIOLOGICAL EMERGENCY RESPONSE

The biological science and medical communities responded to the challenge posed by the fall 2001 anthrax attacks. Los Alamos has been involved in responding to the attacks from the beginning, providing DNA forensics expertise to assist federal law enforcement agencies in the anthrax investigation. Our bioscience experts played an advisory role in the decontamination of the Senate Hart Office Building after the attacks, providing a strategy and advice for decontaminating the building so it could be restored to its regular function.

Although more work and attention is needed in terms of biological emergency response, significant progress has been made through research efforts, many of which reside in DOE NNSA's Chemical and Biological National Security Program (CBNP). For instance, Los Alamos and Lawrence Livermore National Laboratory have been involved in research and development of bio-detection systems as part of CBNP. One such system to detect a biowarfare attack was demonstrated by Los Alamos and Livermore at the Winter Olympics in Salt Lake City. The system, called the Biological Aerosol Sentry and Information System (BASIS), provides public health officials with early warning of a potential bioterrorist attack.

EMERGENCY RESPONSE ISSUES

The following issues related to transferring emergency response authority and responsibility to DHS should be addressed.

- NEST command and control—It must be clear when NEST is under the authority of DHS and when it is under the authority of DOE. For example, continuous monitoring and surveillance looking for threats could be under either department. Once authorities under various options are clear, it will be important to establish joint training to exercise the various options.

- NEST research and development and technology integration—Heretofore, DOE has fulfilled the responsibility for NEST R&D and technology integration. It is important to determine whether this responsibility will continue in DOE or be transferred to DHS. This is the underpinning for the continued and improved effectiveness of NEST. Similarly, the ability to respond to future biological threats depends on synergy with the biological and health sciences.
- NEST legal issues—Legal issues related to liability and indemnification for those that respond to emergency incidents need to be sorted out and resolved. Individuals and contracting entities responding to these incidents at the direction of the federal government need clear legal protections.
- Biological Emergency Support Team (BEST)—The establishment of a national BEST, perhaps modeled after NEST, should be considered. Just as NEST relies on nuclear weapons and threat reduction experts, a BEST will need to maintain close contact with the biological and medical sciences communities.

CONCLUSION

At Los Alamos, we will continue to work with DOE NNSA and the other national laboratories to support the nation's ability to respond to emergencies involving weapons of mass destruction. We will work with the new DHS to ensure the continued effective function of these emergency response capabilities.

Mr. GREENWOOD. Thank you, Dr. Cobb.

Dr. Stringer, for 5 minutes.

TESTIMONY OF LLEWELLYN W. STRINGER, JR.

Mr. STRINGER. Mr. Chairman, members of the committee, thank you for inviting me here today. I was here in October talking about how emergency responders at local, State, and Federal Governments were affected and what we needed. Well, I'm back again today. I'm going to talk about how the homeland——

Mr. GREENWOOD. You did such a good job in October, we wanted a repeat performance.

Mr. STRINGER. Thank you, sir.

I'm going to talk about how this could affect local and State governments, and how it could affect the national medical response system and OEP.

In talking to my local, Federal and State cohorts, I really think we could put all of this together and call it the need for interoperability on a day-to-day basis.

It's part of my job with the State looking at grants, how do we apply for them, trying to get the local and States to understand the stakeholders, what we are going to need to do to get them, and then all the many pots of money that are sort of dangled at us at the State level for locals and State from CDC, OEP, FEMA and many more. They all have different rules. They all have different time tables, they all have different things that we have got to try to understand and then explain to others in the State to be successful in getting the grant and using it. That's a real problem.

In North Carolina, we are trying to develop a single unified terrorist plan, bringing the local and the State entities together to develop strategies for equipping, planning, training and exercising, so we have one plan, wherever it may be in the State, whatever city it may be in. This is very important. The planners in emergency management need one standard set of grant guidelines provided by one unified department for all WMD grants.

I want to compliment the Department of Health and Human Services for the recent bio-state grant program. That was something that was—we could live with and it was really enjoyable to work with, believe it or not. I would recommend that other agencies copy this.

We need funding assigned for program management and equipment maintenance allowances. Most State agencies, local emergency management, and public health agencies are bare-boned. We have limited funds for planning and managing our daily activities, much less new initiatives. I would suggest that 10 to 20 percent of the grant funding be assigned for program management and equipment maintenance. Unfortunately, Federal programs have provided funds for training, planning, and purchasing, but it stops there. If the Department of Homeland Security doesn't follow through with a program that assists the locals and the States with this managing and planning, I'm afraid several years from now it will be like the old civil defense disaster package hospitals, sitting somewhere rotting, unable to be used.

Unfortunately, terrorism is not going to go away, and we need to have continued support to organize a program and to manage it. We need grants that are awarded at 100 percent, not matching funds. I've heard rumors that FEMA's 2003 grants for WMD are going to be on a 75/25 basis. I can tell you that in North Carolina, and I suspect many other States, we can't support this. We are having troubles on a day-to-day basis.

For the National Disaster Medical System and the Public Health Service Office of Emergency Preparedness, it's finally been recognized by Congress in the bioterrorist bill, and I really want to thank you all for that support. It really was greatly needed.

Until recently, NDMS has had little funding, has inadequate staffing and accountability and minimum recognition from DHHS on a regular basis. In years past, some snidely referred to NDMS as the No Damned Medical System. This is no longer true, sir. NDMS responds to help local and State governments when they are overwhelmed with many crises, natural and man-made. Hurricanes, floods, earthquakes, air crashes, animal events, the recent avian influenza, and terrorism. The network of volunteers who step up to the plate and become part-time employees of the U.S. Public Health Service in a crisis has really been helpful.

I have a problem right now, an example with the Federal Team, a WMD issue, which is my team, which is the National Medical Response Team-East, housed in North Carolina, just received one-sixth of our operating budget for 2001/2002. It—to actually get the money appropriated by Congress, I had to get assistance from my Congressman to get HHS to turn the money loose. We were borrowing the money from a non-profit organization to support a Federal team for basic operating expenses. Eight months into the Federal fiscal year, I considered canceling planned training activities because we just could not afford to continue supporting a counterterrorist type team. And if it's not important after 9/11, when will it be?

In closing, you have got to have support for planning and training and maintenance. We need to consider natural and man-made disasters that overwhelm a State or local government. We need not to reinvent a wheel. The FEMA's Federal response plan has been around a while, and it's taken a good while for everybody, including the Federal Government, to fully understand it. It's got a counterterrorist or a terrorist annex since PDD 39 came about, and I believe everybody started going along with it now. Now, if we start something totally new and try to reinvent a wheel, it's going to be another 3 or 4 years at best before it's understood, and we will again have the same problem on a day-to-day basis with interoperability not present; and then in a crisis, whether it be by electronic or face-to-face, we will have a problem.

This needs to be fixed. And I want to thank you for paying attention to it.

[The prepared statement of Llewellyn W. Stringer, Jr. follows:]

PREPARED STATEMENT OF LLEWELLYN W. STRINGER, JR., MEDICAL DIRECTOR, NORTH CAROLINA DIVISION OF EMERGENCY MANAGEMENT, DEPARTMENT OF CRIME PREVENTION AND PUBLIC SAFETY

Mr. Chairman and Members of the Committee, thank you for inviting me here today to discuss the issue of the establishment of a Department of Homeland Security. I am Dr. Lew Stringer, Medical Director of the North Carolina Division of Emergency Management, Department of Crime Prevention and Public Safety. I have a long history of emergency management experience that ranges from services as a local EMS Medical Director for 28 years, Director of the Special Operations Response Team-a disaster organization in North Carolina and involvement with the National Disaster Medical System through the Office of Emergency Preparedness, USPHS since 1990. In October, 2001 I spoke to this committee on WMD issues as it affected the local, state and federal response community.

I am back today to address the issue of how a single homeland security department could affect local and state governments and the Office of Emergency Preparedness/ National Disaster Medical System. During the preparation of my statement and in discussions with my local, state and federal cohorts, this focus become the *issues of "interoperability"*.

I have focused on chemical, biological and radiological response activities, as I know them to be, and have chosen 3 areas of focus: 1. Grants and funding; 2. Preparedness and planning at all levels; and 3. Response efforts.

In my position in North Carolina, I have been involved for several years in the "Grant Process" which includes: applying for grants, explaining the grant requirements to state and local stake-holders, and trying to manage the many different "pots of money" dangled in front of my state by CDC, FEMA, OJP, DHHS and others. (They) all have different requirements, different time tables, different folks to meet with, and different ways to figure out how to be successful. In North Carolina, we are striving to develop a SINGLE, UNIFIED terrorist plan that must bring all the varied state and local agencies together by developing, planning, equipping, training, and exercising strategies into a single unified plan.

Planners in emergency management need a *standard set of guidelines,* provided by one unified department, for all WMD grants. I want to compliment the DHHS on the presentation of requirements for the recent Bioterrorism state grants—others may wish to adopt their guidelines.

We need *funding assigned for program management and equipment maintenance allowances.* MOST state agencies—local emergency management and public health agencies—are "bare boned". We have limited funds to plan or manage our day-to-day activities; much less manage new entities. I would suggest that 10-20 % of the grant funding be assigned for program management and equipment maintenance. Unfortunately, federal programs have provided money for terrorist planning/training and purchase of equipment but have stops there. Otherwise, Homeland Security planning will follow the same path as the old Civil Defense Packaged Disaster Hospital Program—nonfunctional, and useless—if needed in several years. Unfortunately, the need for terrorist preparedness will not go away and support for preparedness must be on going.

We need *grants awarded at 100% and not require matched funding.* I have heard rumors that the 2003 FEMA Domestic Preparedness grants will be at awarded at matching 75/25%. I can tell you that in North Carolina, and I suspect many other states, we can not afford that type of "support".

The National Disaster Medical System (NDMS) and the Public Health Office of Emergency Preparedness (OEP) have recently been officially recognized by Congress in the Bioterrorism bill signed on June 12th. I want to thank you on this committee for your efforts. Until recently, NDSM had little funding, inadequate staffing and accountability, and minimal recognition from DHHS. Some snidely referred to NDMS as *No Damn Medical System*. This is no longer the case. NDMS responds to help state and local governments when the locals become overwhelmed by natural or man-made disasters—hurricanes, floods, earthquakes, air crashes, animal events such as the avian influenza outbreak and terrorist events. The network of volunteer personnel who become temporary employees of the USPHS and respond has been gratifying, especially since September 11.

Let me give you an example of my funding distribution problems:

My Federal WMD team, Nation Medical Response Team-East, housed in North Carolina, has just now received the first ⅙ of our operating budget for 2001-2002. To actually get the money, appropriated by Congress for OEP, I had to request assistance from my Congressman to get the DHHS moving. My Federal team had to use the monies of a non-profit organization, Special Operations Response Team's emergency contingently funds, for a federal team's basic operating expenses. Eight months into the federal physical year, I considered canceling planned training because of the lack of released funding. Since 9/11 certainly, this type of team has never been more needed.

Now that you (Congress) have officially recognized OEP/NDMS and created an Assistant Secretary for Public Health Preparedness, who will direct OEP/NDMS, I am hopeful that such *funding distribution issues* will be resolved. Moving Public Health Preparedness, OEP and NDMS into the Department of Homeland Security should improve these funding distribution issues.

I suspect that others, involved in response, are also looking forward to 100% coordination of efforts—for planning, funding and direction from individuals who are tasked by Congress and our President—to be 100% sure that services are 100% ready to make secure our homeland.

We, in state government, need be confident in knowing that a *coordinated, unified Federal response to natural or man-made disasters will continue* under the Department of Homeland Security. The ground work was begun years ago by FEMA with what is called the Federal Response Plan, (FRP). Federal departments, offices and other Federal entities come under, or are responsible for various emergency support functions, when the Stafford Act is declared. As you know, the Federal Response Plan has a Terrorist Annex since PDD 39, which further defines the functions of crisis and consequence management roles. This plan is fairly well understood by states and is followed by federal, state and local governments.

If the primary consequence offices and agencies, as well as some of the law enforcement entities, are moved from the departments where they now function and are placed under the steadfast management of the Department of Homeland Security, this should improve efficiency, simplify the annual budget process, and reduce redundancies and interagency competition.

A downside for a Department of Homeland Security could be a failure of a service formally provided by the old agency plan not to be honored under the new department plan. For example, when OEP requests from DHHS the temporary assignment of USPHS Commissioned Corp personnel, will that request be honored by DHHS when OEP is no longer under DHHS?

It is critical that in the new department, there must be *a prominent health care focus*. Many of the critical services needed in man-made or natural disaster are health care issues. There needs to be an adequate physician presence-not just a health care administrative presence "to guide the Secretary in health care issues at all levels of department operation.

In closing, I would like to speak about the critical need for *communication "interoperability"*, which has become a "buss word" in Washington and in the state governments. In any emergency, first responders need to be able to communicate with other first responders, i.e. fire with medical, ambulance with police, and all with other agencies who become involved. Mutual aid and the federal response compounds the communication problem by brings more folks who need to talk with each other together. Communications issues have been mentioned in every disaster after-action report I have seen for years. These communication issues involve cost for locals. This will be a huge planning and funding issue that the Department of Homeland Security must address.

I sincerely hope the new Department of Homeland Security will resolve or, at best, improve many "interoperability" issues existing today. The task will be daunting. In these difficult times, the aim should be to make all of us successful.

Mr. GREENWOOD. We thank you, Dr. Stringer, for your testimony again.

Mr. Plaugher.

TESTIMONY OF EDWARD P. PLAUGHER

Mr. PLAUGHER. Good afternoon, Mr. Chairman, and members of the committee. I am Edward Plaugher, chief of the Arlington County Fire Department in Virginia. I appear today on behalf of the Washington, DC Area National Medical Response Team, of which I am its executive agent.

I would like to begin by thanking the committee for having me here today. Issues related to terrorism and related preparedness efforts have taken on a new meaning in our Nation. Our region has been engaged for the previous 5 years prior to the events of September 11 in educating Federal policymakers as to the role of fire and emergency services in mitigating acts of terrorism. The men and women of my fire department were joined by thousands of others from the Washington, DC and New York metropolitan areas in demonstrating that role last fall. I believe we owe it to them and to the public safety good to move forward as quickly as possible in fashioning the most rational and workable national terrorism preparedness policy as soon as possible. The public safety and the memory of 343 fallen firefighter colleagues in New York City demand no less.

Since its inception, the Office of Emergency Preparedness, U.S. Public Health Service, Department of Health and Human Services, has provided an invaluable contribution to the first responder community within our Nation. Creating and supporting the Disaster Medical Assistance Teams, DMATs, the Metropolitan Medical Response System, and the National Medical Response Teams, has provided not only financial support, but leadership and direction in the most critical aspects of disaster response, that is, the aspect of emergency medical care. It goes without saying that without this program, our Nation would not be as prepared as we are today to deal with both man-made and natural disasters.

Long-range relation ships have been developed, and they are vital to the success of the program. As we found on September 11, it is the upfront work that pays dividends during the emergency event.

In addition, the last 6 years has seen the development of both public and local assets under the direction of OEP. These local assets, the Metropolitan Medical Response System, are designed to deal with the consequence of weapons of mass destruction incidents. Each NMRS has an order to get OEP funding that's been required to develop these very critical pre-incident relationships, bringing to the table disciplines who routinely do not work together, but during a disaster or terrorist event must not only work together, but they must do so in a seamless manner. Sacrificing any part of this long-term relationship building and seamless response to medical emergency management must not be allowed to vanish.

Hence, my position on moving OEP to the new Department of Homeland Security is somewhat tied to building upon a well-laid foundation and not allowing this foundation to erode.

I have seen the vast matrix of Federal programs, that is, the good, the bad, and the ugly. Direct relationship-building and financial support for local asset-building has produced outstanding results for emergency medical preparedness. OEP's and NMRS' system has provided this focus and is good.

I have also seen the Department of Defense via the Weapons of Mass Destruction Act of 1996 provide training and exercises but fail to develop lasting relationships within a community or a city. That is bad. And it continues to miss the mark now as a Department of Justice program.

The ugly that I am referring to is the State and local assistance program currently under way at the Department of Justice. Even though well intended by Congress and meaningful in amounts, over $100 million a year since Federal fiscal 1999, almost none of the support has reached the first responder community. Utilizing the States as a funding mechanism has not, and I believe will not, work as intended. As the police chief, my colleague in Arlington County, Edward Flynn, relates, terrorism is a global act with local response.

Back to the concept of transferring OEP to homeland defense. If the transition of the relationship-building cornerstone crumbles, the transition is a giant step backwards. Local response is built on managing a wide array of assets, which is best accomplished in an atmosphere of trust.

On the other hand, if more direct assistance is provided to local first responders with the State in the loop to provide uniformity between States and within States but not as a controlling element or as a barrier to assistance, then homeland defense and OEP could benefit from a single departmental alignment.

Again, I want to thank the committee for giving me this opportunity to testify, and I look forward to your questions.

[The prepared statement of Edward P. Plaugher follows:]

PREPARED STATEMENT OF EDWARD P. PLAUGHER, FIRE CHIEF, ARLINGTON COUNTY, VIRGINIA

Good Morning/afternoon, Mr. Chairman and members of the Committee, I am Edward Plaugher, Chief of the Arlington County Fire Department. I appear today on behalf of the Washington, D.C. area National Medial Response Team (NMRT).

I would like to begin by thanking the Committee for having me today. Issues related to terrorism and related preparedness efforts have taken on new meaning in our nation. Our region was engaged for five years prior to the events of last September in educating federal policy makers as to the role of the fire and emergency services in mitigating acts of terrorism. The men and women of my fire department were joined by thousands of others from the Washington, DC, and New York metropolitan areas in demonstrating that role last fall. I believe we owe it to them and to the public safety to move forward as quickly as possible in fashioning the most rational and workable national terrorism preparedness policy as is possible. The public safety and the memory of 343 fallen firefighters in New York demand no less.

Since its inception, the Office of Emergency Preparedness, U.S. Public Health, Department of Health and Human Services has provided an invaluable contribution to the first responder community within our nation. Creating and supporting the Disaster Medical Assistance Teams (DMATS) and the National Medical Response Teams (NMRTs) has provided not only the financial support but the leadership and direction in the most critical aspect of disaster response emergency medical care. It goes without saying that without this program our nation would not be as prepared as we are to deal with both man-made and natural disasters. Long range relationships have been developed and are vital to the success of the program. As we found on September 11th it is the up front work that pays dividends during an emergency event.

In addition the last six years has seen the development of public and local assets under the direction of OEP. These local assets, the Metropolitan Medical Response Systems are designed to deal with the consequences of weapons of mass destruction incidents. Each MMRS has, in order to get OED funding, been required to develop these very critical pre-incident relationships. Bringing to the table discipline who routinely to not work together but during a disaster of "terrorist" event must not only work together but must do so in a seamless manner. Sacrificing any part of this long-term relationship building and seamless response to medical emergency management must not be allowed to vanish. Hence my position on moving OEP to the new department of Homeland Security is somewhat tied to building upon a well-laid foundation and not allowing this foundation to erode. I have seen with the vast matrix of federal programs "the good, the bad and the ugly". Direct relationship building and direct financial support for local relationship building has produced outstanding results for national medical preparedness without this the ability of the first responder community is greatly diminished. I have also seen the Department of Defense via the "Weapons of Mass Destruction Act of 1996" provide training exercise and expertise but fail to develop lasting relationships within a community or city. That is bad and it continues to miss the mark now as a Department of Justice program. The ugly I am referring to is the State & local assistance program currently under way in the Department of Justice. Even though well intended by Congress and meaningful in amounts, over 100 million a year, almost none of the support has reached the first responder community. Utilizing the states as the funding mechanism has not, and I believe will not, work as intended. As the police chief in Arlington County, Edward Flynn relates "Terrorism is a global act with local response".

Back to the concept of transferring OEP to Homeland Defense; if in the transition the relationship building cornerstone crumbles, the transition is a giant step backwards. Local response is built on managing a wide array of assets, which is best accomplished in an atmosphere of trust. On the other hand, if more direct assistance is provided to local first responders, with states in the loop to provide uniformity between states and within states, but not as a controlling element or as a barrier to assistance, then Homeland Defense and OEP could benefit with a departmental alignment.

I want to thank the committee for giving me the opportunity to testify and look forward to your questions.

Mr. GREENWOOD. The Chair thanks the gentleman. And, for your information, that is not a national alert; that is an indication that we have a series of votes on the floor. For the benefit of the members of the committee, what we will try to do is in the next 15 minutes allow each of the three of us to ask questions, and then we will be able to excuse this panel; and then we will take a brief recess until the next panel comes up.

And let me ask a question, and I would ask, starting with Mr. Plaugher and going to my left, with the exception of Ms. Heinrich, because I have another question for her, this question:

How ready do you think the labs and NDMS teams are today, and the other Federal response assets are, to respond to a true radiological or nuclear incident such as a dirty bomb? Are we sufficiently prepared and adequately organized to handle the threat now? And will the new proposal help improve such preparedness? So if somebody detonated a dirty bomb in Arlington, Virginia tomorrow morning and you had dead bodies and you had people wounded and you had people potentially exposed to radiological materials, how ready are we today, Mr. Plaugher, and how do you see that improving with this legislation?

Mr. PLAUGHER. I think we are very far off the mark as far as for preparedness for a dirty bomb. I think we have focused most of our energy on chemical, and we are now starting to focus on biological. We have yet to begin the preparedness of the nuclear program, and it's just been a matter of assets and resources. We had to start somewhere. I personally chose to focus on chemical attack because of the incident in Tokyo, Japan, and the similarities between our system and their system and what we thought was the likelihood of event.

We have also done a great deal of preparedness for conventional weaponry. So if it's dirty bomb with conventional weaponry, we will have some resources and capability to manage that piece of it. But as far as for the other levels of preparedness, we still have a long way to go.

Mr. GREENWOOD. Does this bill help us get there?

Mr. PLAUGHER. I think this bill will provide us with more focus, which I think is much needed. Coalescing these long-term relation ships, I've heard wonderful testimony today about the NEST teams and about their ability. I do have a relationship with a NEST team in the area, the one out of Andrews Air Force Base.

So there is some capacity and some response capability. But, remember, I'm in the 4-minute business. I've got to make changes in the first 4 to 10 minutes of that incident scene, so I need that equipment and training and capability there immediately.

We just received recently some new radiological monitoring from the Commonwealth of Virginia. So, I mean, we are working in that direction, but we still have a ways to go.

Mr. GREENWOOD. Briefly, Dr. Stringer.

Mr. STRINGER. From a local and State standpoint, we've got a long way to go. As far as the NMRTA is concerned, I think being under the Homeland Security will allow some interopability and get to know the folks better, and I think that should help us in any type of response, bringing in Federal assets to assist a local government.

Mr. GREENWOOD. Dr. Cobb.

Mr. COBB. Two quick comments. One, NEST has been focused since its inception on prior information, and also focused more on the higher-end threat, namely, a stolen or improvised nuclear device. That's one point.

The second point is that while it's recognized that the bolt-out-of-the-blue could happen, and we are moving in that direction, something called the Triage program, I think discussing that capability would be better in a different environment.

Mr. GREENWOOD. Very well.

Mr. Nokes?

Mr. NOKES. Let's see. One answer is the Operation America that Sandia conducted in Portsmouth, Virginia last month, where we had about 100 first responders, including many from the Washington, DC area, teaching them advanced bomb dismantlement techniques. So if the device had not exploded, perhaps the folks who had that training would have an advantage trying to render it safe. If it had already detonated, the effects are variable. They go from almost nothing to very tragic. And so it depends a lot on what the effects were. But, as Don said, the labs have been practicing for the very serious end of that experience, a nuclear weapon, and mostly radiation devices are within that envelope of practice.

Mr. GREENWOOD. Dr. Vantine.

Mr. VANTINE. Mr. Chairman, if an RDD went off in Arlington yesterday, we've already failed. I think the new department can help in two ways: It can help regulate the materials at the source, and it can help detect the materials before this event ever happens.

Mr. GREENWOOD. Ms. Heinrich, very quickly. Do you think we need better coordination between bio, the HHS and this new department in order to be prepared for this kind of an event?

Ms. HEINRICH. I think that the proposal for the most part is broadly stated, and I think that we have to have clarification on, as we have heard here before, the roles and responsibilities. It's not always clear what the control command relationships are. So, I think we need more information.

Mr. GREENWOOD. Thank you. The gentleman from Florida, for 5 minutes.

Mr. DEUTSCH. Thank you, Mr. Chairman. And I'm going to just ask one question, and yield to my colleague from Colorado just not to have to keep you around for about another 45 minutes.

Chief, you seem quite satisfied with your relationships with HHS and FEMA. Except for the possibility of getting more money, is there any reason to move these emergency response activities into a new department?

Mr. PLAUGHER. Well, one of the things that the fire services has said repeatedly to Congress is that we need a national strategy, we need a national focus. And the coalescing of that—and that is all of its subparts—into a single agency has tremendous benefit to first responders in the development of a national strategy. I'd just ask, as this goes forward, that you allow the first responders to have some opportunities for dialog and input into that national strategy. I mean, we are the folks that are going to be there, we are the folks that are going to have to manage the incidents.

So I think it does have some solid purpose and benefit, because we have seen the absence of a national strategy because of the splintered approach to date from the Federal agencies.

Mr. DEUTSCH. Thank you. I'd yield to Ms. DeGette.

Ms. DEGETTE. Thank you very much for yielding.

Mr. Plaugher, let me just comment on what you are saying, because I had a meeting in my district, which is Denver, with the local first responders and the representatives of the State, and they even have a difficult time figuring out who should be giving them directions between the State and the local first responders, much less coordination in urban areas like my district between all the counties that are involved. And I think you are right, there needs to be some kind of directions, so long as it's not, you know, just another bureaucratic layer. I really appreciate what you are saying.

I just have a couple quick questions for Mr. Nokes and Drs. Cobb and Vantine about the labs. First of all, how will the new Department of Homeland Security make the deployment of the technologies that the labs are developing easier to deploy in the field?

Mr. VANTINE. I think what happens is that when DHS starts funding the program and putting it together, they are going to work the whole issue of the systems issue. And so when the technology goes to field, it's going to be already coordinated with local response, regional response, and national response. It's going to be an integrative package. It's going to be vetted at the national laboratories to make sure that it works. It's going to have the best technology. So it will be a package that we put out in the field rather than pieces.

Ms. DEGETTE. And you think under the current structure of the Department of Homeland—or, of what's happening now, it's just in pieces? It's not coordinated?

Mr. VANTINE. I think right now we rely on largesse of other programs. They do R&D in their areas, we take that and try to apply it to this problem, but we don't have the resources to put the technology that we really need on the problem.

Ms. DEGETTE. So you envision that what this department would do, then, would be to take that technology and bring it all together?

Mr. VANTINE. Exactly. That's exactly right.

Ms. DEGETTE. The other two, any additions?

Mr. NOKES. I would make one comment, and that is, right now, as you well know, no one owns the problem and so everybody has a piece of it, and so we have a very tactical fragmented approach of applying technology to the issues. And I would hope that the new department is able to pull together the requirements across the—what are now different agencies and put together a coherent program, so you have good security that is uniform across the country and that would be the best thing.

Ms. DEGETTE. We haven't achieved that yet, have we?

Mr. NOKES. No.

Mr. COBB. Just a quick answer, over the past several months we have been working with NNSA anticipating the possibility that they'd be the lead Federal agency, or they'd have a major role in integrating the technologies. I think much of that is being transferred to the new department, that concept. We now have a lead Federal agency to develop the R&D, so that focus will help.

Ms. DEGETTE. Has this coordination that you all think is so essential, and so do I, and is that part of a specific proposal that you have seen or is that just your hope for what the new agency would show?

Mr. COBB. There has been discussion, but I don't think it is in the framework of a specific proposal. Obviously, the legislation is very broad so the details still have to be worked.

Ms. DEGETTE. Right, and I think that is all of our issues here today. And without, you know—without asking you specific details of how this would work, do you expect you will be consulted on how this coordination can be implemented in a plan?

Mr. VANTINE. I guess I would answer that I think we are in a negotiation stage right now as to how that is going to work. We are trying to talk to Congress and to the different agencies in trying to put together the package of how it is going to work. As you have issues with it, I think we have issues with it too. We don't see the details and I think they will be worked out over time.

Ms. DEGETTE. That is always true when you're talking about a big bureaucracy, the devil is in the details.

Mr. NOKES. I think, one more comment, as I look at the legislation, I see that science and technology is in the infrastructure under the Secretary's office, and the other Under Secretaries don't appear to have a science or technology advocate. So I think you might want to have a chief scientist, or somebody at the top that looks down at all of the technology requirements and makes resource allocation and priority judgments.

Ms. DEGETTE. That was very helpful and now we have to go vote.

Mr. GREENWOOD. The Chair thanks the gentlelady and the Chair thanks each of our witnesses for lending your expertise to this most vital effort and thank you again. You are excused. The Chair would note that we do have series of votes and the committee will recess until 1:35 and then we will bring forward the fourth panel.

[Brief recess.]

Mr. GREENWOOD. The committee will come to order, and we thank our witnesses, and they are Mr. Philip Anderson, Senior Fellow at the Center for Strategic and International Studies, Dr. Ronald Atlas, President-elect of the American Society for Microbiology and Dr. Tara O'Toole, Director of the Center for Civilian Biodefense Studies at Johns Hopkins University and thank each of you for being with us this morning and for your forbearance in waiting for us. You are aware that this is an investigative hearing and that when holding an investigative hearing, it is the practice of this committee to take testimony under oath. Do any of you have objection to giving your testimony under oath? The Chair would also then advise you that pursuant to the rules of this committee and the House, you are entitled to be represented by counsel. Do any of you require or ask to be represented by counsel. In that case if each of you would stand and raise your right hand.

[Witnesses sworn.]

Mr. GREENWOOD. Thank you. You are under oath, and Mr. Anderson we will start with you, and you're recognized for 5 minutes to give your opening statement.

TESTIMONY OF PHILIP ANDERSON, SENIOR FELLOW, CENTER FOR STRATEGIC AND INTERNATIONAL STUDIES; RONALD M. ATLAS, PRESIDENT-ELECT, AMERICAN SOCIETY FOR MICROBIOLOGY; AND TARA O'TOOLE, DIRECTOR, CENTER FOR CIVILIAN BIODEFENSE STUDIES, JOHNS HOPKINS UNIVERSITY

Mr. ANDERSON. Good afternoon, Mr. Chairman, and members of the committee. It is an honor to be with you today to provide my views on vulnerabilities and response capability at the Federal, State and local levels for consideration in addressing the President's proposal to establish a Department of Homeland Security. The slide behind me depicts an area of contamination in the District of Columbia resulting from a detonation of a radiological dispersion device, an RDD, a dirty bomb, detonated on the National Capital Mall area in the area of the Air and Space Museum.

CSIS conducted in-depth research and developed this realistic cross-jurisdictional crisis scenario with the purpose of helping to frame the planning requirement for the Metropolitan Washington Council of Governments, led crisis planning effort by identifying some of the key issues and friction points that needed to be addressed. The exercise portrayed the complexity associated with command control and communications between Federal, State and local government and the private sector and the general public.

The exercise participants included mid to upper level decisionmakers and regional planners from the District of Columbia and other local jurisdictions as well as representatives from FEMA and the FBI. The results of this research effort and the scenario were also presented to the senior leadership of the New York City Police Department. The scenario that was employed involved an explosive dispersal device laced with radioactive Cesium 137. The scenario included expected casualty rates, critical infrastructure damage assessments, and effects across critical key infrastructure.

The addition of a radiological event pushed the recovery portion of the scenario well beyond the scope of the exercise, but it did generate additional thought with respect to future planning. It's important to note that nowhere else in America do the people charged with addressing emergency response and recovery face a more daunting challenge than in the District of Columbia. Nonetheless, the presence of radioactivity was an issue that the participants were clearly not prepared to deal with. This would seem to indicate that the greater Washington region could be prepared for unconventional terrorist attacks involving materials that have the potential of contaminating large areas.

In the absence of well-developed plans and given the complex multi-layer jurisdictions within the greater Washington area, the actions of the Federal, State and local governments could combine to reduce the efficiency and effectiveness of emergency preparedness and response, particularly for unconventional attacks. If you were to ask most Americans to describe their greatest terrorism fears, chances are that they would suggest cataclysmic scenarios involving weapons of mass destruction, nuclear biological or chemical devices.

These views have been reinforced by the media and by the administration's recent spate of gloomy warnings. However, at present, there are significant financial and technical obstacles to terrorists obtaining and deploying effective weapons of mass destruction. There is, however, another category of attack that deserves at least equal attention from government, the private sector and public alike. Not just the high consequence, but very low probability weapons of mass destruction-type attacks or attacks on the opposite end of the spectrum involving a much higher probability, perhaps, a lone shooter or suicide bomber, but yet another category involving attacks that fall somewhere in the middle.

In retrospect, this mid-level space is where September 11 belonged and it is the space in which future terrorists will likely operate. Terrorist attack scenarios in this category typically involve unconventional tactics or weapons that include dirty bombs like that in the scenario we developed. From the terrorist perspective they assume widespread death and destruction is an unattainable goal. So they seek long-term disruption similar to that realized by the September 11 attacks.

Other examples include a well-coordinated attack involving multiple near simultaneous suicide bombings nationwide or targeting unsecured highly visible, nonnuclear aspects of energy infrastructure, very soft targets like oil refineries, petroleum or liquid natural gas terminals or perhaps tanker trunks. These types of unconventional attacks are achievable now and indeed well developed plans along these lines are probably already on the shelf.

Most importantly, although they represent real possibilities, their impact in many cases is far more psychological than real, real in terms of loss of life and injury. Facing up to these threats must not mean giving into fear. Even as a Nation develops defensive technologies from radiation and chemical and biological sensors to bomb sniffing devices, citizens must be equipped with the tools to protect themselves psychologically. An intensive program to create public awareness can help avert the panic and paralysis attacks like these aim to inspire. With the arrest last week of Abdullah al Muhajir, Jose Padilla, the would-be dirty bomber, the importance of educating our first responders and the public in general about the new dangers we face is more apparent than ever.

The response clean-up and recovery effort that would be required following a radiological attack for example, synchronized decisions at the Federal, State and local levels, as well as in the private sector must be fully thought through and incorporated in the comprehensive contingency plans. It is also important that long-term economic recovery plans be developed considering the implications of unconventional attack scenarios. The means to develop greater public awareness and acceptance of risks should be considered. As such scenarios that can be employed in table-top exercises and simulations should be designed and incorporated into the development and testing of plans to address the possibility of unconventional attacks.

While we would all like to believe that the dirty bomb scenario represents a remote possibility, the evidence points to the contrary. How real a possibility that a terrible event like this could happen remains to be seen, but it is clear that adequate preparation for unconventional attack is essential. Addressing all the possible terrorist attacks is a daunting challenge, but it is important to keep in mind that from a terrorist perspective, the challenges are far greater. To kill large numbers of Americans and destroy significant portions of critical infrastructure is extremely difficult. The terrorist must depend on psychological impact to achieve his objectives, disrupting the economy, breaking our spirit and reducing our confidence in our government.

By focusing on the most likely threats, increasing situational awareness and empowering first responders in the public with the knowledge they need, we weaken the terrorist arsenal as we strengthen our own.

Mr. Chairman, over the long term, considering this new and very dangerous environment, the President's proposal must be acted upon to ensure unity of effort and clear lines of authority, responsibility and accountability at every level to effectively address the enormous complexity of securing the homeland. The road ahead remains fraught with challenges yet to be addressed, and we at the Center for Strategic and International Studies are ready and willing to help. Organizing effectively to ensure the security of American homeland is essential to the safety of our country's citizens and to our prosperity as a Nation. We appreciate the committee's leadership on this issue and we look forward to helping in any way we can. Thank you very much.

[The prepared statement of Philip Anderson follows:]

PREPARED STATEMENT OF PHILIP ANDERSON, SENIOR FELLOW AND DIRECTOR, HOMELAND SECURITY INITIATIVE, CENTER FOR STRATEGIC AND INTERNATIONAL STUDIES

I. INTRODUCTION.

Good morning Mr. Chairman—Members of the Committee. It's an honor to be with you today, to present my views on "Creating the Department of Homeland Security: Consideration of the Administration's Proposal . . . focusing on chemical, biological, and radiological response activities proposed for transfer to the Department of Homeland Security." Let me begin by saying that the statement I am about to give represents my views and in no way should be taken as the institutional view of CSIS. Before beginning though, let me provide you with some background on the work we are doing at CSIS.

CSIS has completed a number of homeland security projects both prior to—and since the tragic events of September 11. In January 2001, CSIS released a report on the results of an eighteen-month study, *Homeland Defense: A Strategic Approach.* In June 2001, CSIS co-directed *Dark Winter,* a high-level simulation of a smallpox attack originating in Oklahoma City. In the immediate aftermath of September 11, CSIS convened an internal task force on terrorism, the results of which were published in *To Prevail: An American Strategy for the Campaign against Terrorism.* In March 2002, CSIS completed extensive research on the impact of a "dirty bomb" detonated on the National Capitol Mall. This in-depth research led to the development of a crisis-planning scenario which served as the basis for the Council of Governments led "Greater Washington Crisis Planning Workshop" which was held on March 21, 2002. The results of this research effort and the scenario were also presented to the senior leadership of the New York City Police Department on May 1, 2002.

Currently CSIS is completing a White Paper on the challenges associated with the creation of a Department of Homeland Security that will provide actionable recommendations for decision makers for consideration in this critically important debate. CSIS is also working on a simulation exercise, patterned after our Dark Winter effort, to focus on the vulnerability of U.S. energy infrastructure. Rather than consequence management, this simulation exercise will focus on the less understood—and explored—scenarios in which policymakers must decide on whether and how to act in the case of a credible threat against critical energy infrastructure.

II. OVERVIEW.

With the President's proposal to establish a Department of Homeland Security, there seems to be a renewed sense of urgency in Washington. When considering the number of threats we face from terrorists intent on doing us harm, this would certainly seem appropriate. The Nation is at war—a war that is occurring in many ways beyond the public's view. There can be no greater public recognition of this fact than the President's proposal to establish a Department of Homeland Security.

I was asked to address response capability at the federal, state and local levels for consideration in addressing the President's proposal. In this new and very dangerous environment, it appears that if enacted, the President's proposal would greatly simplify management processes and unify the efforts of the 46 federal agencies that, to varying degrees, have responsibility for Homeland Security. In addition, the President's proposal would seem to represent an effective starting point to ensuring the means to effective communication and coordination between the federal, state and local governments to ensure unity of effort and clear lines of authority, responsibility and most importantly, accountability.

III. THE CHALLENGES.

CSIS conducted in-depth research and developed a realistic crisis scenario to address a plausible—large—cross-jurisdictional crisis in Washington, DC. The overall purpose was to help frame the planning requirement for the Metropolitan Washington Council of Governments (COG) led crisis-planning effort by identifying some of the key issues and friction points to be addressed. The exercise portrayed the complexity associated with command, control and communications between federal, state and local government and the private sector/general public. CSIS facilitated discussions focused on how to resolve lines of communication, authority, and responsibility in an unconventional crisis environment.

The exercise was designed to present participants with a large-scale terrorist attack on downtown Washington, D.C. in order to facilitate discussion and identify questions to be addressed by a coordinated response plan. The exercise participants included mid to upper level decision-makers and regional planners from the COG task force working groups as well as from the District of Columbia and other local governments and from relevant agencies of the federal government to include FEMA and the FBI.

The participant's role was to assimilate the events unfolding and operate within their own committee framework to discuss and determine the actions/recommendations they would take forward to superiors in addressing the regional response to mitigate near term and long-term risks. The exercise was not designed to be a decision driven war game where actions/decisions were analyzed or critiqued against some ideal or textbook solution.

The comprehensive scenario that was employed involved an explosive dispersal device laced with radioactive Cesium 137. The scenario included expected casualty rates, critical infrastructure damage assessments, and effects across key critical infrastructure. The addition of a radiological event pushed the recovery portion of the scenario well beyond the scope of the exercise, but did generate additional thought with respect to future planning. The scenario was presented in three segments with the following questions providing the framework for discussion: What are the key decisions that have to be made? Who will make those decisions? What additional information do you need? Where do you propose to get this information? What are the critical interdependencies? Who will be the authoritative voice for the public? How will you communicate risk to the public?

During the first segment, the participants were not made aware of the radiation associated with the scenario and appeared to be comfortable with near-term response procedures for dealing with a conventional explosion and the resulting crisis. Overall, emergency response procedures and coordination requirements were familiar at this level, due in part to the events of September 11.

The necessity of having coordinated response procedures in place became clearer during the second segment of the scenario that provided the participants with the news that the bomb was in fact a "dirty bomb" that contained Cesium-137. The presence of radioactivity was an issue that the participants were clearly not prepared to deal with. Issues that the participants felt were critical to address at this stage were *whether to shelter in place or evacuate* the city, the requirement for the *President to declare Martial Law,* the possibility that *METRO might be forced to shut down* due to contamination, the *role of the media,* the *presence of radiation, emergency personnel augmentation,* and *protective gear requirements.*

The long-term implications of a radiological attack became increasingly clear and overwhelming as the third segment was introduced. The scenario presented participants with reports of deserted D.C. streets and hotels, workers refusing to return to work, and parents refusing to send their children back to schools that had conducted field trips to D.C. on the day of the attack. These reports were indicative of the deep, long-lasting psychological impact that a radiological attack could have. The public has an inherent fear of radiation, even though there is almost no danger of dying from exposure to this type of isotope—only the potential for long-term health implications in the form of increased cancer and cataract rates. The participants felt that issues associated with *long term economic impact and recovery* were critical to address in advance of this type of attack.

It is important to note that nowhere else in America do the people charged with addressing emergency response and recovery face a more daunting challenge than in Washington, D.C. Nonetheless, the presence of radioactivity was an issue that the participants were clearly not prepared to deal with. This would seem to indicate that the greater Washington region could be unprepared for unconventional terrorist attacks involving materials that have the potential of contaminating large areas. In the absence of well developed plans—and given the complex, multi-layered jurisdictions within the greater Washington region—the actions of the federal, local and state governments could combine to reduce the efficiency and effectiveness of emergency preparedness and response, particularly for unconventional attacks.

IV. GENERAL RECOMMENDATIONS.

At the heart of any effort to establish a Department of Homeland Security is the requirement to address the likely threats. However, defining likely threats in this new environment is problematic in that they will likely derive from multiple sources with different objectives and various means to do us harm. Defining the threat is risky but absolutely necessary for developing plans to organize and allocate resources to address the myriad vulnerabilities that exist.

Later this summer, the White House Office of Homeland Security has said it will unveil a comprehensive national strategy to secure the United States from future terrorist attacks. Governor Ridge has emphasized that the strategy will be guided by a risk management philosophy, "focusing our resources where they will do the most good to achieve maximum protection of lives and property." A risk management approach is essential—but defining the threat, identifying critical vulnerabilities, and developing effective capabilities to address them are a daunting challenge.

With the arrest last week of Abdullah al Muhajir, the would-be "dirty bomber," the importance of educating our first responders and the public in general about the new dangers we face is more apparent than ever. If you asked most Americans to describe their greatest terrorism fears, chances are they would suggest cataclysmic scenarios involving weapons of mass destruction—nuclear, biological, or chemical devices. These views have been reinforced by the media and by the administration's recent spate of gloomy warnings. However, there are significant financial and technical obstacles to obtaining and deploying effective weapons of mass destruction (WMD). But since the consequences of a successful terrorist attack using such weapons would be devastating the government is correct to focus significant resources toward preventing these gruesome possibilities. There is, however, another category of attack that deserves at least equal attention from government, the private sector and the public alike: not just high consequence but very low probability WMD attacks or the less severe consequence but much higher probability suicide bomber attacks, but those attacks that fall in the middle. In retrospect, this mid-level space is where September 11 belonged, and it is the space in which future terrorists will likely operate.

Terrorist attack scenarios in this category are typically unconventional and include "dirty bombs" like the one described herein which employed conventional explosives laced with radioactive Cesium-137—which can easily be found in industry, hospitals and medical labs. Or terrorists could develop a well-coordinated attack involving multiple near-simultaneous suicide bombings nationwide. They could also target unsecured, highly visible, non-nuclear aspects of energy infrastructure—"soft" targets like oil refineries, petroleum or liquid natural gas terminals. These types of unconventional attacks are achievable now; indeed well developed plans along these lines are probably already on the shelf. But although they represent real possibilities, their impact in many cases is far more psychological than real—in terms of injury and loss of life. Facing up to these threats must not mean giving in to fear. Even as the nation develops defensive technologies—from radiation and chemical and biological sensors to bomb sniffing devices, citizens must be equipped with the tools to protect themselves psychologically. An intensive program to create public awareness can help avert the panic and paralysis attacks like these aim to inspire.

The response, clean up, and recovery effort that would be required following a radiological attack for example—that synchronize decisions at the federal, state, and local levels as well as in the private sector—must be fully thought through and incorporated into comprehensive contingency plans. It is also important that long-term economic recovery plans be developed considering the implications of unconventional attack scenarios. The means to develop greater public awareness and acceptance of risks should be considered. As such, scenarios that can be employed in tabletop exercises and simulations should be designed and incorporated into the development and testing of plans to address the possibility of unconventional attacks. While we would all like to believe that the scenario described herein represents a remote possibility, the evidence points to the contrary. How real the possibility that a terrible event like this could happen remains to be seen but it is clear that adequate preparation for unconventional attack is essential.

Addressing all the possible terrorist attack scenarios is a daunting challenge, but it is important to keep in mind that from the terrorist perspective, the challenges are far greater. To kill large numbers of Americans and destroy significant portions of critical infrastructure is extremely difficult. The terrorist must depend on psychological impact to achieve his objectives—disrupting the economy, breaking our spirit, and reducing our confidence in our government. By focusing on the most likely threats, increasing situational awareness and empowering first responders and the public with the knowledge they need, we weaken the terrorist arsenal as we strengthen our own.

V. CONCLUSION

Mr. Chairman, over the long term, considering this new and very dangerous environment, the President's proposal must be acted upon to ensure unity of effort and clear lines of authority, responsibility and most importantly, accountability at every level to effectively address the enormous complexity of securing the homeland.

Mr. Chairman, the road ahead remains fraught with challenges yet to be addressed. The Center for Strategic and International Studies is ready and willing to help. Organizing effectively to secure the American Homeland is essential to the safety of our country's citizens and to our prosperity as a nation. We appreciate the

Committee's leadership on this issue, and we look forward to helping in any way we can.

Mr. GREENWOOD. Thank you, Mr. Anderson.

Dr. Atlas for 5 minutes. You need to push the button on your microphone, sir.

TESTIMONY OF RONALD M. ATLAS

Mr. ATLAS. Chairman Greenwood, members of the subcommittee, we would like to thank you for inviting the American Society for Microbiology to testify on issues related to the administration's proposal to create the Department of Homeland Security. The ASM has submitted a written statement which I will briefly summarize. The ASM, which has particular expertise in biomedical research and public health protection, supports the establishment of a Department of Homeland Security that would have oversight, coordination and leadership functions for biodefense activities. We agree that the Department of Homeland Security should be established to serve the important function of integrating threat analysis and vulnerability assessments and to identify strategic priorities for preventative and protective steps that can be taken by other Federal agencies.

We believe that the Department of Homeland Security would be able to work with the Department of Health and Human Services and the National Institute of Allergy and Infectious Diseases to pursue highly managed rapid paced and even classified research and development projects, which are needed to defend against the threat of biological weapons. ASM thinks that having a strong science and technology component within the Department of Homeland Security is essential and would help provide critical linkage among the numerous mission agencies charged with science development.

By having a strong science component, the Department of Homeland Security would be able to play a vital role in coordinating, reviewing and evaluating scientific and technical programs relating to human animal and plant life. We need to recognize, however, that biodefense research is part of the continuum of the breadth of biomedical research aimed at protecting the Nation aimed at infectious diseases. This field is different than many other areas because of its duality and the high degree of overlap with the public health and biomedical research activities of the Nation.

We do not want to create a duplicative system. Rather, we want a seamless, integrated and highly coordinated biodefense response system. Therefore, ASM believes that it is critical that a scientific health organization, namely HHS, continue to prioritize and conduct Federal research relating to civilian human-related, biological, biomedical and infectious diseases. We feel it is important to distinguish between policy and planning guidance, which would be well served by the Department of Homeland Security and the responsibility and/or authority for the direction, control and conduct of scientific research, which should remain within HHS.

HHS and the National Institutes of Health are best qualified to establish biomedical research and development programs and to prioritize scientific opportunities and research. The National Institute of Allergy and Infectious Diseases bring to bear all aspects of biomedical research and full capability of science to achieve scientific advances and biodefense. The ability to build on the body of scientific knowledge underpins the capability of the United States to combat bioterrorism.

Because it is difficult to distinguish an introduced infectious disease from a naturally occurring one, the strategies to protect against either event in terms of new scientific and technical approaches are the same. Since 9/11, National Institute of Allergy and Infectious Diseases has rapidly accelerated work to protect the Nation against the threat of bioterrorism. This acceleration has occurred across the spectrum of scientific activities from basic research in microbial biology to the development of vaccines and therapeutics to research related to diagnostic system.

We fear that the proposal to transfer responsibility for biodefense research to the Department of Homeland Security could create unpredictability and loss of momentum for these research programs, would very likely divert money from research and would not be the optimum way to obtain the integrated work of the best scientific minds. It is clearly not the aim of the administration's proposal. We, therefore, feel that the HHS, in consultation and coordination with the Department of Homeland Security, should retain primary responsibility for accelerated biodefense research and development programs.

ASM also would leave primary responsibility for planning for such emergencies for the Centers of Disease Control and prevention. We do not want to create a separate public health system for biodefense. A public health emergency arising from biological causes public health authorities must determine the nature of the organism, distinguish between a bioterrorism attack and a natural event, and respond rapidly to the health threat.

Regarding the select agent registration program, the administration bill would transfer this and the enforcement programs of HHS to the new department. HHS currently has the scientific and institutional knowledge and expertise relating to dangerous biological agents, biosafety and biosecurity to administer the program, and ASM continues to believe that the CDC should be responsible for the select agent registration program, which is key to the development of the Nation's biodefense capability.

Further, the administration bill does not appear to transfer the select agent registration and enforcement programs newly assigned to the Department of Agriculture. ASM believes that coordination and the registration programs for agriculture and human agents is critical as was recognized in H.R. 3448. The proper administration of the select agent program must balance public concern for safety with the need to not unduly encumber legitimate research and diagnostic testing. We need an integrated program that adds protection against misuse of microbial resources.

Therefore, ASM is recommending that an interagency group with the involvement of scientific societies address the advisability of removing the select agent program from HHS authority. Finally, ASM's full testimony touches upon a number of other specific issues. These issues include management and oversight of the National Pharmaceutical stockpile and response to infectious disease outbreaks, be they natural or intentional and provisions relating to research programs and activities of the USDA and DOE.

Each of these specific areas merits careful review by this committee. In closing, I want to reaffirm ASM's commitment to working with the administration and the Congress to achieve the most effective and most efficient system in the world for research control and response to the threat posed by biological agents.

[The prepared statement of Ronald M. Atlas follows:]

PREPARED STATEMENT OF RONALD M. ATLAS, PRESIDENT ELECT, AMERICAN SOCIETY FOR MICROBIOLOGY

INTRODUCTION

The American Society for Microbiology (ASM) is pleased to testify before the House Energy and Commerce Subcommittee on Oversight and Investigations hearing on creating the Department of Homeland Security: Consideration of the Administration's Proposal with a focus on chemical, biological and radiological response activities proposed for transfer to the Department of Homeland Security (DHS). The ASM is the largest life science society with over 40,000 members and its principal goal is the study and advancement of scientific knowledge of microbiology for the benefit of human welfare.

The ASM has worked with the Administration, the Congress and federal agencies on measures to protect against biological weapons and bioterrorism. Most recently, ASM provided expert advice on provisions to expand the Biological Weapons Statute in the USA Patriot Act and on Title II of the Public Health Security and Bioterrorism Preparedness and Response Act of 2002, which expands controls on certain dangerous biological agents and toxins. ASM members are involved in research and public health initiatives aimed at eradicating the scourge of infectious diseases, which daily end the lives of thousands of Americans and tens of thousands around the world. Infectious diseases remain the major cause of death in the world for those under the age of 45 and particularly for children. They are the third leading cause of death in the United States.

The ASM considers it critical that the proposed DHS build upon existing science and technology programs that hold promise in the defense against bioterrorism and in the effort against deadly infectious diseases. We would like to focus our comments on issues that Congress should consider on how best to achieve this goal.

THE ROLE OF THE DEPARTMENT OF HOMELAND SECURITY

1. Role of science and technology in Homeland Security is Critical

The terrorist events of September 11 and the anthrax biocrimes reveal the need and complexity of homeland defense. The ASM, therefore, supports oversight, coordination and leadership for biodefense activities in a Department of Homeland Security (DHS). Given that science and technology will play a vital role in the biodefense of the nation, the ASM believes it is essential to establish a strong science and technology function in the DHS. This science component will provide the necessary linkage between the Secretary of Homeland Security and the numerous mission agencies charged with science and technology development.

2. The Department of Homeland Security has an important role to play in defending the nation against biological threats.

The DHS will have an important role in developing the nation's defenses against, and responses to, biological threats. The role of DHS should be to integrate threat analysis and vulnerability assessments and to identify priorities for preventive and protective steps to be taken by other federal agencies to protect the American public. The DHS can coordinate, review, and evaluate scientific and technical programs related to human, animal, and plant life. The DHS will be a proper governmental vehicle to coordinate and to integrate the expanded roles of mission agencies in bioterrorism related research. The important role of the United States Army Medical Research Institute for Infectious Diseases (USAMRIID) should be recognized and strengthened and it should interface with the proposed DHS.

It will be important to define the boundaries between DHS and the mission agency with major responsibility for protecting the nation's health, HHS. An appropriate coordination office or position should be established within DHS. One approach, for example, would be for DHS to establish a position or appoint a person with the appropriate scientific background who would report to both the DHS Secretary and the

HHS Secretary. That person would also work with the National Institutes of Health (NIH) and National Institute of Allergies and Infectious Diseases to ensure integration of threat and vulnerability analysis about bioterrorism. The goal, of course, would be mutually agreed upon research priorities that address threatening biological agents.

Other mechanisms and/or functions may be needed for HHS and DHS to serve the vital role of coordinating the pursuit of an integrated research and development agenda for counter-terrorism, including highly directed, high risk, fast-paced, classified projects, and to manage between research results and applications to develop and evaluate specific technologies and for procurement. For example, NIH/NIAID has already accelerated basic and clinical research related to bioterrorism to focus on "Category A" agents considered by CDC to pose the highest threat. Last fall, the NIAID conducted a study to show that existing stocks of smallpox vaccine could be diluted at least 5-fold to provide immediate protection in case of a smallpox attack. NIAID also accelerated screening of antiviral compounds for activity against smallpox and related viruses and accelerated development of a "new generation" bioengineered anthrax vaccine and a promising Ebola virus vaccine. It has launched seven new fiscal year 2002 initiatives to expedite biodefense research.

3. *ASM recommends that HHS continue to be responsible for the prioritization, direction, and conduct of federal research efforts related to civilian, human, health-related biological, biomedical, and infectious diseases.*

Pathogenic microbes pose a threat to national security whether they occur naturally or are released in a bioterrorism attack. Biodefense research is part of the continuum of biomedical research aimed at protecting the nation and the world against infectious diseases. The capability to develop countermeasures and interventions is directly related to information generated by biomedical research on pathogenic microbes and the host response to these microbes. Therefore, it is critical that federal research efforts related to civilian human health-related biological, biomedical, and infectious diseases should be prioritized and conducted by, and at the direction of, the Department of Health and Human Services (HHS).

It is important to distinguish between oversight functions such as policy and planning guidance and coordination, which would be served by the DHS and the responsibility and authority for the direction, control and conduct of scientific research. ASM recommends that HHS, a public health and biomedical research agency of unparalleled success, should continue to be responsible for the conduct and direction of scientific research.

The Administration's Bill recognizes the necessity that HHS conduct the research and development programs related to infectious diseases. Section 303(a)(1) of the Bill provides that the Secretary of DHS shall carry out responsibilities related to civilian human health-related biological, biomedical, and infectious diseases through HHS and the Public Health Service "under agreements with the Secretary of Health and Human Services, and may transfer funds to him in connection with such agreements." Section 301(2) of the Administration's Bill, however, gives DHS primary authority and responsibility for the conduct of national scientific research including "directing, funding, and conducting research and development" related to biological threats. Additionally, at Section 303(a)(2), the Bill provides that DHS, in consultation with HHS, "shall have authority to establish the research and development program, including the setting of priorities." The ASM believes that the proposed restructuring of program authorities in the Administration's bill will create unpredictability for research programs, will divert monies from research and will not be the best approach to achieving the goal of civilian biodefense, which requires the involvement of the best scientific minds and the support of excellent science based on merit review.

The HHS, the federal agency with the major mission for protecting the public health, is best qualified to establish biomedical research and development programs, identify scientific opportunities and the research approaches for ensuring that biodefense needs are met in the best way possible. The National Institute of Allergy and Infectious Diseases (NIAID) is best able to bring together all aspects of biomedical research and the full capability of science to ensure breakthroughs and advances of high quality for biodefense. The ability to build on the body of scientific knowledge underpins the capability of the United States to combat bioterrorism. For example, the national response mounted by NIH/NIAID to AIDS demonstrates the capability of science to respond to a threat. The response was based on years of accumulated scientific knowledge and biomedical research that had been well supported by Congress. The response to bioterrorism will require the same long-term dedication of financial resources and scientific talent.

The NIAID, working with the DHS, has the knowledge about scientific capabilities to respond to threats and vulnerabilities related to the biological sciences. It can identify the science and infrastructure relevant to the most pressing issues and take advantage of the most highly leveraged opportunities for research that can contribute to counter-terrorism solutions. Because it is difficult to distinguish an introduced infectious disease from a naturally occurring one, the strategies to protect against either event in terms of new scientific and technical approaches, including surveillance, prevention and response, are the same. There will be dual benefits for public health in that investment in research to develop new therapeutics, vaccines, antivirals, genomics, diagnostics, sensitive detection devices and innovative surveillance approaches for biological agents will carry over to public health breakthroughs for all infectious diseases.

The nation has already seen the ability of HHS to respond to bioterrorism. In the months since September 11, 2001, the NIAID has rapidly accelerated work to protect the nation against the threat of bioterrorism. This acceleration has occurred across the spectrum of scientific activities from basic research in microbial biology to the development of vaccines and therapeutics to research related to diagnostic systems. It is critical that this work continue to develop rapidly and efficiently without delay, disruption or loss of momentum.

A scientific health agency, HHS, rather than the nonscientific, nonpublic health DHS should have the principal authority for developing and prioritizing scientific and health related programs. Essentially, therefore, the ASM suggests reversing the responsibilities identified in Section 303(a)(2) of the Administration's Bill. HHS, in consultation and coordination with DHS, should retain responsibility for accelerated research and development programs, including prioritizing such projects

THE PUBLIC HEALTH SYSTEM FOR BIODEFENSE

The ASM is also concerned that the nation not create a separate public health system for biodefense. Therefore, the ASM would leave primary responsibility for planning for public health emergencies arising from biological causes with the HHS and Center for Disease Control. At the earliest possible moment after the outbreak of a contagion, it is critical to determine the nature of the organism and to distinguish between a bioterrorism attack and a natural event. Then, public authorities must respond rapidly and appropriately to the health threat that either one would present. The ASM believes CDC should be charged with these tasks.

Section 505(a)(2) of the Administration's Bill requires DHS to carry out these functions under agreement with HHS. Again, the ASM believes the important and appropriate role for DHS is to coordinate planning and development of programs and to lend technical assistance to the responsible agency. It is entirely appropriate for HHS to coordinate and consult with DHS. As with the direction and control of research, however, the primary duty and authority should remain with the scientific agency with the existing knowledge, experience, and expertise to fulfill the critical mission. A scientific person within the DHS with the appropriate public health background and reporting to both the DHS Secretary and HHS Secretary could work closely with the CDC Director to achieve mutually agreed upon public health priorities for bioterrorism preparedness and response.

ADMINISTRATION AND ENFORCEMENT OF THE PROGRAM FOR REGISTRATION FOR POSSESSION AND USE OF SELECT AGENTS

Agriculture, the food supply, and the environment are potential targets of bioterrorism along with humans. It is important, therefore, to integrate and coordinate programs related to human, animal, and plant agents. Section 302(a) of the Administration Bill transfers to DHS the select agent registration and enforcement programs of HHS. However, it does not transfer the select agent registration and enforcement programs of the Department of Agriculture to the DHS. Subtitle C of the Public Health Security and Bioterrorism Preparedness Act of 2002 mandated coordination of activities of HHS and the Secretary of Agriculture regarding "overlap agents"—that is, agents that appear on the separate lists prepared by HHS and Agriculture. Without doubt, such coordination must occur. Bioterrorism research and surveillance extends and applies to infectious disease and select agent research. The ASM believes that integration of the select agent registration program inevitably will assist in the creation of an efficient registration process thereby expediting registration.

The proper administration of the select agent program is key to the development of the nation's biodefense capability and response and must balance the concerns for public safety with the need to not unduly encumber legitimate scientific research and laboratory diagnostic testing. The ASM continues to believe that HHS has the scientific and institutional knowledge and expertise related to dangerous biological agents, biosafety, and biosecurity in microbiological and biomedical laboratories and that it is best qualified to achieve the goal of protecting the public health and safety without interfering with research, and clinical and diagnostic laboratory medicine. Transferring this program to DHS raises many questions with regard to the administration of this program which must be carefully considered by Congress, which recently enacted new legislation and additional requirements for select agents. The ASM, therefore, requests that a review be done by an interagency group with the involvement of scientific societies to assess the advisability of removing the select agent program from HHS authority.

EACH TRANSFER OF A SCIENIFIC FUNDTION SHOULD BE SPECIFICALLY REVIEWED

Some additional specific measures in the Administration Bill require further consideration and comment by the ASM. The ASM continues to study the Administration Bill to evaluate the best approach to achieving expedited research that advances the defense against bioterrorism but does not dilute the continuing, critical battle against naturally occurring infectious diseases. The ASM suggests expeditious review of the appropriateness of each transfer of a facility or responsibility related to biological organisms from an existing agency. Similarly, the proposed transfers within the USDA should be carefully reviewed, in particular the justification should be considered for transferring Plum Island which addresses animal diseases but not incorporating the equivalent functional unit that addresses plant diseases.

For example, as noted above, the defense against bioterrorism must be fully integrated into the nation's public health system that is led by the Centers for Disease Control and Prevention. Currently, CDC would use the national pharmaceutical stockpile in response to infectious disease outbreaks—both natural and intentional. Sections 501(3)(B) and 502(6) would transfer the Strategic National Stockpile to DHS. Such transfer should be reviewed carefully during further consideration of the Bill. HHS should be responsible for developing the materials in the stockpile. Therefore, it seems appropriate for HHS to continue management of the stockpile. The ASM, however, understands the coordination and oversight function envisioned for DHS, and the final resolution of the management of the stockpile ultimately must depend upon the resolution of the scope and role of DHS responsibilities and activities. At this time, we also recommend that there be an external review of the CDC to ensure optimal preparedness for public health emergences and bioterrorism and to ensure appropriate integration with existing programs.

CONCLUSION

We appreciate the opportunity to present this testimony. The ASM is committed to working with Congress and the Administration to achieve the most efficient and effective system in the world for research, control, and response to the threat posed by biological agents.

Mr. GREENWOOD. Thank you, Dr. Atlas.

Dr. O'Toole for 5 minutes.

TESTIMONY OF TARA O'TOOLE

Ms. O'TOOLE. Thank you, Mr. Chairman. I am a physician and a public health professional by training, so I am going to restrict my remarks to those aspects of the proposed new agency related to bioterrorism activities. First, I would like to say that I support the President's call for a new agency dedicated to homeland security. We are also extremely admiring of the President's and the administration's initiatives on bioterrorism, particularly over the past year. I think that the R&D initiative situated in NIH as well as the funds now going to State and local health departments for public health preparedness reflect the President's recognition of the importance of the bioterrorism threat as well as the unique nature of this threat and the necessary response.

That said, however, I think the proposed reorganization as it pertains to bioterrorism functions raises several serious issues, and I would like to suggest some of them to you today. As I look at the proposed new agency, it appears to be a tiny island of bioscience, public health and medical functions around bioterrorism concerns within a very large ocean of more traditional national security and law enforcement functions. This worries me.

First of all, my understanding is that the rationale for consolidating many of these other border security type functions into a single agency is to improve coordination, cooperation and collaboration amongst similar functions and to get them all under one roof. In the case of bioterrorism programs, however, we would not be consolidating public health and bioscience research functions, we would be splitting them out to a new agency. This raises the specter of either, as Dr. Atlas suggested, having to create redundant parallel programs in homeland security in order to have enough leadership to figure out what to do in these areas and do it properly, or leaving one of the other agencies, either HHS or homeland security with insufficient robustness and expertise to carry out these important and difficult tasks.

The second problem that is raised by the proposed reorganization is the question of talent. The Hart-Rudman report talked quite eloquently about the crisis of competence that the Federal Government is facing, and it noted that in particular, we have failed to attract people with science and technology backgrounds into the Federal service. This is a problem we should confront now. Whoever belongs to this new agency, I think the Congress would do the country a great service if you could figure out ways to attracting young people, particularly with scientific backgrounds into Federal service.

I don't see that in the new bill as of yet. I would like to list five things that I think are essential elements of any department, whatever we name it, or whatever it is that has to lead the Nation's biodefense and biopreparedness efforts. First of all, as I said, they must have adequate expertise and personnel. I believe the crisis of competence is already afflicting the Federal agencies. And whether or not bioterrorism programs move to homeland security, we must, we must hire many new, I would say, at least 100 professionals to deal with bioterrorism programs in CDC, HHS or homeland security. This is for the medical and public health parts of bioterrorism.

Second, one of the critical aspects of success in bioterrorism prevention and preparedness is liaison with local authorities. The core of bioterrorism response is going to reside in hospitals, in clinics and in State health agencies. The Federal Government has to enable those entities to work properly. I am concerned that the programs already underway, particularly the public health preparedness programs initiated by the administration in February, are going to be disrupted with this move or even the threat of this move. These programs are getting started. The money is on the ground in the States.

Whatever we do, however we do it, we should ensure that that progress is not thwarted. We will also, if we create a homeland security agency as the home for bioterrorism preparedness programs, be creating a two-stop shopping problem for local authorities. They will go to CDC for traditional infectious disease help and guidance. They will go to homeland security, should we move the bioterrorism programs there. I understand we are going to try to have tight coordination between those agencies, but again, we seem to be splitting rather than consolidating functions. That could be a real problem for local agencies which have limited resources to interact with the Federal Government.

Third, I am worried about sustained support. However we configure bioterrorism programs, we are going to have to put resources into these programs for many years to come. If we move bioterrorism preparedness programs into homeland security, we may lose the opportunity to build dual use programs, that is bioterrorism response capabilities and systems that serve routine organizational purposes in the medical and public health field. That is not necessarily the case, but again, moving it to a new agency threatens to create parallel systems rather than one integrated system.

Fourth, we have to have a robust biodefense research and development program. The President recognized this by granting NIH the greatest budget increase in history in the past year. We ought not to build this program from scratch, but we are starting from very far back in the field. NIH does not now do production and development of technologies. No one in the government does it well. However we go forward with biodefense R&D, we must engage the universities and the private sector in this enterprise. That is where the real talent in bioscience research lies in this country, not in the government.

The government, aside from NIH, actually has very few bioscientists who are expert in R&D. So the new agency, whoever it is, has got to be able to engage industry and the university researchers and biodefense R&D.

Finally, I think that it is critically important that bioterrorism and biodefense be seen as a top national security priority. Coming from the Hopkins Center for Biodefense Strategies, I am, as you might imagine, deeply worried about this threat. I believe that this threat will grow considerably in the next few years because the power and the diversity of biological weapons is linked to advances in the life sciences and these advances which will have great booms for human kind are moving very, very quickly. That said, should we decide to leave bioterrorism preparedness programs in HHS, we must make sure that those programs don't get left behind and left out of the national security policies planning and strategies. It has taken several years to get national security experts to recognize that it is essential to have public health and medical people at the table making decisions about these issues, and we should not lose that progress in the new move to the new agency should we decide to go in that direction. Thank you, Mr. Chairman.

[The prepared statement of Tara O'Toole follows:]

PREPARED STATEMENT OF TARA O'TOOLE, DIRECTOR, CENTER FOR CIVILIAN BIODEFENSE STRATEGIES, JOHNS HOPKINS UNIVERSITY

Mr. Chairman, my name is Tara O'Toole. I am a physician and public health professional by training, the Director of the Johns Hopkins Center for Civilian Biodefense Strategies, and a faculty member of the Bloomberg School of Public Health. From 1993-97 I served as Assistant Secretary of Energy for Environment Safety and Health, and prior to that was a senior analyst at the Congressional Office of Technology Assessment. It is a privilege to come before you today to discuss the implications of President Bush's proposed bill to create a Department of Homeland Security. I shall confine my remarks to those aspects of the bill which deal with bioterrorism preparedness and biodefense activities.

I strongly support the formation of a federal department of Homeland Security as outlined by the U.S. National Commission on National Security in the 21st Century (the "Hart-Rudman report"). It makes great sense, as President Bush has advocated, to consolidate some of the many departments and agencies that share similar functions pertaining to border security, customs procedures, etc. in order to achieve greater collaborative power, efficiency and accountability.

There are some potential advantages to be gained from placing bioterrorism preparedness and biodefense research and development activities in a new federal agency. The activities dealing with the biodefense mission are profoundly important to the nation's security and deserve the attention and support the new agency is likely to command in the coming years. If biodefense activities do not reside in the Homeland Defense Department, there is some peril that these crucial functions will be neglected. It is also important that the operational public health and medical biodefense functions are integrated with national security objectives and that biodefense experts be full participants in national security policymaking and strategic planning.

I do, however, have serious concerns about the implications of moving bioterrorism preparedness programs and biodefense activities into the new agency, at least in the form presently envisioned.

A bioterrorist attack would be unlike any other type of terrorist assault. This would not be a "lights and sirens" event with firefighters, police and emergency rescue teams rushing to the scene of attack. We will know we have been attacked with a biological weapon when victims become ill and report to doctors' offices and emergency rooms. The "first responders" to bioterrorism will be physicians and public health professionals from state and local health agencies. The center of action will be hospitals, clinics and laboratories. Bioterrorism response activities—which will involve actions needed to treat the sick and perhaps stem the spread of contagious disease—are quite different from the emergency response to other types of catastrophic terrorism or to natural disasters.

Allowing for the inevitable transition period of confusion and adjustment, it is likely that the new agency will be more successful in instilling work habits of cooperation and collaboration to the extent that the agency's mission is coherent and tightly interconnected. It is not clear to me how or whether simply combining highly diverse functions from dozens of existing agencies under a single department results in better coordination or operational accountability. The description of the new department seems to envision an agency that is largely dedicated to security functions—border protection and control, vulnerability assessments of critical infrastructures, etc. The bioterrorism related programs and the scientific research and development aspects of the proposed department seem strikingly different from everything else the agency would handle.

President Bush exercised admirable leadership this winter when he greatly increased funding for bioterrorism preparedness programs in Centers for Disease Control and Prevention (CDC) and initiated a significant investment in bioterrorism research and development to be administered through the National Institutes of Health (NIH). The anthrax attacks of 2001 revealed that considerable improvement is needed in the nation's ability to respond to such attacks. In the past six months, notable progress has been made by the DHHS Office of Public Health Preparedness (OPHP). The OPHP has set sound goals for upgrading local medical and public health response capabilities, and the "critical benchmarks" it has demanded state health authorities achieve will provide clear indications of progress. We should consider disassembling and transferring this successful effort to the new department only after careful deliberation of what might be lost in the process. A recent poll reports that most Americans would seek and trust the advice of CDC during a public health emergency. It is unclear if such public confidence would transfer to the new department.

Part of the rationale behind the formation of a Homeland Security agency, as I understand it, is to combine similar functions—such as border control, customs services and immigration policy, etc.—within a single department, thereby enhancing program focus, fostering cooperation and collaboration and improving operational effectiveness. Yet moving bioterrorism programs from the Department of Health and Human Services (DHHS) to the proposed new agency will likely impede all these goals. Instead of consolidating similar programs, the proposed agency would split bioterrorism preparedness programs from the related but more encompassing mission of public health protection which is DHHS' main objective.

Rather than producing organizational coherence the proposed move would require that parallel capacities be created in both DHHS and the new agency. Homeland Security could not hope to lead the development of an effective bioterrorism response capability unless it were staffed with health officials and scientists having considerable expertise and experience in infectious disease, epidemic control, laboratory diagnosis, etc. Again, the country would be forced to create parallel workforces: one in Homeland Security for bioterrorism preparedness and another in DHHS for "normal" public health functions.

Moving bioterrorism programs to Homeland Security would disturb the existing relationships between DHHS bioterrorism programs and the state and local public health departments and health care facilities which are the central core of bioterrorism response. This is an especially important consideration right now, when the federal grants to state health departments are just hitting the streets and programs to upgrade response capacities at the city, county and state level are getting started. Changing the federal partner for these path-breaking grants will almost inevitably slow progress in this critical arena.

Moving bioterrorism preparedness and response activities out of DHHS may also sacrifice opportunities to construct dual use programs. Ideally, one would design bioterrorism response systems that also serve routine organizational purposes. There is a real danger that by sequestering bioterrorism programs in Homeland Security, they will be treated as "emergency use only" functions or seen as such, reducing the efficiency of preparedness efforts, and quite possibly compromise response effectiveness.

Bioterrorism is, arguably, the type of terrorism with which the country is least familiar and for which the United States is least well prepared. A bioterrorist attack could be calamitous, killing many thousands of people in the initial assault. The consequences would be sustained and the crisis could continue for weeks or months, especially if the weapon used were a contagious disease. The economic and social disruption would be significant—as was seen in the aftermath of the 2001 anthrax attacks when only 22 people were infected with a disease treatable with antibiotics. According to the Defense Science Board, we currently have countermeasures of some effectiveness (vaccines, drugs) for only 13 of the 50 pathogens most likely to be used as bioweapons. In addition, the institutions and infrastructures which would be at the core of bioterrorism response—health care organizations and the public health system—are financially frail, highly stressed, and have almost no capacity to contend with a sudden surge in demand for care.

These factors make it imperative that we make significant headway quickly in our capacity to manage bioterrorist threats. If one looks at the description of the proposed department, bioterrorism-related activities appear to be a tiny island of bioscience, medical and public health functions within a gigantic ocean of security and border control operations. I am skeptical that such an odd coupling can be made to work, particularly in the short term when there is such need for rapid progress.

I am especially worried about the fate of science and technology within the proposed department. Although there is clearly value in linking national security needs to research and development priorities, it is a very tall order to ask a single agency to develop national security strategy and implement operations on the scale envisioned for Homeland Security AND create a sophisticated scientific research and development capability over a broad range of disciplines and technologies.

Furthermore, we should have no illusions that creating a viable biodefense R&D capability is merely a matter of transferring or consolidating existing capabilities and programs. Regardless of how biodefense R&D programs are structured, the US government will have to build its capacity in these areas far beyond our present state. This nation has tremendous talent in bioscience and biotechnology—but the majority of talent lives in universities and the private sector, not in government. Any successful biodefense strategy must find ways to engage top scientists and young scientists in these sectors. Creating a robust biodefense R&D capability should be a top national security priority however we eventually design the architecture of biosecurity functions.

Bioterrorism must be considered a special category of terrorist threat. The potential power of bioweapons is easy to lose sight of in the aftermath of the thankfully limited anthrax attacks of 2001. But it is important to keep in mind that bioterrorism occupies a special category of terrorist threat that deserves careful scrutiny. The Hart-Rudman Commission noted in its first volume of analysis that

> *"...the most serious threat to our security may consist of unannounced attacks on American cities by sub-national groups using genetically engineered pathogens." [US Commission on National Security/21st Century, Sept. 15, 1999]*

As we design programs to prevent and respond to bioterrorist attacks we must proceed carefully, especially so since these weapons are largely unfamiliar to policy experts. However we decide to proceed in organizing federal bioterrorism activities, the nation's ability to respond to mass casualty situations and to effectively contain spread of contagious disease remains a grave concern. We must use our prodigious talent in bioscience to create the vaccines and therapies needed to respond to the bioweapons of today and of the future. We cannot afford a pause or loss of momentum in accomplishing these tasks.

Mr. GREENWOOD. Thank you, Dr. O'Toole.

The Chair recognizes himself for 5 minutes for inquiry. Dr. Atlas, in your testimony on page 3 you said the role of DHS should be to interrogate threat analysis and vulnerability assessments and to identify priorities, and I underscore priorities for preventative and protective steps to be taken by other Federal agencies to protect the American public. DHS can coordinate, review and evaluate scientific and technical programs relating to human animal and plant life. It seems to me you support some role for the new Secretary with respect to public health R&D and preparedness grants, including in some instances having the Secretary set the priority for such activities. Can you explain the distinction you are proposing and some alternative models such as dual reporting?

Mr. ATLAS. Yes, in a couple of ways. We see a very important strategic role for the new Secretary. The new Secretary will bring more of the intelligence community of the overall government perception of threat to human health and services for incorporation into the Nation's R&D plan. We could well imagine that the Assistant Secretary that has been discussed by your subcommittee today having a dual reporting responsibility, and I know that is normally very difficult, but we are dealing with such a complex issue with such duality, such overlap that we think that perhaps such a unique solution of having an individual with the health background that we need being able to assist both the Secretary of HHS and the Secretary of DHS in this area.

Mr. GREENWOOD. Let me ask a question of Dr. O'Toole. I understand that you support the increased flexibility in the administration's proposal for personnel-related decisions. You talked about the need to bring young scientifically trained people in the government, and to do it as quickly as possible. Why is it necessary, in your opinion, for there to be this civil service rule flexibility for this new agency?

Ms. O'TOOLE. Well, I think it wouldn't be necessary if we were allowed to hire several thousand new FTEs into the Federal Government. But absent that, in order to get a new skill mix into the government, it has been my experience that it was necessary to be able to move people in and out in ways that were not permitted by the civil service regulations.

Mr. GREENWOOD. Mr. Anderson, what additional measures to coordinate the Federal, State and local response to a nuclear attack have been implemented subsequent to the Air and Space Museum exercise?

Mr. ANDERSON. In following conversations I have had with local first responders, public safety, public health folks, there seems to be a great deal of momentum. I am not convinced that we are anywhere near close to being able to solve this problem and address it effectively. But at least the situation awareness exists that didn't exist previously, and I think that is going to lead to effective processes and procedures and hopefully equipment procurement, and all of the coordination that has to occur between the 17 surrounding jurisdictions in order to effectively deal with this.

Mr. GREENWOOD. Did this exercise exclusively look at the consequences to the post explosion, or did you look at questions at all as to the access to the Cesium, for instance?

Mr. ANDERSON. We developed the back end of this thing completely. We selected Cesium as the radioactive material, simply because it is readily available and there's enough of it missing or unaccounted, for according to the NRC, that it's reasonable to believe that it could have fallen into terrorists hands right here in this country.

Mr. GREENWOOD. What are the sources of it?

Mr. ANDERSON. Medical research, cancer research, cancer treatment. It has industrial applications with various types of diagnostic equipment. It is out there in large amounts. It would take a pound and-a-half to do what the slide depicted. And that was a DOD model that just plugged in 1,043 curies of Cesium 137, or a pound and a half and 4,000 of TNT. We absolutely believe—well, when we began, we thought it was a very remote possibility.

We only selected a dirty bomb because we were looking for a cross-jurisdictional crisis that would help in their planning effort. When we finished the research—and again it included all the back-end stuff like where are you going to get the materials and where are you going to get a school bus and all the rest. We absolutely believe this a real possibility. How real remains to be seen, but real enough that we need to think it through in terms of how we are going to respond.

Mr. GREENWOOD. My time has expired. The gentleman from Florida for 5 minutes.

Mr. DEUTSCH. Thank you, Mr. Chairman. Dr. O'Toole, obviously you listened to Mr. Anderson's testimony in terms of the threat of biological and chemical, which is not something that he discounts, but is describing as very far away. Would you characterize those the same way he did?

Ms. O'TOOLE. No.

Mr. DEUTSCH. Do you want to elaborate on that?

Ms. O'TOOLE. I think it is quite possible there could be a large bioterrorism attack in this country. It is very easy to do. The materials are at least as available as those necessary for a dirty bomb, although I agree with Mr. Anderson that a dirty bomb is also quite feasible. It's also possible that several small or a medium-sized biological attacks could be levied upon the United States and we would have a very difficult time figuring out what was going on and how to respond to it. As we saw with the anthrax attacks, which is only 22 cases, it doesn't take thousands and thousands of people being killed in order to cause disruption and economic loss.

Mr. DEUTSCH. Let me follow up with the issue at hand which is our responsibility in terms of the proposal of the Department of Homeland Security. You have expressed grave doubts about this new department that it will have the capacity to address bioterrorism and infectious disease outbreaks. How would you envision if you were writing the legislation, how would you structure the public health research and response system?

Ms. O'TOOLE. Well, as I said, first of all, I would hire more people into the system with appropriate expertise. I think that we need to build a much more operational Federal public health capacity that is able to go into the field, figure out the epidemiology.

Mr. DEUTSCH. Let me make my question clearer. The people on the appropriation side, we are the subcommittee that authorizes, but will be involved at a direct level in terms of actually structuring. The issue that we have talked about this whole day so far is how much is HHS doing now in basically biologicals with the component—and, you know, I have spent some time at CDC and talked to them and they seem to have an incredible, sophisticated, well-run operation now, but the concept is to take that out of HHS and CDC and bring it into homeland security.

So the issue in front of us is there seems to be some that's coming out and some that's staying in. I mean, would your advice be keep it in one place, whether it is HHS or bring everything over to homeland security, or Dr. Atlas suggested—I will be honest with you. I heard what you said. I don't think it is possible. The whole point of doing this is you have two people responsible and no one is going to be responsible as creative as you want to be. So I think—and Dr. Atlas, I would be happy for you to respond. But Dr. O'Toole, you can just respond specifically about that issue.

Ms. O'TOOLE. If I had a magic wand in hand, I would keep it in HHS and make it more robust. I would, however build in mechanisms to both coordinate activities between HHS and homeland security as well as to ensure that bioterrorism gets appropriate notice and someone is accountable for bioterrorism and HHS. I think the bioterrorism functions are basically medical and public health functions. It's going to be really hard to transplant them into this new security agency. It's possible maybe 10 years from now, it will be highly desirable. But in the near term, as I said, we run the risk of disrupting our capabilities in this area with this transplant.

Mr. DEUTSCH. Would that mean there is no advantages of thinking about the sort of public health response from a terrorism basis or just a naturally occurring event basis. Do you gain anything about that in sort of the discovery process or treatment process or prevention process?

Ms. O'TOOLE. No. What you would gain is focus and attention directed toward bioterrorism. It would clearly be a national security priority, and it would be funded that way. And the people dealing with bioterrorism would be part of the national security inner circle. Public health is not now at the Federal or State level on the hot water circuit. That presumably would change to some extent. But again, I think you would lose a lot of functional capacity, at least in the near term with the move.

Mr. DEUTSCH. This is obviously a decision that Congress is making over the next several months, and I think your perspective—and all three of your perspectives are unique in terms of the panels we have had up to this point. Hopefully—I know our staffs are interacting with you. But clearly, the direction of everyone up to you and the direction that most of us are at least coming from, and I think we still have open minds and this is what this whole process is about is that we are really talking about taking it away or setting up a two tiered system. And again having some experience no where in the orders of magnitude your experience with this and

I know you're being sincere with your belief and based on your background, I think it's something we need to take very seriously.

If you could just work with us because all of us are trying to get to the same goal, but the opinions you are expressing really are a minority opinion which might be the correct opinion but I think if I could encourage you to interact with our staffs and with members directly because you know, I think we have the ability to influence it and shape it the correct way.

And as strongly as you can be—this is not politically driven, it's not anything driven. It's just trying to come to the best outcome. And I hear exactly what you're saying. Dr. Atlas, if I could give you the opportunity to respond.

Mr. ATLAS. I suggested what may be even more complicated, one individual, an individual who could serve that coordinating function, that integration between two secretaries. Like Dr. O'Toole and the testimony clearly indicated that HHS should retain the authority over the biomedical research and the public health response, but clearly, there is this new function of homeland security. There is a real need for it. It brings other assets of the government to bear and it is somehow linking those that we are, I think, debating as we are going back on this, and I do see the possibility that appropriate high level individual who can walk back and forth between the two with great freedom would be a valuable asset in homeland defense.

Mr. DEUTSCH. Mr. Chairman, if I could just, for 15 seconds, kind of follow up with my last comment. But having spent some time at CDC I think, Dr. O'Toole, what you are specifically talking about where there are people there who basically made their career there and they seem to be able to attract the best and brightest within their little world of doing this biological. And I agree with you completely, this is really an individual basis. I mean, you need some incredibly bright—the brightest of the brightest people in the world looking at this to understand it. And if we are going to create a culture where we are not going to be able to attract and keep those people, it is going to be a failure. There is going on within HHS. Is there is at least, from a laymen's perspective, there is a culture that has been able to attract the brightest of the bright, even if the salary structure is not as good as it could be, and even if we can do a lot more. But we have got some people there who really are the best of the best, and I guess my real concern, which I hear you saying a little bit is, if we move this over to a new agency without any history, without any culture without any understanding how—you just can't move the whole function and move it over.

Ms. O'TOOLE. Could I clarify just a minute, Mr. Chairman. I think you can move it but you better prepare that ground. I also think that you have to significantly revamp the CDC operation and bring a lot more people than had been coming into CDC in to do bioterrorism work in the near future. Either way, I think that the bioterrorism functions deserve a lot of attention and consideration. But what you don't want to do is break the operations that are now beginning to work out of HHS. They are young. They are like new chutes. If you transplant them too early into hostile soil, it's not going to work.

Mr. GREENWOOD. Chair thanks the gentleman. The gentlelady from Colorado is recognized for 5 minutes.

Ms. DEGETTE. Thank you, Mr. Chairman. Dr. O'Toole, I really empathize with what you're saying, and I think part of the problem we have since the details of this proposal aren't completely fleshed out, it is hard for us to exactly see what would happen. But here's something that I kind of wanted to throw back at you, and if the gentlemen would like to answer it, that would be great, too.

The problem with biological and also chemical warfare is that it really cuts across many agencies and many disciplines and the GAO's testimony today talked about—really highlighted the types of problems when you're dealing with competing authorities among different agencies. With the recent anthrax events that you referred to, for example, local officials were complaining that the FBI and the public health officials had competing priorities about handling specimens, and this proved problematic because the FBI was briefing FBI officials, and local health departments didn't know what was going on and first responders.

I saw some of this in a town hall meeting—wasn't a town hall meeting, but a meeting with first responders talking about anthrax in Denver and the Postal Service employees in Denver got into a big argument with the State and with the FBI local offices there because the Postal Service couldn't get the FBI to test questionable specimens, and the Post Office didn't know what to do with them.

And I hear what you're saying, but I wonder, does the solution of hiring more people really help resolve issues of how do you prioritize and how do you deal with these interdisciplinary issues, and maybe you have some idea and I would like to hear it, because I think it would help all of us.

Ms. O'TOOLE. Hiring more people doesn't solve all problems. But I think some of the problems you mentioned would be at least alleviated to some extent if we had more realistic exercises so the people got used to working together and they had a better sense of what the protocols would be in an actual crisis. That was part of the problem in the anthrax response. It was sheer confusion. It was also lack of expertise. You had person A saying A and person B saying something else.

So getting our acts together actually is going to be a real challenge, no matter where the bioterrorism functions lie within the Federal bureaucracy. So hiring people is not a one-size-fits-all solution, but if you had more people, you could run more exercises and train more people. I think it would help.

Ms. DEGETTE. How would you deal with the interdisciplinary issues that are such a problem right now?

Ms. O'TOOLE. The interdisciplinary issues are always going to be there.

Ms. DEGETTE. If you have one per—if you had a correctly structured agency where one person was in charge of saying here's the protocol for who's notified.

Ms. O'TOOLE. You can do that no matter how you structure the organization. The problem is anticipating that we're going to have to be dealing with the Post Office, okay. There is an infinite number of scenarios that one can imagine for these nontraditional attacks and we need to create organizations that are expert enough and inventive enough and nimble enough to respond appropriately to things we never imagined before.

Now all of the literature and all of the experience of emergency disaster personnel and scholarship shows that planning is the one thing that seems to help get people ready for the next unexpected disaster, not because you put together plans that you use, not because you generate protocols that you snap into place, but because people know each other and they work better, particularly when they have to invent things on the run if they do know each other.

Ms. DEGETTE. You know, I think you're right, but we just had Operation Top-Off in Denver, which was a year ago, which was exactly this, planning for a biological attack. And yet that experience, which involved all the very same agencies that I was just talking about didn't help them even deal with an anthrax threat, much less a real incident.

Ms. O'TOOLE. Top-Off was 2 years ago. There is still no public analysis of what we learned in Top-Off, in part because of a personnel deficiency. I think it did help, but I think the failings in Top-Off are an indication of how hard this is and how far we have to go. We need to be careful of silver-bullet solutions. This reorganization is not going to be a solution. It may be one step toward an ultimate solution, but it could also be a step backwards. We need to be very thoughtful about that.

Ms. DEGETTE. I think those are very wise words, and thank you, Mr. Chairman.

Mr. GREENWOOD. The Chair thanks the gentlelady from Colorado. The Chair thanks the panel, Mr. Anderson, Dr. Atlas, Dr. O'Toole. We appreciate the good service you offered us today to help us with this really important work. Thank you again. Panel is excused and the committee is adjourned.

[Whereupon, at 2:30 p.m., the subcommittee was adjourned.]

CREATING THE DEPARTMENT OF HOMELAND SECURITY: CONSIDERATION OF THE ADMINISTRATION'S PROPOSAL

TUESDAY, JULY 9, 2002

HOUSE OF REPRESENTATIVES,
COMMITTEE ON ENERGY AND COMMERCE,
SUBCOMMITTEE ON OVERSIGHT AND INVESTIGATIONS,
Washington, DC.

The subcommittee met, pursuant to notice, at 9:15 a.m., in room 2123 Rayburn House Office Building, Hon. James C. Greenwood (chairman) presiding.

Members present: Representatives Greenwood, Bilirakis, Gillmor, Burr, Whitfield, Bass, Tauzin (ex officio), Deutsch, Stupak, Strickland, and DeGette.

Also present: Representatives Shimkus, Wilson, Buyer, Green, Capps, and Burton.

Staff present: Tom DiLenge, majority counsel; Amit Sachdev, majority counsel; Ray Shepherd, majority counsel; Peter Kielty, legislative clerk; Brendan Williams, legislative clerk; Chris Knauer, minority investigator; Jonathan J. Cordone, minority counsel; Edith Holleman, minority counsel; David R. Schooler, minority general counsel; and David Nelson, minority investigator.

Mr. GREENWOOD. The committee will come to order. The Chair recognizes himself for 5 minutes for the purpose of making an opening statement.

Billboard ads for the movie "Sum of All Fears," based on the Tom Clancy thriller, are emblazened with these chilling words: 27,000 nuclear weapons. One is missing." while the phrase is classic Hollywood promotion, in the post-September 11 world, we find ourselves asking are we now at a point in our history when we have to be prepared for even such a doomsday?

The threat of a terrorist attack involving nuclear weapons, or, more likely, radiological materials mixed with conventional explosives, the so-called dirty bombs, are more tangible than any of us could have ever imagined in the cold war period. The International Atomic Energy Association has documented 18 cases of trafficking since 1983 involving highly enriched uranium or plutonium, the key ingredients for an atomic bomb. These cases represent only those instances where the perpetrators were caught. A recent Washington Post article reports that the Intelligence Community believes that al Qaeda could already control a stolen Soviet-era tactical nuclear warhead or enough weapons-grade material to fashion a functioning, if less efficient, atomic bomb.

And what about the so-called dirty bomb? Experts estimate that the loss of life would not approach that of an atomic bomb, but the economic consequences could be just as devastating.

In testimony before the Senate Foreign Relations Committee, the Federation of American Scientists President Henry Kelly said that if the proper ingredients were used, a dirty bomb explosion could spread enough radioactive material to contaminate all of Manhattan, making it uninhabitable for 40 years, and leading to the potential destruction of $2 trillion of real estate.

The consequences of failing to safeguard our Nation and our people against such nuclear and radiological threats are simply too horrific to ignore.

We must take the steps needed to prevent weapons of mass destruction or the materials used to make these devices from being shipped into and around our country. Fortunately, good work is now underway to reduce the threat of nuclear terrorism. U.S. Customs service, for example, is taking bold long-term steps in the right direction. Its Container Security Initiative seeks to secure shipments into America by requiring that containers bound for U.S. ports will be examined at their ports of origin. While this is a valuable security measure, it will take time to fully implement.

For that reason, we must do more in the immediate future to protect our Nation from the agents of terror. To date, government agencies have been slow to take all available steps needed to protect our Nation's borders. Yet other Nations, including Germany, Ukraine, Slovakia, and Italy, currently use state-of-the-art technologies like portal monitoring systems to examine vehicles at border crossings, and for the past 10 years vehicles seeking entry into Poland have had to pass through a similar radiation detection system.

None of this is unavailable to our own Federal agencies. Ironically, the U.S. Department of Energy has been working closely to install these devices at Russian border crossings. Indeed, DOD officials estimate that there are already 100 such vehicle monitoring devices in Russia right now.

Clearly, the American people and this Congress feel a sense of urgency. So the question becomes why are we so far behind in this critical area, especially when the technology, much of it U.S. technology, exists to protect our seaports and our mail and express package delivery system infrastructure?

This committee's 10-month investigation suggests two principal answers:

First, the Federal Government has not provided sufficient guidance and assistance to the governmental and private sector entities at the front lines of homeland security on how to identify, evaluate, and implement currently available technologies that could measurably reduce the threat of nuclear smuggling. Indeed, we have been unable to find any Federal agency that believes it has the responsibility to do so.

Second, not surprisingly, the Federal Government's research and developmental efforts in this area have not been sufficiently focused and coordinated. Much of the work is redundant, with numerous agencies contracting with various laboratories to conduct similar R&D activities again. This occurs because of the lack of organizational clarity. Up to this point, no one agency has been charged with developing a strategic plan for such research for its direction, funding, coordination, and implementation.

All of this brings us to today's continuation of the hearing this subcommittee begun 2 weeks ago to consider the Bush administration' proposal to create a new Department of Homeland Security. Besides examining how this proposal may help to alleviate the two core problems described above, we will also review two other important aspects of the President's proposal at today's hearing: how public health research and development may be affected by the proposed transfer of certain authorities for terrorism-related biomedical research to the new Department; and how the critical infrastructure assessment and other related activities of the new Department may help improve our Nation's protection of key assets such as the energy and telecommunications grids and our food and drinking water supplies.

We have many panels and witnesses to hear from today as we embark on this very serious undertaking. I appreciate the patience of our members and our witnesses as we proceed through what will undoubtedly be a long day.

For purposes of information, let me provide a quick outline of the day before us. The first two panels will focus on the public health research and development activities potentially affected by title III of the administration's proposal. The third and fourth panels will focus on critical infrastructure protection issues based on title II of the administration's proposal, including discussion of public access to such information.

The final two panels will discuss those aspects of title III that relate to research and development of nuclear, chemical, and biological detection technologies and other related programs at the Department of Energy currently proposed for transfer to the new Department.

Based on my consultation with the committee minority staff, I expect that the subcommittee will approve a motion to close to the public the last two panels of today's hearing due to the sensitive nature of that discussion.

Before I recognize the ranking member for an opening statement, I would like to point out that Thermo Electron Corporation and Sandia National Laboratory will conduct equipment demonstrations throughout the day. Members and staff are encouraged to use this opportunity to assess the capabilities of currently available detection devices.

Thermo's equipment is set up in the chairman's meeting room, right next to the members' lounge near the hearing room, and Sandia's devices will be in the meeting room attached to the minority lounge. Thermo's demonstrations consist of a human portal device capable of detecting both metals and radiological material. The company will also have a live demo of the detectors used in vehicle portal systems, which will include a software display showcasing a graphical representation of the information collected by the detector. Thermo has also set up a model of a radiological detection device that can be used on cranes.

Sandia will display various bomb disassemblement devices. While these devices are used to disassemble common explosive devices, they can also be used to deactivate dirty bombs. In addition, Sandia will have a nuclear detection device, but it will not be active. So you can take your nuclear devices right past it.

The members are reminded that this is a continuation of a previous hearing. Opening statements will not be required, but they will be tolerated, and the Chair recognizes the ranking member Mr. Deutsch.

Mr. DEUTSCH. Thank you, Mr. Chairman. Before I make a very brief comment, there are three organizations that have contacted us that want to be able to provide testimony, and without objection, we can allow that: National Association of City and County Health Officials; the American Public Health Association; and the Association of State and Territorial Health Officials.

Mr. GREENWOOD. Without objection.

Mr. DEUTSCH. Thank you, Mr. Chairman. As you mentioned, this is a continuation of our previous hearing. I look forward to working with you. I think this is really an issue which I described as working shoulder to shoulder, heart and soul together, to create this new Department. But as we're doing that, I think our job—and I think both of us would agree—is working out some of the details, particularly within areas of our jurisdiction, like HHS and NIH. And just a concern that we had expressed at the prior hearing, that some of the incredibly significant jobs that they do now not be put into a second and third or fourth place under an agency that clearly will have the most significant task that our government is facing.

So with that, I would yield back the balance of my time.

Mr. GREENWOOD. The Chair thanks the gentleman and recognizes the chairman of the full committee, the gentleman from Louisiana, Mr. Tauzin.

Chairman TAUZIN. Thank you, Mr. Chairman, and I too submit statements for the record. But I want to just mention a couple of things I think are critical as you begin an extraordinary day. I think you've got 26 witnesses ahead of the committee today as we work on a very short deadline to produce for the President and for the House our recommendations on this new Cabinet-level Department of Homeland Security. Our deadline is July 12, and we're going to meet it. So this hearing is critical to put on the record all of the investigation that is going to help us formulate those final recommendations before the end of this week. So we thank you, for all of you who participate today.

I wanted to mention a couple of things, Mr. Chairman. You mentioned the striking lines from the movie, the "Sum of All Fears," and it occurred to me that the attacks on our country may well take very different forms. Just last week we learned of an all-points bulletin announcement, announcing an effort to retrieve or recover a tanker truck that was stolen from a locked yard in Florida somewhere, the tanker truck containing chemical waste material. Obviously we've seen some of the reports indicating that those types of low-technology attacks are being discussed by al Qaeda and by some of the cells and networks that exist in this very country.

We learned in the newspaper this week how when Mohammed Atta appeared before a USDA official seeking a loan to buy a crop duster, trying to get Federal taxpayer dollars to buy a crop duster that obviously was intended in his mind for a terrorist attack, that he went absolutely ballistic when he saw the beautiful aerial map of Washington, DC in the office there, and that he put cash on the table, trying to buy that beautiful map from the representative of the USDA because that map represented to him, obviously, a source of information upon which he might plan or his friends might plan an attack on this city.

It calls to our attention the importance in this legislation of amending the Freedom of Information Act to make sure that road maps, vulnerability assessments of assets, both public and private, other road maps of sensitive installations and sensitive places in this country are not so easily available to people who might have improper motives, such as Mr. Mohammed Atta, in using those road maps to hurt this country or its people. Balancing the needs of freedom of information in this country against the concerns that I think of all that the USDA office represents is going to be a difficult challenge of this committee and the Congress, but I know that this committee will be up to it as we make our recommendations to the House.

And one final thought, and that is that we've got some people we need to perhaps congratulate today. I want to single out the Port of Norfolk, Virginia, which is a private port authority which on its own set up a radiological detection system on one of its cargo planes with almost no help from the Federal Government. We've since come in and assisted them, but that kind of initiative is to be recognized, and I want to thank the Port of Norfolk, Virginia and the other ports of America who are doing things like that on their own.

I have nine ports in my district, and we are working with every one of them. One of them is Port Fourchon, where 16 percent of the oil that serves our country enters this country. I don't believe it is yet on the list of ports that receive Federal assistance in security, but we're working on that. Making sure that all of these ports are more adequately protected is going to be a critical component of this new Department.

I want to congratulate the Customs Service pilot project that is underway at the Detroit/Windsor border where a single vehicle portal system is being tested, and we're very interested in learning the results of that test.

I want to thank you, Mr. Chairman, and our staff for the work you did with FedEx and UPS. These two organizations are now planning to install radiation detection systems at their overseas hubs, and they intend to achieve 100 percent coverage of all packages that move through these organizations. That is a big step forward, and I want to thank all of you, Mr. Chairman and members of the staff, who have worked with these organizations in ensuring that.

But we need to point out there is a lot of work to be done. We've got six Dewey labs, for example, that are currently working on research related to the detection of radiological and nuclear material, but little coordination behind their efforts. We hope this new Department will begin to coordinate those very important efforts to make sure that port authorities and customs services and other private entities who want to use technologies like this know what is the best technology and what works and what doesn't work. Now, that is going to be part of the recommendations.

Finally, Mr. Chairman, if anyone wants to know when the next attack on America is going to occur, the answer came in a radio program today as I was driving into work. The next attack on America will occur today, every day. There are 30 cyber attacks that we can identify on sensitive government entities in this country, at least 30. There are days when there are hundreds, and there are days when there are thousands of such attacks, coming in from places as far away as China. Some of those are just business espionage attacks on private entities and Web sites. Some of those are probing attacks on very sensitive cyber systems that exist, that operate and that protect this country, that operate sensitive installations and protect this country in many important ways. Thirty identified attacks every day; today, tomorrow, every day. If we don't make sure the Homeland Security Department is prepared in this critical area of cyber security, we will have failed in our duty, and I hope this committee approaches that issue very seriously as we move forward with our recommendation.

Thank you, Mr. Chairman.

Mr. GREENWOOD. I thank the chairman.

Are there members of the minority who wish to make opening statement? Beginning with Mr. Stupak.

Mr. STUPAK. Thank you, Mr. Chairman, and I'll be brief. I look toward to today's hearing. I believe this is the second hearing now we have had on the President's proposal to make Homeland Security a Cabinet-level position.

I'm concerned that we're under a time line here of July 12, that this homeland security must be completed by then and passed. It is too important of an issue to put a time line out there and say you have to pass it. In my 10 years' experience when we do things under time line, that we don't have much rhyme or reason, usually we rush things through and questions are not answered. And then after the fact we are saying, why did we do that?

So I would hope that we could slow this process down and give this President's proposal the due consideration it deserves, because I accept the principle that homeland security is so important, that it demands a Cabinet-level position. However, I'd like to see a clear chain of command in whatever new structure we approve, and I don't see that in the President's authority. And when we had the discussions last week with Director Ridge, he told me that this Cabinet-level position, after explaining—we were talking about ports and radiation, how it went from Customs and DOE to Sandia, and we still don't have an answer—and he said that is the way the Federal Government works.

Well, if that is the way the Federal Government works, I don't want to pass a Cabinet-level position to have the Federal Government in a horizontal chain of command when a decision is never made; and once the decision is made, no one accepts the responsibility, but points to another person as the person who made that decision. So I would like to see a clear chain of command in whatever new structure we'll approve. We need to know there is a vertical authority and people will accept responsibility for their actions. We need to know where the buck stops, if you will.

Portions of our homeland security are being neglected. Again, last week I pointed out where hundreds of miles of international coastland, about 700 miles, which is basically my district along Canada there, is currently without any surveillance or security measures whatsoever. And I think we need to know who we would go to talk to to get this thing fixed. That is not clear in the President's proposal.Reorganization will come, but it needs to be better than the piecemeal structure we see throughout the Federal Government today.

So, again, while I'm supportive of the idea, I want to see a clear level of command here. I just don't want merely a shuffling of chairs at the table and calling it a Cabinet-level position and somehow we cured this ill that we have called homeland security by July 12.

So I look forward to today's hearing. I understand it is going to be a long hearing, so I would yield back the balance of my time and I look forward to hearing from the witnesses. Thank you, Mr. Chairman.

Mr. GREENWOOD. The Chair thanks the gentleman.

Does the gentleman, Mr. Whitfield, care to make on opening statement?

Ms. DeGette?

Ms. DEGETTE. Yes.

Mr. GREENWOOD. She's recognized for 5 minutes.

Ms. DEGETTE. Thank you, Mr. Chairman. I'd like to echo some of the sentiments of my colleague, Mr. Stupak. I think that it is essential that we have a Homeland Security Office with real authority, with a Cabinet-level authority. What I don't want to see is more layers of bureaucracy layered on what we have right now and no clear decisionmaking process. And I don't think any of us wants to see that, but if we rush this through in the way that it is envisioned, I think we could face a lot more problems when we face terrorist attacks, either from abroad or domestically. And with those sentiments, I yield back.

Mr. GREENWOOD. The Chair thanks the gentlelady.

The gentleman from New Hampshire, Mr. Bass, does not care to make an opening statement. Mr. Green, do you have a brief opening statement?

Mr. GREEN. Mr. Chairman, just following up my colleagues on both sides of the aisle, I hope the new Department would be able to bring together what our goals are, but mainly the established collaborative relationships between all these agencies, that—maybe that is the goal of the administration, but think there may be some lacking in the actual language of the bill. But I support the Homeland Security Cabinet-level position, but also hopefully we'll give it the authority it needs to be able to protect us. Thank you.

Mr. GREENWOOD. The Chair thanks the gentleman.

Does the gentleman, Mr. Shimkus, wish to make an opening statement?

Mr. SHIMKUS. I just want to thank the Chair for allowing other members from the committees come and sit in on the hearing and I look forward to the testimony.

Mr. GREENWOOD. Delighted to have you with us.

And that brings us to our first panel. Welcome, Mr. Hauer. Mr. Hauer is the Director of the Office of Public Health Emergency Preparedness in the Department of Health and Human Services. We thank you and we welcome you. Sir, you're aware that this committee is holding an investigative hearing and when we do so, it is our custom to take testimony under oath. Do you have any objection to giving your testimony under oath?

Mr. HAUER. None whatsoever.

Mr. GREENWOOD. So hearing that, you are also advised that pursuant to the rules of this House and committee, you are free to be represented by counsel. Do you choose to be represented by counsel?

Mr. HAUER. No.

Mr. GREENWOOD. In that case, would you stand and raise your right hand?

[Witness sworn.]

Mr. GREENWOOD. Thank you. You are under oath, and you are recognized for 5 minutes for your opening statement.

TESTIMONY OF JEROME M. HAUER, DIRECTOR, OFFICE OF PUBLIC HEALTH EMERGENCY PREPAREDNESS, DEPARTMENT OF HEALTH AND HUMAN SERVICES

Mr. HAUER. Good morning, Mr. Chairman, and thank you. I thank you, too, members of the committee, for giving me the opportunity to appear before you today on behalf of Secretary Thompson to discuss the proposed Department of Homeland Security.

The Secretary strongly supports the reorganization initiative, and, as the President announced earlier this month, the threat of terrorism in its myriad forms are becoming an ever-present part of our daily lives. The new Department will enable us to make further significant advances in protecting the American people from those who are bent upon inflicting death, destruction and social disorder, and to achieve their ideological goal.

We are pleased that the Congress is giving the President's proposal prompt and thorough attention. Secretary Thompson and I look forward to working with this and other committees to ensure passage of the legislation for the new Department.

The President's proposal deals with certain terrorist-related activities that currently are the responsibility of the Department of Health and Human Services. Some of these HHS activities would be transferred to the Department of Homeland Security. For other relevant public health and medical activities, DHS would assume responsibility for setting goals and providing strategic direction, but would rely on HHS to implement and operate on a day-to-day basis.

My written statement focuses on all activities being moved from the Department of HHS to the Department of Homeland Security. I'll focus today on two examples of those in the transfer.

First, the Select Agent registration program. Within HHS, the Center for Disease Control and Prevention currently regulates the transfer of certain dangerous pathogens and toxins, commonly referred to as "select agents" from one registered facility to another. These agents are widely used in research laboratories across America. Examples are the bacterium that causes anthrax, the bacterium that causes plague, and the virus that causes Ebola, a lethal hemorrhagic fever. Select agents are prime candidates for use by would-be terrorists, and thus when used in research must be kept constantly under safe and secure conditions.

The recently enacted Public Health Security and Bioterrorism Preparedness Response Act of 2002 authorized HHS to promulgate and enforce regulations concerning the possession and use of select agents as well as air transfer. While CDC has done its best to manage this program, CDC is a public health agency and not a regulatory body. We believe that the new Department, with its strong multipurpose security and regulatory infrastructure, will be well suited to prevent nefarious or other irresponsible uses of select agents. HHS will be prepared to provide DHS with whatever scientific expertise and other technical assistance it may seek to help to manage this program. Under the administration's bill, the Secretary of Homeland Security would administer the Select Agents program in consultation with the HHS Secretary, and HHS would continue to make key medical and scientific decisions, such as which biological agent should be included in the select agents list. Certain specific program-level details and administrative choices are still being studied in order to ensure the most seamless transition.

Let me focus now on civilian human health-related biological and biomedical infectious disease defense and research programs.

The President's proposal provides that the new Department's civilian human health-related biological, biomedical, and infectious disease defense research and development work shall, unless the President otherwise directs, be carried out through HHS. Under the President's proposal, the Secretary of Homeland Security, in consultation with the Secretary of Health and Human Services, shall have the authority to establish the R&D program that will be implemented through HHS. Thus, as the agency responsible for assessing threats to the homeland, DHS, in consultation with HHS, will provide strategic direction regarding the Nation's biological and biomedical countermeasure research priorities.

The President's proposal provides that the new Department shall, unless otherwise directed by the President, carry out through HHS certain health-related activities such as programs to enhance the bioterrorism preparedness of State and local governments and nonFederal public and private health care facilities and providers. The object of this provision is to continue the important role that HHS plays in assisting State and local governments and the hospital and public health community in preparing for and responding to large-scale public health emergencies. As with the research program, the Secretary of Homeland Security, in consultation with the Secretary of Health and Human Services, will establish the Nation's antiterrorism preparedness and response program and priorities, but the implementation of the public health components of that program will be carried out largely through HHS.

Mr. Chairman and members of the committee, our Nation needs a Department of Homeland Security. The Department of Health and Human Services strongly supports the President's proposal, and we look forward to doing whatever is necessary to effect a smooth and swift transition of responsibilities and operations.

We believe that the President's proposal strikes the right balance. It plays to the strengths of HHS and recognizes this Agency's core mission, the protection of our Nation's public health, while capitalizing on the strategic and logistical strengths of the new Department of Homeland Security. We will ensure that HHS fulfills its obligations to the new Department and provides it with whatever public health, medical and scientific expertise it may require.

I thank you for the time, and I'd be happy to answer any questions.

[The prepared statement of Jerome M. Hauer follows:]

PREPARED STATEMENT OF JEROME M. HAUER, ACTING ASSISTANT SECRETARY OF PUBLIC HEALTH EMERGENCY PREPAREDNESS, DEPARTMENT OF HEALTH AND HUMAN SERVICES

Thank you, Mr Chairman and members of the Committee for giving me the opportunity to appear before you today on behalf of Secretary Thompson to discuss the proposed Department of Homeland Security. The Secretary strongly supports the reorganization initiative that the President announced earlier this month.

The threat of terrorism in its myriad forms has become an ever-present part of our daily lives. The new Department will enable us to make further significant advances in protecting the American people from those who are bent upon inflicting death, destruction, and social disorder to achieve their ideological ends. We are pleased that the Congress is giving the President's proposal prompt and thorough attention. Secretary Thompson and I look forward to working with this and other Committees to ensure passage of the legislation for the new Department.

The President's proposal deals with certain terrorism-related activities that currently are the responsibility of the Department of Health and Human Services (HHS). Some of these HHS activities would be transferred to the Department of Homeland Security (DHS). For other relevant public health and medical activities, DHS would assume responsibility for setting goals and providing strategic direction but would rely upon HHS to implement and operate the activities on a day-to-day basis.

I will discuss examples from each group of activities in turn.

EXAMPLES OF ACTIVITIES PROPOSED FOR TRANSFER FROM HHS TO DHS

HHS functions conveyed to the new Department in the President's proposal include:

- The Select Agent registration enforcement program;
- The Office of the Assistant Secretary for Public Health Emergency Preparedness; and
- The National Pharmaceutical Stockpile.

Select Agent Registration Program

Within HHS, the Centers for Disease Control and Prevention (CDC) currently regulates the transfer of certain dangerous pathogens and toxins—commonly referred to as "Select Agents"—from one registered facility to another. These agents are widely used in research laboratories across America. Examples are the bacterium that causes anthrax, the bacterium that causes plague, and the virus that causes Ebola, a lethal hemorrhagic fever. Select Agents are prime candidates for use by would-be bioterrorists and thus, when used in research, must be kept constantly under safe and secure conditions.

The recently enacted Public Health Security and Bioterrorism Preparedness and Response Act of 2002 authorized HHS to promulgate and enforce regulations concerning the possession and use of Select Agents, as well as their transfer. While CDC has done its best to manage the Select Agent program, CDC is a public health agency and not a regulatory body. We believe that the new department, with its strong multi-purpose security and regulatory infrastructure, will be well-suited to prevent nefarious or other irresponsible uses of Select Agents. HHS will be prepared to provide DHS with whatever scientific expertise and other technical assistance it may seek to help it manage the program. Under the Administration bill, the Secretary of Homeland Security would administer the select agents program in consultation with the HHS Secretary, and HHS would continue to make key medical and scientific decisions, such as which biological agents should be included in the select agents list.

Office of the Assistant Secretary for Public Health Emergency Preparedness

The Public Health Security and Bioterrorism Preparedness and Response Act of 2002 created the HHS Office of the Assistant Secretary for Public Health Emergency Preparedness for which I serve as Acting Assistant Secretary. The responsibilities of this new office include the supervision of the Office of Emergency Preparedness, the National Disaster Medical System, the Metropolitan Medical Response Systems, and related HHS emergency management functions. This cluster of activities is a logical and proper candidate for transfer to DHS—thereby enabling seamless integration of national public health and medical emergency management assets with the Nation's new preparedness and response infrastructure at DHS. The Public Health Service Officers and other HHS employees who have faithfully performed disaster relief work over the years have done a wonderful service for our Nation. They are a credit to HHS as they surely will be to the new Department.

Strategic National Stockpile

CDC currently manages 12 "push packages" of pharmaceutical and medical supplies and equipment strategically located around the United States; additional lots of pharmaceuticals and caches of medical materiel are maintained by manufacturers under special contractual arrangements with CDC. You may recall that one of the push packages was dispatched to New York City on September 11th and that elements of the stockpile were used to respond to the anthrax attacks. CDC has done an exemplary job managing what is now called the Strategic National Stockpile and this fine work has set the stage for integration of the Stockpile with other national emergency preparedness and response assets at DHS.

The President's proposal is designed to achieve this integration by tapping the strengths of DHS and HHS in a precisely coordinated way. Thus, the Secretary of Homeland Security will assume responsibility for continued development, maintenance, and deployment of the Stockpile—making it an integral part of the larger suite of federal response assets managed by FEMA and other future DHS components—while the Secretary of Health and Human Services will continue to determine its contents. The arrangement will ensure effective blending of the public health expertise of HHS with the logistical and emergency management expertise of DHS.

DHS FUNCTIONS TO BE CARRIED OUT THROUGH HHS

Certain specific program level details and administrative choices are still being studied in order to ensure the most seamless transition, and to give the greatest possible levels of efficiency and effectiveness to our fight against the threat of biological warfare and to protect the public health. However, the President's proposal clearly designates the following two activity areas that the Secretary of Homeland Security will carry out through the Department of Health and Human Services:

1. Civilian Human Health-Related Biological, Biomedical and Infectious Disease Defense Research and Development

The President's proposal provides that the new Department's civilian human health-related biological, biomedical, and infectious disease defense research and development work shall—unless the President otherwise directs—be carried out through HHS. Under the President's proposal, the Secretary of Homeland Security, in consultation with the Secretary of Health and Human Services, shall have the authority to establish the research and development program that will be implemented through HHS. Thus, as the agency responsible for assessing threats to the homeland, DHS, in consultation with the HHS Secretary, will provide strategic direction regarding the Nation's biological and biomedical countermeasure research priorities.

2. Certain Public Health-Related Activities

The President's proposal provides that the new Department shall—unless otherwise directed by the President—carry out through HHS certain public health related activities (such as programs to enhance the bioterrorism preparedness of state and local governments and non-federal public and private health care facilities and providers). The object of this provision is to continue the important role that HHS plays in assisting state and local governments and the hospital and public health community in preparing for and responding to large-scale public health emergencies. As with the research program, the Secretary of Homeland Security, in consultation with the Secretary of Health and Human Services, will establish the Nation's antiterrorism preparedness and response program and priorities, but the implementation of the public health components of that program will be carried out largely through HHS.

CONCLUSION

Mr. Chairman and members of the Committee, our Nation needs a Department of Homeland Security. The Department of Health and Human Services strongly supports the President's proposal and we look forward to doing whatever is necessary to effect a smooth and swift transition of responsibilities and operations. We believe that the President's proposal strikes the right balance: it plays to the strengths of HHS and recognizes this agency's core mission—the protection of our Nation's public health—while capitalizing on the strategic and logistical strengths of the new Department of Homeland Security. We will ensure that HHS fulfills its obligations to the new Department and provides it with whatever public health, medical, and scientific expertise it may require.

At this time, I would be happy to answer your questions.

Mr. GREENWOOD. Thank you, Mr. Hauer, we appreciate your presence and your testimony.

The Chair recognizes himself for 5 minutes for questions.

As you know, sir, many in the public health community have expressed concern about sections 301 and 303 of the administration's proposal, which seem to grant this new Department of Homeland Security direction over the conduct of traditional public health research activities, albeit ones involving potential terrorist weapons such as anthrax and smallpox. The question is why did the administration propose this change, and what advantages do you see flowing from it, if adopted, and how do you respond to the concerns that have been raised about it?

Mr. HAUER. Well, first of all, we have heard some of the concerns, but at the end of the day, I think that some of it is misinformation that has—or misunderstanding of the direction in which the new Department will play and the role the CDC will continue to play in addressing public health research. CDC will continue its normal course of public health research. What will happen is the new Department of Homeland Security will take responsibility for ensuring that certain aspects of the research, those related to threats that impact public health, are coordinated through them, because as with other types of R&D, this new Department will have access to information, will be able to coordinate research, and will ensure at the end of the day the focus of public health research when it comes to dealing with threats that this country now confronts is well coordinated by one Agency at the Federal level.

Mr. GREENWOOD. Does that in your mind have the effect and is it the intent that if the Secretary of Homeland Security says to CDC, I want you to conduct the following research projects and I want you to do it right away and I want you to do it with this level of intensity and so forth, that given all these finite resources of government, that would trump and take priority over CDC's other projects at that time, and that the Department of Homeland Security would have the ability to sort of push some of CDC's agenda off the table temporarily while that——

Mr. HAUER. Depending on the nature of the threat, I would envision that if in fact there is a threat or a threat or concern that Homeland Security feels is a high priority that they would work with CDC. CDC does ongoing biological and nuclear and chemical research currently, and it consistently rearranges its priorities based on things that are going on. One good example is West Nile.

In 1999, when West Nile first was recognized in New York City, CDC had to reshift its priorities fairly quickly to understand what was going on.

Mr. GREENWOOD. But I think we understand that, and that says it ought to be—I think the concern that has been raised here that needs to be clear in all of our minds is that on a given day if the Secretary of Homeland Security says to CDC, I want you to devote X resources and personnel to studying an Ebola in some capacity, and the Secretary of HHS says, no, that is—we don't need to do that. I'm much more interested in West Nile right now. Keep the folks on West Nile. And there is a disagreement about that, how is that resolved?

Mr. HAUER. Well, the disagreements—I think that there will continue to be an Office of Homeland Security in the White House reporting to the President, and disagreements of those natures can be resolved at the Office of Homeland Security. But I would envision that if Homeland Security—the new Department of Homeland Security feels that the needs are that great, that HHS and in fact CDC will do everything possible to try and work with the new Department to ensure that research is done as expeditiously as possible.

Mr. GREENWOOD. In your testimony you note that the new Secretary will have the ability to, quote, establish the research and development program for public health threats of a terrorist nature that will be carried out by HHS, which is what the President's proposal language says. Yet later you infer that this only means that the new Secretary will provide, quote, strategic direction regarding priorities for research. Is that all that is meant by the statutory language proposed or does the administration envision other functions for this new Secretary in this R&D area?

Mr. HAUER. Well, I think the new Secretary, one, will set some of the research and development priorities for the chemical, biological, nuclear threats that we face. Clearly, the ongoing research at NIH already focuses on some of those; and in working with the new Department I think there will be a significant synergy so that as the new Secretary begins to set research priorities the new Secretary will clearly have to work with the center directors at NIH and at CDC. And the new Secretary—I don't envision the new Secretary actually conducting or in any way getting into the research business but working through NIH, working through CDC to actually conduct the research, set some of the priorities, to fund some of the research, but not to get into the actual research business in and of itself within the new Department.

Mr. GREENWOOD. The Chair recognizes the ranking member, Mr. Deutsch, for 5 minutes.

Mr. DEUTSCH. Thank you, Mr. Chairman.

The Centers for Disease Control is one of the really international experts in infectious diseases and select agents. In fact, it is probably the most complete laboratory used for infectious disease identification research and surveillance. It works with researchers around the world in these diseases. However, when Governor Ridge testified before us in June, he said the new role of the Centers for Disease Control would be working with the maternal health, smoking and immunizations. Was that an accurate representation of

CDC's new role? Why are we putting money to making their laboratory more secure and capable of working on bioterrorism if they are not working with these agents?

Mr. HAUER. Well, in fact, CDC is working on threat agents. CDC has a very aggressive program in working with threat agents, and the new Department of Homeland Security will continue to work with the CDC. The new Department of Homeland Security will be—the intelligence component of the new Department will help determine what the focus of the new research is on threat agents, because they will able—through the intelligence work be able to determine what threat agents we confront at that point in time. CDC will continue their ongoing research in dealing with these kinds of threat agents, as will the NIH in looking for new ways, new vaccines, new ways to treat these types of threat agents, new ways of dealing with these types of threats.

Mr. DEUTSCH. The Public Health Security Bioterrorism Act of 2002 also gave CDC the responsibility for improving public health communication facilities and networks. Where does that task go under this new structure?

Mr. HAUER. The HAN and NEDS, which are the infrastructure that are being developed nationwide to allow CDC, State health departments and city and county health departments to communicate, will stay at CDC, as will NEDS, which is an infrastructure in development at CDC to allow hospitals and health care providers to communicate with local health departments for either data mines, surveillance and other types of surveillance systems. That will again take place in the traditional public health activities—they will stay at CDC as they are now.

Mr. DEUTSCH. If CDC is going to take on these additional responsibilities, how will that be funded and who will be——

Mr. HAUER. Those responsibilities are at CDC right now and are funded through the Bioterrorism Act of 2002.

Mr. DEUTSCH. Does Homeland Security have a clear understanding of the difference between law enforcement first responders, communications such as were needed on September 11 and the public health surveillance network which is for public health and medical staff which may not be first responders?

Mr. HAUER. Well, I don't think there is any question that there is a very clear understanding in the Homeland Security—in the Office of Homeland Security right now and the people that I deal with that the responsibility for public health emergency is significantly different than the response to chemical or nuclear energies. One is clearly a type of response and a public health emergency—the first responders is a completely different community, and the detection of a public health emergency is one that will occur over time. It is not an immediate—I don't think there is any question that that is clearly understood and as part of a new structure of homeland security is integrated into that structure.

Mr. DEUTSCH. Who is going to maintain responsibility to improve the capacities of the State and local laboratories?

Mr. HAUER. That will be done through the Centers for Disease Control, through the laboratory response network. That will be an ongoing process. While Homeland Security will have some oversight of that, the function will remain with CDC.

Mr. DEUTSCH. Has HHS failed with its recommendations about how to improve the preparedness of the public health system?

Mr. HAUER. No. In fact during the last 8 months—I've worked in this environment for almost 8 years now and in fact had the first surveillance system in the country when I was still a commissioner in New York City and was extremely frustrated with HHS over time because HHS was not moving forward. During the last 14 months or so, the Department has made incredible strides.

Secretary Thompson, even before being confirmed, recognized that bioterrorism was an issue that we would have to confront, and we did briefings for the Secretary early on in the administration. Since September of last year, we have accelerated our programs in large part because of the support we've gotten from Congress and the money we've received.

We have done—the original plan, just one example, was to have 40 million doses of smallpox vaccine by 2005. When the Secretary took over, that was the plan. We are now going to have 260 million new doses of new vaccine by the end of this year.

We have seen enormous strides at the State and local level on the ability to receive the national pharmaceutical stockpile, the training that is going on at the State and local level, the exercises we're seeing at the State and local level. We're finally seeing hospitals talking with their local health departments. We are finally seeing strides that heretofore have never occurred, and we expect as this gets integrated into the Department of Homeland Security that there will be even closer coordination with the—at the State and local level.

Mr. DEUTSCH. If I can indulge in just one very short follow-up question, all that good future planning that you described, how will that be affected in terms of the switching of responsibilities to Homeland Security?

Mr. HAUER. I don't see any effect. In fact, because of so much of what goes on in emergency response at HHS, it is so closely coordinated with FEMA, as this becomes integrated into the new department, I would imagine that the synergies will be even stronger and that we will clearly see more coordination both at the State and local level.

Mr. GREENWOOD. The Chair thanks the gentleman and recognizes the chairman of the full committee, Mr. Tauzin.

Chairman TAUZIN. Thank you, Mr. Chairman.

Mr. Hauer, as you're well aware, in the bioterrorism bill we've beefed up the select agent's program at CDC, basically making sure it is not only a registration but also a tracking system on select agents that may affect human health, and I understand that that is—that function is scheduled to be transferred to the Homeland Security office. But we also at the same time, as you know, put together a program at USDA, a similar program for tracking and registering not only the transfer or possession of select agents that may affect animal health and we also in the bioterrorism bill encouraged those two agencies to coordinate so that we end up with eventually a single registration and tracking system. After all, a select agent that can hurt an animal could well likewise hurt a human being.

The question I have for you is, is it good for us to be transferring this function from CDC and at the same time leaving the other function at USDA, or should both functions be transferred simultaneously or neither one? What is your recommendation?

Mr. HAUER. Well, let me give you the perspective from HHS. I'm not sure I can address the USDA perspective, but the—from HHS's perspective, CDC has never really been a regulatory agency. So having that function within CDC I think is probably not the best spot to locate it. CDC is quite good at determining what agents should be on a select agent list, but I think moving it into Homeland Security, where they have more of a coordination with law enforcement and intelligence, again provides a better synergy——

Chairman TAUZIN. But if that is correct, isn't it also true that the USDA function should move with it?

Mr. HAUER. I don't know why at this point in time the USDA component remains where it is. I can't answer that.

Chairman TAUZIN. So you can't answer that. We can't either, and we don't understand that. It seems to us, at least—I would love your thoughts on this. But if you're going to move the CDC program because of the fact that it now connects the registration and tracking system to the coordination of law enforcement efforts in the area, that it would be logical to do the same thing for the USDA program as well.

Mr. HAUER. Well, at first blush I would envision that the USDA in fact does do regulatory work and does have the capability to go out and do an inspection and track these kinds of things, where CDC and HHS have historically not. That would—that would be my——

Chairman TAUZIN. Nevertheless, the coordination with law enforcement personnel in materials of the investigatory outreach of those agencies in tracking and identifying perhaps the mishandling or mistransfer or improper possession of these agents would seem to compel an argument that those functions ought to all be coordinated out of the same office, does it not?

Mr. HAUER. I certainly understand your thinking on this, but I will ask the folks at Homeland Security who are working on this what their thinking was, and I'll get back to you on it.

Chairman TAUZIN. Please do.

Finally, we've got a lot of work to do today, and I don't want to keep you, but we watched with great interest the shooting at LAX Airport this last week. Interestingly enough, when the shooting occurred they immediately began a debate as to whether or not it was a terrorist attack. We noted local law enforcement—local guards, first of all, El Al Air and the local guards at LAX did their job, apparently, well. Local police came in and effectively did their job well. FBI was called in to find out whether it was a terrorist attack. Are there some parallels here to the questions of how we manage the very close similarity and features between a bioterrorism attack and a naturally occurring infectious disease in our Nation, and are their lessons there for how this new Homeland Security will work with other agencies that deal with everyday disease and research, et cetera?

Mr. HAUER. Yeah. I think that is an excellent comparison, because early on in the evolution of an outbreak one might not know whether it is a naturally occurring outbreak, whether it is, you know, a bunch of kids who have had something bad to eat at school or—in a lunchroom or whether in fact we've had something a little more nefarious occur.

It's only—you know, there are certain assumptions at this point in time. If we were to see anthrax or smallpox we could assume that it probably is not something innocent, but some of the other agents, some of the other things that could potentially be used could create some confusion——

Chairman TAUZIN. Although even with anthrax you have animals spreading the——

Mr. HAUER. In the Southwest we see——

Chairman TAUZIN. The point I'm making is that the way in which we structure this new department, we had best be careful not to remove from certain agencies in the health community their ordinary capacity to be the first on the scenes and to do their job in terms of assessing an outbreak before you know whether it's a terrorist attack or a naturally occurring process. Right?

Mr. HAUER. I think that is absolutely right. I think at the end of the day this Department, first of all, will not have impact on the way State and locals do outbreak investigation, other than to potentially strengthen what goes on at the local level and to be a catalyst for better coordination with law enforcement. But public health at the local level will continue as it is, as will public health activities at the Centers for Disease Control; and we have to ensure as the new Department evolves that the coordination between the new Department and HHS is a solid relationship so that those things that could evolve very innocently, at the end of the day that might not be—are escalated to the new Department for——

Chairman TAUZIN. Yes. We can draw another really far-out analogy. There was a horrific story a few years ago of the Boy Scouts, you know, trekking around the mountains here in America only to have one of the kids get lost. A helicopter located him, but nobody seemed to be—have the authority to tell the helicopter to land, pick him up, and to leave the kid out there overnight. Parents knew he was out there alone at night lost in the woods. The helicopter couldn't get authority to land to pick him up.

That's our worst fear as we make these changes. Please know that. We hope—work with us on that. Our worst fear is that some bureaucrat is sitting around trying to figure out whether this is a bioterrorism attack or whether this is naturally occurring and so nobody moves until that decision is made. Our worst fear is that we take away authorities that currently would respond quickly, regardless of what caused the problem and begin to deal with it rapidly and effectively. Now I would hope again that be central in all of the considerations and the transfers of authority that take place.

Thank you, Mr. Chairman.

Mr. GREENWOOD. The Chair thanks the gentleman.

The gentlelady from Colorado, Ms. DeGette.

Ms. DEGETTE. Thank you, Mr. Chairman.

I'd like to follow up on Chairman Tauzin's questions, because I think they are important. You see, when a gunman goes into the L.A. Airport and shoots somebody, you know a crime has been committed. So local law enforcement officers respond to the crime, and they do what they need to do. Then we come and say, was it a terrorist attack or something else? But with biological warfare it is not so clear-cut, and I think that is the issue we've got.

Because what happens is there's an outbreak of something, not anthrax or smallpox. I mean, we're going to make that same assumption just like when you have a gunman going into the L.A. Airport. But let's say you have some other kind of outbreak, and what would happen—I mean, right now, the CDC has responsibility for researching naturally occurring public health issues, issues like flu, food-borne illnesses, new and emerging infectious diseases. So we're now going to know—we're not going to be able to say if there's an outbreak of one of these diseases, well, this could be bioterrorism. It's probably likely going to be naturally occurring, unlike a shooting or some clear-cut crime. And so the question—so what would happen right now is there's an outbreak of Legionaire's disease or something else. CDC begins to investigate it. How are we going to be able to separate those normal functions of the CDC from biological warfare, and how is the new agency going to do that without totally subsuming the CDC's routine functions?

Mr. HAUER. Well, let's start at the local level, because all terrorism really starts at the local level. I very much like your analogy of the law enforcement officers and the way they respond. An outbreak at the local level usually starts with a local or county health department, and it's escalated usually to the State health department and then to CDC. None of that would change. Much as at the local level the law enforcement agencies call the FBI in, local health departments will call in either the State health department or CDC.

If in fact the outbreak or the incident is suspected based on what they see, based on the patterns to be something intentional rather than a naturally occurring outbreak, then the Department of Homeland Security clearly would be brought in early on so that the coordination with law enforcement and with other agencies is begun as quickly as possible.

Ms. DEGETTE. Let me stop you right there. Who makes the determination to bring them in?

Mr. HAUER. Well, that would be part of—as the information becomes available, HHS would notify Homeland Security very quickly that an outbreak investigation that they are looking at looks to be other than a naturally occurring outbreak; and as a matter of course, the new Department would probably be notified anytime there's a large outbreak or something suspicious as a matter of courtesy.

Ms. DEGETTE. Okay. So I'm just trying to follow this, because I think it's important.

Denver health, which actually—we have, as you know, one of the few coordinated departments. They see something they suspect is a big outbreak of something. They bring in the State health department. The State health department brings in CDC. Is it the CDC through HHS that then decides whether or not to bring Homeland Security in? Are there going to be Homeland Security people looking at that, too?

Mr. HAUER. No. I would envision the HHS would notify Homeland Security that there's an outbreak of some kind, that it's being tracked. The other——

Ms. DEGETTE. Okay. And then what's Homeland Security going to do that CDC doesn't do now?

Mr. HAUER. Well, it depends on the nature of the outbreak, depending on what assets are needed, if the national pharmaceutical stockpile is needed. A lot of that requires infrastructure from other Federal agencies, air assets, mobilization of people, of volunteers, depending again on the nature of the outbreak, if we have to do a mass prophylaxis of people or a vaccination. All of that requires coordination from other Federal agencies, and that is better coordinated at the Department of Homeland Security.

Ms. DEGETTE. Right. But I'm really talking about disease research and identification. In our last hearing, some of the medical and research experts said that they thought that transferring some of the responsibility for research in biological countermeasures and identification is simply going to add bureaucracy and it's going to put scientific decisions out of the hands of scientists and into the hands of bureaucrats. I'd like to know your response.

Mr. HAUER. Yeah. But we've got to separate outbreak investigation or a response to a potential incident to the actual research and development activities that are going on on a daily basis. And the R&D activities that are currently being conducted by NIH and CDC would continue as they are——

Ms. DEGETTE. Those would not be supervised by the new Department?

Mr. HAUER. The new Department would coordinate and would oversee the bioterrorism, chemical terrorism and nuclear terrorism-related activity to ensure that at the Federal level we have a better coordinated approach. We have numerous R&D activities going on at the Federal level. By bringing them all together in one department we will have a more effective and more efficient R&D program.

Ms. DEGETTE. Just one last question. I would ask unanimous consent for one additional——

Mr. GREENWOOD. Without objection. The Chair would just remind you it's going to be a long day.

Ms. DEGETTE. I know.

But the Department of Homeland Security, we sort of have—it would say to CDC and to NIH, we want you to do this research, because we think it's important.

Mr. HAUER. Yes, I believe that is——

Ms. DEGETTE. That is——

Mr. HAUER. The question is one the chairman had earlier on.

Ms. DEGETTE. Thank you, Mr. Chairman.

Mr. GREENWOOD. The Chair recognizes the gentleman from New Hampshire, Mr. Bass, for 5 minutes.

The gentleman from North Carolina, Mr. Burr.

Mr. BURR. Thank you, Mr. Chairman.

Welcome, Mr. Secretary. I really want to focus on the office that you're Acting Secretary of that we created in the bioterrorism bill, because I think that was debated heavily within the ranks of this committee. We saw a tremendous need for it, and I would take for granted you still see a tremendous need for it.

Mr. HAUER. Yes. I think that there needs to be some kind of a coordinated function within HHS to continue to coordinate HHS's activities; and, again, the need for an Assistant Secretary's position is one that I would defer to Congress to decide about.

Mr. BURR. Well, the great thing is you're here and we get to ask you. For that reason, I would ask you, do we need in the newly configured world of a shift of some responses over to Governor Ridge and Homeland Security to still create an Assistant Secretary slot at HHS that's configured much the same way?

Mr. HAUER. Well, I think again there needs to be some coordinated function at HHS to be interfaced with the new Department and to coordinate the multiple activities within HHS that will remain with the Department once components are shifted over to Homeland Security.

Mr. BURR. And the components that currently are under debate to shift are pretty obvious ones—the National Medical Response Team, the Metropolitan Response Teams, who also have significant roles as it relates to training, precautionary deployments because of national and international conferences that we feel that they may be needed for, which you could make a tremendous argument that they ought to stay over in HHS because their hopeful deployments are more for practice than they are for the actual threats.

But we do leave at HHS the responsibilities, as I understand it, for bioterrorism preparedness, emergency preparedness, a number of things that are still over at HHS.

I think the reason that this committee specifically created this Office of Assistant Secretary was we saw that without an Assistant Secretary there was an inability to focus on the preparedness that was needed. So I am asking you if we only shift a few of the functions and we've still got this huge slice that deals with our ability to prepare, don't we still need an Assistant Secretary level at HHS to drive that focus?

Mr. HAUER. Yeah, the functions that are being shifted over to Homeland Security, or I should say the most obvious ones, are those that have response functions to either natural or intentional incidents, OEP, NDMS. Those functions because of the work they do with FEMA, with the Secret Service, with the FBI would fit quite naturally in the new department. The funds that will be coordinated and remain with HHS do need to have some coordinating focus within the Secretary's office. And how that's structured, you know, until the bill was passed we had a Director of the office, a Special Assistant. Any structure would work but we do need to have a coordinated focus within HHS for the Department of Homeland Security to deal with.

Mr. BURR. I make the claim that there's a greater tendency today on the part of any agency to say now that we have an Office of Homeland Security it's their responsibility to make sure that the infrastructure for preparedness, whether it's national, whether it's State, whether it's local, is in fact in place, that they've the correct training, that they've got the right equipment, that they're prepared. Now we all understand it's the Office of Homeland Security that would make the notification in the event of a threat. But clearly the way we've designed it downstream, the equipping, the training will stay as a responsibility of HHS, am I correct?

Mr. HAUER. That's correct.

Mr. BURR. I guess what I would ask you is given that we were there before and even though we knew we needed this and this was from administration to administration to administration yet we didn't make a lot of progress, I'll ask you again don't you need an Assistant Secretary level to drive the type of focus within HHS regardless of who's there to make sure that downstream we have the components in place to be able to adequately address the call from Homeland Security that says we have to mobilize?

Mr. HAUER. Yes, we definitely need somebody in the Secretary's office who reports to the Secretary, has the weight of the Secretary, to get and to maintain coordination of all the operational divisions within HHS.

Mr. BURR. I might also add that we saw this office as a key component to our ability to rebuild our public health infrastructure in America. We cannot lose focus of that opportunity that we have. Even as we talk about how to split up these things we cannot miss the opportunity to rebuild that public health infrastructure, and that's through CDC.

Mr. HAUER. It's within the focus of HHS and I think it deserves input from the Assistant Secretary level person.

Mr. GREENWOOD. The Chair thanks the gentleman. Mr. Stupak is recognized for 5 minutes.

Mr. STUPAK. Thank you, Mr. Chairman. Following up Mr. Burr's question, whether we need an Under Secretary, the bill we have before us would give the current Secretary the authority to establish guidelines for State and local government efforts to develop and implement countermeasures to chemical, biological and other threats. But I believe the bill is ambiguous in conferring this authority. Is it your understanding that the guidelines are mandatory or are they merely suggestions?

Mr. HAUER. The guidelines for State and local governments?

Mr. STUPAK. Yes.

Mr. HAUER. Right now most of the money comes from the Federal Government.

Mr. STUPAK. But the bill. The bill's language, suggestive or mandatory?

Mr. HAUER. Subjective of what, the guidelines?

Mr. STUPAK. Conferring the authority. Who's going to have the authority to do this?

Mr. HAUER. It would be the Under Secretary of the Department of Homeland Security.

Mr. STUPAK. That's mandatory; they would have to do that?

Mr. HAUER. That they would establish guidelines.

Mr. STUPAK. But yet it appears the bill doesn't really grant that authority. Do you believe these guidelines will be effective and will help to improve our State and local emergency preparedness if the Federal Government does not assist State and local units in acquiring the necessary technology? You can do the guidelines, you can say it's mandatory, but how do you assist them in acquiring the necessary technology to actually implement these guidelines to make us more secure?

Mr. HAUER. At the State and local level? In fact, I think the Department of Homeland Security would fix one of the greatest problems for State and local responders, and that's been this multiple department approach. And, again, coming from the local level and the State level, I was always quite frustrated dealing with numerous departments here in DC.

Mr. STUPAK. How does this fix it?

Mr. HAUER. It has one department that is coordinating units, one department that is coordinating technology, one department that is sending out a single message.

Mr. STUPAK. I won't have to go to HHS or DOE; I can go right to this department?

Mr. HAUER. Correct.

Mr. STUPAK. Good. As you may be aware, the administration issued its homeland security proposal in advance of the joint House and Senate Intelligence Committee's completion of its analysis of what happened on September 11, how it might be prevented and how, moving forward, we might effectively respond to another attack. In your experience, isn't this analysis key to understanding what is needed to effectively prepare and to respond to an attack? Basically we're pushing this administration proposal, the chairman said it's going to be done by July 12. Shouldn't we really have the analysis of the joint committee before we push through a proposal? Don't we have the cart before the horse here?

Mr. HAUER. In fact, I think that the proposal as it stands shows a fair amount of insight into the problems that we've confronted as a government.

Mr. STUPAK. A fair amount, but not all the insight, correct?

Mr. HAUER. Poor choice of words. I think as it was drafted it really addresses many of the problems that I as a local responder for many years——

Mr. STUPAK. Many, but not all. See, the point I'm trying to say, if we've got these joint committees doing the work, shouldn't they do the work before we push through this proposal? I don't want to push this thing through and get it on the floor because we have this end of month recess, we're going to August, and say, jeez, we did this and we find out we did it half right or many things right but not all.

Mr. HAUER. I think the longer we wait to try to resolve some of these problems, the longer we wait in trying to address the creation of this department, the longer we maintain some of the vulnerabilities that we have at the State level, the local level and at the Federal level. I think that's why the President at this point is anxious to move forward so aggressively to address this, because I think they see and understand exactly what the issues are.

Mr. STUPAK. But with all due respect, in our aggressiveness we shouldn't be blinded to the real needs that this country needs in homeland security.

Mr. HAUER. Agreed. But I think that as the proposal stands, I think it addresses those issues. I think it addresses the needs.

Mr. STUPAK. Let's move on. I'd like to see the joint committee issue a report, their analysis, before we jump forward with this bill and not under this arbitrary time line. But the bill would also grant DHS the authority to develop interoperative communications technology in helping to ensure that emergency response providers acquire such technology, yet the bill does not include the authority for DHS to provide grants to assist State and local units of government to equip the first responders with the so-called interoperative communications technology. So how are we going to do this? How will the communications interoperability be improved without the authority?

Mr. HAUER. I believe through the FEMA grant program and through the Department of Justice grant program there is the capability for State and local governments——

Mr. STUPAK. But you told me earlier we were going to do it all under Homeland Security for one-stop shopping.

Mr. HAUER. But that will all be moved under Homeland Security.

Mr. STUPAK. So FEMA won't be in any more?

Mr. HAUER. In combining those grant programs, one of the biggest frustrations at the local levels is all these multiple grant programs coming from FEMA, Justice and trying to figure out guidelines and who can do what, in trying to streamline that particularly, and I think you're absolutely right, interoperable communications has been a problem. It was a problem at the World Trade Center. And you know in many of the incidents I've managed over the years, interoperable communications has been a need. This new department will allow a better funding stream, a more coordinated funding stream and a more efficient funding stream so that things like interoperable communications can be funded.

Mr. STUPAK. So FEMA and Justice would be out of it?

Mr. HAUER. They are moved into the new department and the money that they currently flow to the State and local governments would be better coordinated.

Mr. GREENWOOD. The time of the gentleman has expired. Chair recognizes the gentleman from Illinois, Mr. Shimkus.

Mr. SHIMKUS. Thank you, Mr. Chairman. I'll try to be brief, too. To my friend from Michigan I would just say that it's—he needs to talk to his minority leader. One of the driving emphasis to the movement so quick is the September 11 date set by the minority leader. I, too, am concerned about the speed at which we're moving. It's a bipartisan rapid movement to a finish line. I would—I like the aspect of this registrar to track and get information down or up the chain through the community health departments, the other Clancy book where they have the ebola virus spreading all over the place. The President at that time just stopped all flow of individuals within the country to try to contain the spread. I think that's part of it, the vision of what would happen if you could gather information of sporadic outbreaks and if you stopped the movement of people then you could possibly stop the spread, and I think that's a very positive thing.

The question I have is why is the scope of the research authority given to the new Secretary limited to civilian efforts of the Federal Government and to what extent is the DOD research in these same areas, which I understand is quite extensive, being incorporated into the new department's authority?

Mr. HAUER. I think when you look at the research that DOD does versus the research that we do at NIH and CDC, the research that DOD has historically done is combat environment research.

It's based on military constructs and the needs of the military. And very often the military research does not transfer that easily to the civilian environment. We saw that in 1996 when a lot of DOD equipment was given to cities as part of the original Nunn-Lugar funding and a lot of cities really could not integrate military equipment into their response. The civilian research, however, is pretty much focused on the needs of the civilian responders, whether it's for chemical attack, nuclear or biological.

So I think keeping that separation is a good one, I think, though it doesn't prohibit nor does it in any way obstruct the ability for the researchers to share data, to share what they're doing, but the military focus is really just a different focus. Ideally what you can do is take some of that military research that's going on, and this in fact occurs now, and hand it over to the civilian environment but to just, to take the military research per se, their focus has historically just been a little different.

Mr. SHIMKUS. Does that sharing go on now?

Mr. HAUER. Yes. In fact, we've got an ongoing program with DOD and we work closely with DOD. But again we work with them, a lot of their focus—and we're doing one project now on bio detection with them, but they look from the military perspective and how it would work in a combat environment. We have to translate that into how it works in the civilian environment.

Mr. SHIMKUS. Being from the military environment, I understand there's great differences.

I yield back my time.

Mr. GREENWOOD. The Chair thanks the gentleman. The gentleman from Texas, Mr. Green, for 5 minutes.

Mr. GREEN. Thank you, Mr. Chairman. We heard in our last hearing and again today we'll hear from medical and research experts that they believe transferring the responsibility of research and development in biological countermeasures to the Homeland Security is inefficient and adds additional bureaucracy and puts scientific decisions in the hands of the nonscientific agency. We haven't heard a response from the administration. Do you have a response to that concern about the adding bureaucracy, then taking the decisionmaking away to a nonscientific agency?

Mr. HAUER. First of all, I don't think it takes any decisionmaking away from the researchers of NIH or CDC. I think at the end of the day having a central agency centrally coordinate the needs for research and then helps coordinate the direction of research based on the needs of terrorism will provide more efficiency with the use of our research dollars and I think more effectiveness. But it in no way inhibits what's going on in the scientific community. It in no way inhibits what is going on right now. It does not do anything to undermine the basic science research at NIH or the public health research at CDC.

Mr. GREEN. You're the Acting Assistant Secretary of Public Health Emergency Preparedness. Your entire office would go over to the Homeland Security. Does that mean that HHS does not need to be in the public health and emergency preparedness program and would HHS be out of the preparedness business?

Mr. HAUER. No. As I mentioned earlier, I think there's still a need to have somebody on the Secretary's staff who would be coordinating the activities of HHS.

Mr. GREEN. And they would coordinate, I assume, with the——

Mr. HAUER. With the department.

Mr. GREEN. Would you explain the relationship between the select agent registration and lab security program with that of the lab bio safety programs already in operation under the proposed select agent registration lab security program.

Mr. HAUER. The new select agent program is one that allows the new Department of Homeland Security to track and monitor agents that are being used. Mainly the pathogens that are being looked at in research labs around the country right now are pathogens that could be used as bioterrorist weapons. The new legislation, the legislation that was passed and would now be transferred to Homeland Security, allows the Department of Homeland Security to inspect labs, to trace and track the shipment of select agents, because clearly these agents could potentially be problematic if there's no control.

Mr. GREEN. Who's going to take over the bio safety inspection program and will there be coordination?

Mr. HAUER. Clearly there will be coordination. There's no question about it. CDC and NIH will continue to give guidance and technical assistance to the department re the lab safety and security issues.

Mr. GREEN. Who's going to actually do it? Will it be the Homeland Security?

Mr. HAUER. Yes, the Department of Homeland Security will oversee the select agent registry. The CDC and NIH will have input as to what agents should be on the select agent list.

Mr. GREEN. Okay. Will the new Homeland Security also take over the bio safety regulation inspection program or will that still be under the CDC?

Mr. HAUER. That would be under, if I'm correct, under the CDC.

Mr. GREEN. Are you going to have dual inspectors? Will there be collaboration between the two?

Mr. HAUER. I would envision there would be good collaboration between the two.

Mr. GREEN. So there wouldn't be dual inspections by two Federal agencies?

Mr. HAUER. I would not envision it at this point. That level of detail needs to be worked out as the department evolves.

Mr. GREEN. Thank you, Mr. Chairman.

Mr. GREENWOOD. The Chair thanks the gentleman, recognizes the gentlelady from New Mexico, Ms. Wilson, for 5 minutes.

Mrs. WILSON. Thank you, Mr. Chairman. I appreciate your having this hearing today. I appreciate your having so many good witnesses. I wanted to start out with an observation, listening to the questioning that's going on, and then ask a couple of questions. It seems to me that kind of getting away from some of the detail and getting up to a more general strategic level, there are two strengths that the United States can buildupon to ensure homeland security. One is the ability to collect and integrate information. That's not within the realm of this committee's jurisdiction, but it is critical. The second is technological superiority. Without leveraging those two strengths, this department is not going to be successful, I don't think. Even the draft legislation is weak in both of those areas, as evidenced by the confusion about the R&D structure and the lack of an R&D structure in the bill itself. In the draft bill there's only one Under Secretary that has any kind of responsibility for R&D. It's underemphasized in the bill. It's unclear. And that's been demonstrated amply by the questions and answers we've gotten this morning where it needs to be clarified and we need to know how we're going to pursue research, development, tests, and evaluation, because if we don't we won't be able to leverage those strengths long term.

I don't think the answer is necessarily moving more cells back and forth between a new department and old department in regard to particular research. I also think we should avoid the temptation to designate a particular group or a particular laboratory as the one that does research on homeland security, because every Under Secretary is going to have different needs. If you're the Under Secretary who's responsible for biological events you will be wanting to call on NIH and CDC and the pharmaceutical agency and national laboratories and universities, depending on the nature of the problem, of the alligators that are after you today. At the same time there needs to be a longer term focus. The guy who is responsible for emergency preparedness has problems and operations he needs to conduct today and will be looking at applied research. But there is nothing in the bill that gives the Department of Homeland Security that broad look and that long-term look for both basic research and applied research that will give us the things that all of those guys are going to be asking for 10 years from now or 20 years from now. It's a critical weakness in the bill, and I think we're going to have to remedy it in part in this committee as well as other committees that are looking at it.

Let me ask a couple of questions. What is your understanding in the base bill of where the budget authority lies? Who gets the money and manages the money for bioterrorism research? Is it you, or is it HHS or is it Department of Homeland Security?

Mr. HAUER. It's the Department of Homeland Security.

Mrs. WILSON. So they are the ones who will determine where the dollars get farmed out to, but they can call on the CDC lab or NIH or whatever?

Mr. HAUER. Or one of the national labs. Again, one of the reasons you have better coordination of funding is by having a central focus for all the R&D on these different threats.

Mrs. WILSON. You currently do work for others within HHS? I mean, do you have a Department of Defense or other Federal agencies come and say we've got this piece of work to do and there's a laboratory within HHS that has the expertise to do it?

Mr. HAUER. On occasion, yes.

Mrs. WILSON. Do those relationships—are they difficult to perform or are they—I mean, is it hard to do or is it a fairly seamless process for that money to be——

Mr. HAUER. It depends on the nature of the research.

Mrs. WILSON. Okay. Do you do cooperatively research with different agencies or even private industry and HHS laboratories-funded research?

Mr. HAUER. Yes.

Mrs. WILSON. How do those work?

Mr. HAUER. It depends again on the nature of the research. We have ongoing research with other agencies, with universities; depends on whether it's the biologic area where we have strengths or in the nuclear area where we work with some of the national labs on some of the R&D programs. We have certain needs, but they might have certain strengths. Again it depends on the research.

Mrs. WILSON. I know that your position is new and you're just getting up and running, but have you got in progress a strategic plan or an R&D road map for setting priorities in both short-term and long-term research in the biological area?

Mr. HAUER. Absolutely. We're looking at a number of areas, including new vaccines, new antibiotics. We're working with NIH on the research strategy, both short and long term, on immunomodulations to try and avoid using antibiotics. We're again working both within HHS and with outside experts to help formulate some of the agenda.

Mrs. WILSON. Thank you.

Mr. GREENWOOD. The Chair thanks the gentlelady; recognizes for 5 minutes the gentlelady from California, Ms. Capps.

Mrs. CAPPS. Thank you, Mr. Chairman, for holding this hearing and to you, Mr. Hauer, for your testimony. The publication, weekly publication, Mortality and Morbidity Report, from CDC was very useful to me in the years that I was a school nurse working in public health, as I did in my career before coming to this place. When I visited with Mr. Deutsch at the CDC a few months ago, I was struck by how intimately involved they are with health officers and health facilities on the local level, probably more than about any other Federal agency that I'm aware of. To me that is a real skill.

To cite another example, shortly after 9/11, I, as did many of my colleagues, went back home to the various local entities that had come together around this event and put together disaster teams of preparedness using the local structure. I asked them what they needed. They told me they felt the public health and other infrastructures was already stretched before 9/11 at the local level. We know that any event happens locally, witness the LAX event, whatever that means, which happens at a precise place.

So what we have both in CDC and NIH, in my opinion, is that ability with professional expertise to go up and down the line from the local research interests to the national ability to gather together, coordinate and so forth. If we were stretched before 9/11 in our infrastructure, which I and others believe we were, how can whatever is being put together now through Homeland Defense mitigate those real needs that are there and not be seen, as some at home have said to me, where's the money for the first responders. We know what we need to do to prepare. We don't even have vaccines on hand for flu or various other things. That's where I'm coming from today.

Mr. HAUER. The issue of the money getting to the first responders, HHS, and in probably record time for the Federal Government, the Secretary, once the President signed the bill in January, told us that we had a very short window to get money to the local and State public health departments, and in fact we did that. Thirty days after the grants came in they were reviewed. We had over 114 grants reviewed within 30 days from the four cities and the States and Territories, and they were reviewed and the money was committed and got it out to the States.

Mrs. CAPPS. I don't want to interrupt you, but right at that point that was before this legislation. This new department that we are struggling with and asking questions about is going to be yet another layer. I mean, the immediacy that you just described is what is needed. What will this agency do?

Mr. HAUER. I think a couple of things. First of all, this department, you mentioned earlier on that local responders continue to wonder where the money is. Having been one of them for many years, I understand those frustrations. I think what the President is trying to do is ensure that there is an efficiency in getting that money out as well as a rapidity in getting money out to State and local responders, but ensuring at the same time that there's not duplication of programs, as there is now, which is confusing local responders and making it very difficult for them to understand training guidelines. We have training programs that sometimes have contradictory information. We are giving out equipment, hand-held assets for detecting biological equipment, which should not be used by local responders.

Mrs. CAPPS. I see the yellow light. I want to pin this down because I appreciate the fact that you have had very local experience and so you know what this is. It is very easy to be cynical and say this is just one more bureaucracy. Precisely at the area you're describing, coordination some of us are hearing, is that a transfer of authority? And I don't think we're—I for one am not even on the subcommittee but I'm very interested in this topic. I need to be convinced more clearly that the authority of professionals in the field, in the area of research, whatever, is not going to compromise that integrity.

Mr. HAUER. I don't think there's any question that what this bill envisions is allowing from an R&D perspective those folks doing R&D, whether it's the national lab, NIH or CDC, to continue their focus, but when it comes to the threats that confront the country right now, whether it's chemical, biological or nuclear, to have a more coordinated effort in researching those types of remedies so that we don't have one agency giving something out to first responders that might not work while another agency is saying don't use them. And that's where we stand right now. I think that's what the President is trying to do.

Mrs. CAPPS. Thank you.

Mr. GREENWOOD. The Chair thanks the gentlelady. Mr. Hauer, thank you so much for your testimony, for answering our questions and for your service. You are excused.

The Chair would call the next panel, consisting of Ms. Jan Heinrich, who is the Director of Health Care and Public Health Issues at the General Accounting Office; Dr. Gail Cassell, Vice President, Scientific Affairs, Distinguished Research Scholar for Infectious Diseases at Eli Lilly and Company; and Dr. Margaret

Hamburg, Vice President, Biological Programs, Nuclear Threat Initiative.

Thank you for being with us. This committee is holding an investigative hearing. When doing so, it is our practice to take testimony under oath. Do any of you have qualms about testifying under oath?

In that case, I should also inform you that pursuant to the rules of the committee and the House, that you are entitled to be represented by counsel. Do any of you wish to be represented by counsel?

Thank you. If you will stand and raise your right hand.

[Witnesses sworn.]

Mr. GREENWOOD. Thank you very much. You are under oath. Ms. Heinrich, you are recognized for 5 minutes.

TESTIMONY OF JANET HEINRICH, DIRECTOR, HEALTH CARE AND PUBLIC HEALTH ISSUES, GENERAL ACCOUNTING OFFICE; GAIL H. CASSELL, VICE PRESIDENT, SCIENTIFIC AFFAIRS, DISTINGUISHED LILLY RESEARCH SCHOLAR FOR INFECTIOUS DISEASES, ELI LILLY AND COMPANY; AND MARGARET A. HAMBURG, VICE PRESIDENT, BIOLOGICAL PROGRAMS, NUCLEAR THREAT INITIATIVE

Ms. HEINRICH. Mr. Chairman and members of the subcommittee, I appreciate the opportunity——

Mr. GREENWOOD. I think your mike is not on yet.

Ms. HEINRICH. Is that better?

I appreciate the opportunity to be here today to discuss one component of the proposed creation of the Department of Homeland Security. My remarks will focus on the potential effects of reorganization of biomedical research under Title III.

The proposed department is tasked with developing national policy for and coordinating the Federal Government's civilian research and development efforts for all threats, biological, radiological and nuclear. The new department could improve coordination of the biomedical research efforts, most of which is sponsored or conducted at the National Institutes of Health. The President's proposal could help improve coordination by giving one person the responsibility for a single national research and development agenda.

In the past, we have recommended the creation of a unified strategy to reduce duplication and leverage resources, and suggested that the plan be coordinated with Federal agencies performing research as well as State and local authorities. Such a plan would help to ensure that research gaps are filled, unproductive duplication is minimized, and that individual agency plans are consistent with the overall goals.

Interagency coordination will remain essential under the proposal. It should be noted that the legislation focuses on civilian efforts only. The new department will also need to coordinate with DOD because it also conducts biomedical research geared toward protecting service members, but applicable to the civilian population as well.

NIH and DOD currently collaborate on a number of projects, such as a shared data base to compare the sequences and functions of pox virus genes and testing of new vaccines. This coordination needs to continue.

It also includes four academic centers, CDC, USAMRID, DARPA, and American Type Culture Collection.

Despite these positive aspects of the proposal, we are concerned about the implications of the proposed transfer of control and priority setting for dual-purpose research. The President's proposal would transfer the responsibility for biomedical defense research to the new department, but the programs would continue to be carried out by NIH. These programs include a variety of efforts to understand basic biological mechanisms of infection and to develop and test rapid diagnostic tools, vaccines, and antibacterial and antiviral drugs.

The research on biologic agents that could be used by terrorists cannot be readily separated from research on emerging infectious diseases. For example, research being carried out on antiviral drugs for biodefense research is expected to be useful in the development of treatments for hepatitis C. Research to expand our knowledge of factors that play a role in determining antibiotic resistance, virulence and invasiveness, as well as factors influencing the severity of disease, are critical to emerging infectious diseases as well as biodefense research.

In addition, the proposal would allow the new department to direct, fund and conduct research on its own. This raises the potential for duplication of efforts, lack of efficiency, and an increased need for coordination with other departments that would continue to carry out relevant research. It is inefficient to build and duplicate the expertise and facilities that already exist in the current Federal laboratories that are needed to conduct this work.

In conclusion, better coordination of research efforts could reduce wasteful duplication and increase efficiency. We are concerned, however, with the President's proposal to transfer broad control of biomedical research to the new department. Although there is a need for a strategic plan for research, there is also a need for maintaining the synergy of biodefense, emerging infectious diseases, and basic biomedical research efforts.

The R&D funding and priority setting needs to be vested at the program level best positioned to understand the benefits of both the basic and applied research efforts. If disagreements arise over priorities for biomedical research, there may need to be a mechanism for resolution within the Office of the President or in the Congressional appropriations process.

Mr. Chairman, this concludes my remarks. I would be happy to answer any questions.

[The prepared statement of Janet Heinrich appears at the end of the hearing.]

Mr. GREENWOOD. Thank you, Ms. Heinrich.

Ms. Cassell.

TESTIMONY OF GAIL H. CASSELL

Ms. CASSELL. The establishment of a new Federal Department of Homeland Security can potentially achieve greater efficiency, effectiveness and accountability regarding many aspects of terrorism. However, there are unique characteristics of bioterrorism that deserve special consideration and suggest the need to address them in a manner differently than that proposed by the administration's bill. I will limit my comments this morning to those that directly relate to research leading to countermeasures.

There is no simple counter to bioterrorism, no magic bullet. Instead, development of an integrated set of strategies is required. Such efforts must include preventing countries from acquiring bioweapons in the first instance, dismantling existing programs and capabilities where proliferation has already occurred, deterring the use of biological weapons and ultimately putting in place countermeasures that can rapidly detect and effectively defend against such use. It is the latter that requires special consideration with respect to the proposed role of DHS.

In the long term the only way to defend against bioterrorism is through a combination of constant surveillance, accurate diagnostics to identify threats as early as possible, and availability of high quality vaccines and drugs that can be useful against any attacks that do occur. Research related to bioterrorism is inextricably linked to that of naturally occurring infectious agents and development of the new antibiotics, antivirals diagnostics and vaccines. Thus, the research and development of technologies for biodefense should be synergistic and duplicative.

The diversity of existing biological weapons and the ever increasing possibilities preclude simple therapeutic countermeasures to bioterrorism. Currently our countermeasures are very limited, even for known threats. This is a very important consideration. There are 13 viruses on the select agent list today, yet there is only one antiviral and this is for smallpox, and it must be administered intravenously. There are no truly broad spectrum antivirals. We have only a limited number of antivirals for a few naturally occurring viruses.

The situation is somewhat better, but still worrisome with respect to antibiotics. There has only been one new class of antibiotics developed in the past 30 years. The Russians are known to have constructed bioweapons resistant to current antibiotics. While there are currently 23 antibiotics in Phase I through III clinical trials today, there are few new classes and importantly no new broad spectrum antibiotics, only more quinolones like Cipro. In short, our antibiotic armamentarium is limited and there is growing concern about an increase in resistance to existing antibiotics, exemplified by two different bills introduced within the past few months.

An idea of the problem of resistance, in fact it's now known that the E. coli strains occurring, 90 percent of these are—or 50 percent of these actually are resistant. Thus it seems clear that no public health response to bioterrorism is likely to prove effective without addressing the overall problem of existence and the technical challenges of drug discovery and development.

Development of effective countermeasures will depend on interdisciplinary research ranging from basic research into the mechanisms by which the agents cause disease, how the body responds, and how the agents are transmitted. This new knowledge then must be used to develop innovative vaccines, antibiotics, and antivirals and immunomodulators. Equally important will be to benefit from knowledge gained in previous failures in development efforts. It is important to realize while development of a new bioweapon only takes months, development of a single new drug or vaccine on average, based on many years of experience, requires anywhere from 8 to 10 years. Thus, meeting time lines and research goals are vital to our defense given our current situation.

The magnitude of this challenge cannot be underestimated. Success will require involvement of the very best scientific and medical talent in government, academia and the private sector. Likewise, in order to achieve success in a timely manner the United States must be able to capitalize upon the existing infrastructure for product development. Engaging the full spectrum of private industry from the smallest biotech to the largest pharmaceutical companies in the search for solutions will not only greatly raise the chances of success, but also significantly lower the total cost to taxpayers, augmenting public appropriations with private capital investment.

The NIH, specifically NIAID, is uniquely positioned to lead this effort. The NIH recognizes that significant advances occur when they often are unforeseen. These advances expand the experimental possibilities. It must be recognized that not all research problems are equally approachable no matter how urgent and important to public health. Development of countermeasures for bioweapons is not like designing a new tank. Research and development of countermeasures will be a long-term endeavor.

There is always uncertainty about where the most valuable discoveries can be made, but NIH is best placed to identify scientific opportunities and applications that are relevant to the most pressing issues. NIAID is unrivaled in its track record of bringing the right scientists and rigorous peer review and oversight of funded research. Indeed, many of the best investigators have already been funded by NIAID and have recently made major advances in determining the mechanism of action of the anthrax toxin and the molecular mechanism by with the Ebola virus induces death.

As evidenced by mechanisms put in place early in the AIDS epidemic, NIAID has a positive track record of working with the private sector from early phase discovery to clinical development. NIAID can quickly mobilize the entire research community. Last fall NIAID conducted a study to show that existing stocks of smallpox vaccine could be diluted at least fivefold to provide immediate protection to a larger number of individuals should the need arise. Within 3 months post-9/11 a comprehensive biodefense research agenda was developed with broad input from the scientific and medical communities, including those from industry. Over 20 initiatives already have been launched to expedite biodefense research by NIAID.

This impressive efficiency is in part based upon the synergy which is derived by driving the biodefense research in parallel with all other infectious disease research. Separation of these two efforts could result in failure due to missed opportunities, failure to apply the latest technologies or knowledge gained from the study of other infectious agents. Therefore, I strongly recommend that the Department of Health and Human Services continue to be responsible for the prioritization, direction, and conduct of Federal research efforts related to the development of countermeasures for bioterrorism.

Although the administration's bill recognizes the necessity that HHS conduct the research and development programs, the bill provides that DHS in consultation with HHS shall have final authority to establish the research and development program, including the setting of research priorities. The proposed transfer of program and funding authority in the administration's bill gives ultimate control of research spending and priorities to DHS, a nonscientific, non-public, health-based agency. I will submit that you cannot wisely set research priorities without being actively engaged in research and with wise medical input from the medical and scientific investigators.

The bottom line is that DHS could under the current administration's bill change priorities midstream and by budget allocations. To create the appropriate scientific and medical infrastructure in DHS would result in loss of momentum and unpredictability of new and ongoing research programs within HHS. There is no time to reinvent the wheel. Rather it should capitalize on the solid infrastructure that already exists in the infectious disease research in this country.

It is not clear which activities by the DHS would duplicate, supplant, or replace existing programs conducted by HHS and create increased and recurring costs. One of the most critical determinants of success in biodefense research will be support and oversight of excellent science based upon peer review and merit. As stated earlier, NIH-NIAID has an unparalleled track record of success based upon external peer review. A scientific health agency, HHS, rather than the nonscientific, non-public health DHS, should have the principal authority for developing and prioritizing scientific and health related programs.

The role of DHS should be to integrate threat analysis and vulnerability assessments and research agenda. This could be accomplished by appointment of an Assistant Secretary who would have dual reporting to HHS and DHS and to work closely with NIAID. The desired outcome would be mutually agreed upon research priorities that address threatening biological agents, whether they be intentionally released or naturally occurring.

Last, regulation and oversight measures for work with infectious agents must be balanced so as not to impede legitimate research, diagnosis, and treatment of naturally occurring infectious agents. I recognize that there's concern, and I share those concerns, about pathogenic microorganisms being used as biological weapons by nations or individuals. As these concerns are addressed, however, I urge that there be careful review of possible measures that might be taken to establish appropriate safety and enforcement measures. HHS has the best scientific and institutional knowledge to provide oversight of select agent registration and to develop rational enforcement programs.

Thus, I believe the program for select agents should remain within HHS at the CDC. To transfer it to DHS will result in a delayed implementation, which could considerably slow down implementation of the biodefense research agenda. And as currently structured, I would just point out that minimal regulations are being put in place now so there's opportunity to change those obviously by DHS further down the road. As a result of this, more importantly, DHS could result in undue tension within the research community. Inappropriate policy measures and regulations to prevent terrorists from acquiring pathogens could have unintended consequences for research aimed at developing the very countermeasures that could eventually remove agents from the select agent list. There needs to be careful balancing and public concern about safety and security and the need to conduct legitimate research to protect the public.

I thank you for the opportunity to share my concerns with you this morning, and I'm happy to answer questions.

[The prepared statement of Gail H. Cassell follows:]

PREPARED STATEMENT OF GAIL H. CASSELL, VICE PRESIDENT FOR SCIENTIFIC AFFAIRS AND DISTINGUISHED RESEARCH SCHOLAR IN INFECTIOUS DISEASES, ELI LILLY AND COMPANY

Mr. Chairman and members of the Committee, thank you for the opportunity to participate in this hearing. Before sharing my views, some comments about my background may be helpful. My name is Gail Cassell. I am a microbiologist currently serving as Vice President for Scientific Affairs and Distinguished Research Scholar in Infectious Diseases, Eli Lilly and Company. Prior to my arrival at Lilly five years ago, I was the Charles H. McCauley Professor and Chairman of the Department of Microbiology of the University of Alabama Schools of Medicine and Dentistry, Birmingham, Alabama. My background is that of a research scientist in infectious diseases working in laboratories of both industry and a research-intensive university as well that of a Director of large training programs for pre- and postdoctoral students in molecular genetics, virology, and immunology. I have served on the Advisory Committee of the Director of the National Institutes of Health (NIH) and on the Advisory Council of the National Institute of Allergy and Infectious Diseases of the NIH and as Chair of the Board of Scientific Councilor's of the National Centers for Disease Control and Prevention (CDC). I am currently a member of the Director's Advisory Committee of the CDC. Over the years, I have participated in reviews of the biomedical research programs (including bioweapons defense research) in the Department of Defense. Of particular relevance to the discussions today, I have been actively involved in issues related to biodefense for over the past decade, as a past President of the American Society for Microbiology (ASM), Chair of the Public and Scientific Affairs Board of the ASM and a member of ASM's Task Force on Biological Weapons. I served as Co-chair of the committee that reversed the decision to destroy the U.S. stocks of smallpox and as a member of the Advisory Committee to establish the first unit in the U.S. military to address a bioweapons attack on U.S. soil. In addition, I continue to serve on a number of committees in the National Academy of Sciences (NAS) dealing with bioweapons, including a Russian research advisory committee. Most recently I served on the Bioweapons Subcommittee of the NAS Committee on Science and Technology for Countering Terrorism.

The events of September 11, and the anthrax incidents which followed, have proven the vulnerability of the United States to terrorism and the complexities of preparedness. The need to strengthen planning, coordination, implementation and oversight of homeland security is obvious. The establishment of a new federal Department of Homeland Security, at a cabinet level, can potentially achieve greater efficiency, effectiveness and accountability regarding many aspects of terrorism. However, there are unique characteristics of *bioterrorism* that deserve special consideration and suggest the need to address them in a manner differently from that proposed by the Administration's Bill. These characteristics include: (1) inadequacy of existing countermeasures and urgent requirement for *interdisciplinary* research; (2) indistinguishable features of bioterrorism and naturally occurring infectious diseases; and (3) the nature and extent of the bioterrorism threat and the need to balance public safety and legitimate research in regulation and oversight measures.

1. INADEQUACY OF EXISTING COUNTERMEASURES AND URGENT REQUIREMENT FOR INTERDISCIPLINARY RESEARCH

There is no simple counter to bioterrorism, no "magic bullet." Instead, development of an integrated set of strategies is required. Such efforts must include preventing countries from acquiring bioweapons in the first instance, dismantling existing programs and capabilities where proliferation has already occurred, deterring the use of biological weapons, and, ultimately, putting in place countermeasures that can rapidly detect and effectively defend against such use. It is the latter that requires special consideration with respect to the proposed role of DHS.

In the long term, the only way to defend against bioterrorism is through a combination of constant surveillance, accurate diagnostics to identify threats as early as possible, and continuous innovation to provide high quality vaccines and drugs that can be useful against any attacks that do occur. Research related to bioterrorism is inextricably linked to that of naturally occurring infectious agents and development of new antibiotics, antivirals, diagnostics and vaccines. The research and development of technologies for biodefense should be synergistic and not duplicative.

The diversity of existing biological weapons and the ever-increasing possibilities preclude simple therapeutic countermeasures to bioterrorism. Furthermore, response possibilities are limited even for known threats. Although there are 13 viruses on the current select agent list, there is only one antiviral, which is for smallpox and must be administered intravenously. There are no truly broad-spectrum antivirals, and only a limited number of antivirals for routine pathogens like influenza, herpes, hepatitis B, and HIV. The situation is somewhat better but still worrisome with respect to antibiotics. There has only been one new class of antibiotics developed in the past three decades. The Russians are known to have constructed antibiotic resistant bioweapons. In short, our antibiotic armamentarium is limited, and there is growing concern about an increase in resistance to existing antibiotics. It seems clear that no public health response to bioterrorism is likely to prove effective without addressing the overall problem of antimicrobial resistance and the challenges of drug discovery and development. Finally, the best deterrent against the use of a biological weapon of mass destruction may be a constant stream of new, innovative antibiotics, antivirals, and vaccines. Knowledge of such commitment and successful developments would surely dissuade the efforts of our enemies in such an arena.

Development of these countermeasures will depend on interdisciplinary research ranging from basic research into the mechanisms by which the agents cause disease, how the body responds, and how the agents are transmitted. This new knowledge then must be used to develop innovative vaccines, antibiotics, antivirals, and immunomodulators. Equally important will be to benefit from knowledge gained in previous failures in countermeasure development efforts. Given the long lead-time necessary for development of vaccines and drugs (average 8-10 yrs), achieving timelines and goals are critical.

The magnitude of the challenge to develop effective countermeasures is great. Success will require involvement of the very best scientific, medical, and pharmaceutical talent in government, academia, and the private sector. Likewise, in order to achieve success in a timely manner, the United States must be able to capitalize upon the expertise of and existing infrastructure for product development that resides in the pharmaceutical industry. Engaging the full spectrum of private industry, i.e., from the smallest biotech to the largest pharmaceutical companies, in the search for solutions to infectious diseases, will not only greatly raise the chances of success, it can also significantly lower the total cost to taxpayers, augmenting public appropriations with private capital investment. Thus, it is critical to recruit these organizations into the biodefense effort and assure effective alignment between government, academia and industry.

NIH/NIAID is uniquely positioned to lead the effort. The NIH recognizes that significant advances occur when they are often unforeseen. These advances expand the experimental possibilities and open new pathways for research. It must be recognized that not all research problems are equally approachable no matter how urgent and important to public health. Research and development of countermeasures will be a long-term endeavor. There is always uncertainty about where the most valuable discoveries can be made but NIH is best placed to identify scientific opportunities and applications that are relevant to the most pressing issues that will yield solutions. NIH/NIAID is unrivaled in its track record of bringing together the brightest scientists and rigorous peer review and oversight of funded research. Indeed, many of the best investigators have already been funded by NIAID and have recently made major advances in determining the mechanism of action of the anthrax toxin and the molecular mechanism by which the Ebola virus induces death. As evidenced by mechanisms put in place early in the AIDS epidemic, NIAID has a positive track record of working with the private sector from early phase discovery to clinical development. They can quickly mobilize the research community. Last fall, the NIAID conducted a study to show that existing stocks of smallpox vaccine could be diluted at least 5-fold to provide immediate protection to a larger number of individuals should the need arise. Within three months a comprehensive Biodefense Research Agenda was developed with broad input from the scientific and medical communities, including those from industry. Over 20 initiatives already have been launched to expedite biodefense research. This impressive efficiency is in part based upon the synergy, which is derived by driving the biodefense research in parallel with all other infectious disease and immunology research. Separation of these two efforts could result in failure due to missed opportunities—failure to apply the latest technologies or knowledge gained from the study of other infectious agents. Therefore, I recommend that the Department of Health and Human Services (HHS) continue to be responsible for the *prioritization, direction, and conduct* of federal research efforts related to development of countermeasures for bioterrorism.

Although the Administration's Bill recognizes the necessity that HHS *conduct* the research and development programs related to infectious diseases, Section 303(a)(2) of the Bill provides that DHS, in consultation with HHS, *shall have final authority to establish the research and development program, including the setting of priorities.* The proposed transfer of program and funding authority in the Administration's Bill gives ultimate control of research spending and priorities to DHS, a non-scientific, non-public health based agency. To create the appropriate scientific infrastructure in DHS would result in loss of momentum and unpredictability of new and ongoing research programs within HHS. There is no time to "re-invent the wheel" rather we should capitalize on the solid infrastructure that already exists in infectious disease research in this country. It is not clear which activities by the DHS would duplicate, supplant, or replace existing programs conducted by HHS and create increased and recurring costs. One of the most critical determinants of success in biodefense research will be support and oversight of excellent science based upon peer review and merit. As stated earlier, NIH/NIAID has an unparalleled track record of success based upon merit review.

In summary, a scientific health agency, HHS, rather than the non-scientific, non-public health DHS should have the principal authority for developing and prioritizing scientific and health related programs. The role of DHS should be to integrate threat analysis and vulnerability assessments into the research agenda. This could be accomplished by appointment of an Assistant Secretary that would have dual reporting to HHS and DHS and to work closely with NIH/NIAID. The desired outcome would be mutually agreed upon research priorities that address threatening biological agents.

2. INDISTINGUISHABLE FEATURES OF BIOTERRORISM AND NATURALLY OCCURRING INFECTIOUS DISEASES

While bioterrorism poses grave threats, the human race has been ravaged by infectious diseases throughout its history. The emergence of new infectious diseases (notably HIVAIDS) has decimated entire societies, while infectious agents such as influenza can turn unexpectedly virulent, e.g. the 1918 influenza pandemic killed tens of millions of people. In this broader context of emergent and resurgent infectious disease, the victims of a bioterrorist attack pose an indistinguishable set of public health challenges from any number of foreseeable natural outbreaks. Since well over 30 previously unknown infectious agents (including several new hemorrhagic fever viruses and new highly virulent strains of streptococci) have been identified since 1973, it is imperative that our public health infrastructure and surveillance systems be structured to recognize both naturally occurring and intentionally released infectious agents. CDC should have this responsibility. Section 505(a)(2) of the Administration's Bill requires DHS to carry out these functions under agreement with HHS. A separate public health system for biodefense should not be created. The primary duty and authority should remain with CDC, which has the existing knowledge, experience, and expertise. Again, an Assistant Secretary with dual reporting to HHS and DHS could coordinate planning and development of programs and lend technical assistance. Working closely with the CDC Director mutually agreed upon public health priorities for bioterrorism preparedness and response could be achieved in an efficient manner.

3. THE NATURE AND EXTENT OF THE BIOTERRORISM THREAT AND THE NEED TO BALANCE PUBLIC SAFETY AND LEGITIMATE RESEARCH IN REGULATION AND OVERSIGHT MEASURES

Biological weapons have varied characteristics. High potency, substantial accessibility, and relatively easy delivery characterize the most fearsome agents. Humans, animals, and plants are potential targets for bioterrorism. Many of these agents-bacteria, viruses, and toxins—occur naturally in the environment. Thus the agents and much of the technology required to produce them are available for civilian or military use in many countries. Regulation and oversight measures for work with infectious agents must be balanced so as not to impede legitimate research, diagnosis, and treatment of these naturally occurring infectious agents. I recognize that there is public concern about pathogenic microorganisms being used as biological weapons by nations or individuals. As these concerns are addressed, however, I urge that there be careful review of possible measures that might be taken to establish appropriate safety and enforcement measures. The response taken should be carefully weighed and it should be balanced to avoid over regulation and intrusive schemes that could interfere with the flow of research activities in academia and industry. Any resulting harm to research could deprive society of the benefits of research advances. Scientific research must not be discouraged by unreasonable restrictions. To do so would not serve the public interest.

In reviewing the possible risks and options for responses, we should consider emulating the process used in overseeing recombinant DNA research. This experience is an example of where a technical problem was recognized and a balanced analysis and an appropriate mechanism were set in place for overseeing activities. The NIH Recombinant DNA Advisory Committee developed a rational approach to regulatory oversight of recombinant DNA. The NIH Guidelines were developed by a committee of experts and an oversight regime was designed with an understanding of the issues and risks. We should use the same model to construct a reasonable method that will not impede research or result in unnecessary costs. Institutions must take a proactive role in assuring that hazardous agents are brought into or shipped from their facilities and used in compliance with applicable regulations. The most effective approach to adequate oversight and record keeping is for institutions to monitor possession, transfer and use of select agents. Placing responsibility at the level of individual institutions for compliance with Title II of HR 3448 will be the least inhibitory to research.

It is important to coordinate programs related to human, animal, and plant agents because some of the threats for each are the same. Section 302(a) of the Administration's Bill transfers to DHS the select agent registration and enforcement programs of HHS. However, it does not transfer the select agent registration and enforcement programs of the Department of Agriculture to the DHS. Subtitle C of the Public Health Security and Bioterrorism Preparedness Act of 2002 mandated coordination of activities of HHS and the Secretary of Agriculture regarding "overlap agents"—that is, agents that appear on the separate lists prepared by HHS and Agriculture. Title II of that legislation expands the current select registration program to include mandatory registration of possession of select agents. Mr. Chairman, the Energy and Commerce Committee is to be congratulated for their role in this important legislation. Indeed, integration of the select agent registration program will undoubtedly result in a more efficient registration process thereby expediting registration.

Coordination among agencies that have regulations for infectious substances is important. Better compliance can be achieved if regulations are clear and coherent, streamlined and integrated, based on real risks, and effectively communicated to individual researchers. Emphasis must be placed on education, guidance and dissemination of information to research investigators, who must clearly understand their role and responsibilities. Institutional Biosafety Committees can be strengthened and there should be qualifications and training for institutional biosafety officers. Laboratory scientists and safety managers in institutions must have input into the rule-making procedures and work to assure that regulations are realistically applied with minimal intrusiveness.

The core elements of a regulatory regime are already in place in 42 Code of Federal Regulations Part 72 and in the Biosafety and Microbiological and Biomedical Laboratories (BMBL) Manual. Appendix F includes guidelines for Laboratory Security and Emergency Response for Microbiological and Biomedical Laboratories. Although it is currently nonspecific, it is a reasonable basis for the development of biosecurity requirements. It should be possible for HHS to modify its current regulatory regime to govern registration for possession and build on the BMBL guidance to provide for threat and risk based regulations. Security for select agents should be based upon risk levels.

HHS has the best scientific and institutional knowledge to provide oversight of select agent registration and to develop rational enforcement programs. The scientific communities, both in universities and in the private sector, are accustomed to self-regulation in use of radioactive materials, chemicals, and infectious agents. This service is provided by institutional Biosafety Offices. Likewise, review of protocols and inspection and accreditation of facilities are the norm for use of laboratory animals in research. Again, implementation of regulations related to select agents is reminiscent of the oversight put in place with the advent of recombinant DNA technology. In short, once the regulations have been established, implementation can be achieved through use of a system modeled after Biosafety Office Programs already in existence.

I believe the program for select agents should remain within HHS. To transfer it to DHS will result in a delay to implementation, which could considerably slow down implementation of the biodefense research agenda. More importantly, housing it within DHS could result in undue tension with the research community. For example, it is unclear whether the regulations to be put in place within the next 180 days will be changed taking on more of a criminal approach rather than one based upon scientific knowledge and insights into the biomedical research process utilizing infectious agents. The Administration's Bill states that interim regulations will be put in place thereby leaving freedom following the transfer of authority to DHS for other regulations to be drafted.

In summary, I support Title II and its protections for the legitimate and critical performance of research and diagnostic testing. Security for biological facilities is different from security for nuclear and chemical facilities and must take into account the unique aspects of work with biological agents. Inappropriate policy measures and regulations to prevent terrorists from acquiring pathogens could have unintended consequences for research aimed at developing the very countermeasures that could eventually remove agents from the select agent list. There needs to be a careful balancing of public concern about safety and security with the need to conduct legitimate research to protect the public. Because of the enactment of HR 3448, which again the Energy and Commerce Committee and this Subcommittee had direct responsibility, the United States is in a leadership position with regard to the establishment of reasonable controls on select agents. However, we should not have a false sense of security since no other country in the world has adopted similar legislation, which will be necessary. Ultimately, successful oversight will depend upon the integrity of the personnel who have access to select agents and on local institutional commitment.

CONCLUSION

Again, I appreciate having been given the opportunity to share my views and concerns with you. The inadequacy of our current public health infrastructure and existing biomedical defenses against a range of possible bioterrorist attacks has become clear. This inadequacy has, moreover, served to underscore the already well-documented need for better and more varied antimicrobials, vaccines, and other agents to detect, prevent or treat infectious diseases. One likely outcome from increased attention to bioterrorism threats will be the development of more comprehensive public health measures and countermeasures to threats posed by naturally occurring infectious diseases. I believe the recommendations I have made today provide the greatest chances for success.

Mr. GREENWOOD. We thank you. Thank you for being with us. Ms. Hamburg for 5 minutes.

TESTIMONY OF MARGARET A. HAMBURG

Ms. HAMBURG. Thank you for the invitation to discuss the proposed Department of Homeland Security, and the implications for public health and bioterrorism.

I strongly support current efforts to give greater authority and accountability to our homeland security program, including the creation of a new department with cabinet level authority. Yet we should move forward carefully as you are doing. Realistically, a reorganization of this magnitude requires a strategic framework for action, one that defines critical goals and objectives and how best to achieve them and one that defines the relative roles and responsibilities of the different entities involved.

The opportunities for greater efficiency, effectiveness and accountability through a new department is fairly evident in realms of overlapping security, such as border security, customs procedures and aspects of emergency response. How best to organizationally address the activities related to bioterrorism prevention, preparedness and response is a more complicated question.

Bioterrorism is fundamentally different from other security threats we face. Meaningful progress against the biological threat depends on understanding it in the context of infectious and/or epidemic disease. It requires different investments and different partners. Unless we recognize this, our Nation's preparedness program will continue to be inadequately designed, the wrong first responders will be trained and equipped, we will fail to build the critical infrastructure we need for detection and response, the wrong research agendas will be developed, and we may miss important opportunities to prevent an attack from occurring in the first place.

And before a major reorganization of the agencies and activities involved in biodefense, we would be well advised to examine our recent experience with the deadly dissemination of anthrax for lessons learned. It is stunning and disappointing that we have not done this. An independent, comprehensive analysis of the anthrax episodes and response should be undertaken. Looking within and across the relevant agencies of government, levels of government and at the relationships with private sector organizations, an informed analysis with identification of gaps and preparedness would be of enormous value. There may be certain real advantages to consolidating programs within the new Department of Homeland Security.

The biological threat and the public health programs required to address it is of profound importance to our national security. By residing within this new department it may command more priority attention and support. It may help ensure that experts in biodefense and public health preparedness are full partners at the national security table.

However, including biodefense and public health programs in the new department has some serious drawbacks. A fundamental concern is that you will loss program focus and organizational coherence by combining biodefense activities which deal largely with infectious disease medicine and public health into a department devoted mainly to a very different set of security functions and concerns. These biodefense activities could well be swallowed up in this huge new agency, which will likely lack the expertise and technical leadership necessary to plan and direct vital bioterrorism preparedness functions.

In addition, most of the public health activities required for bioterrorism are just as important for day-to-day functions of public health and medical care. In the months since 9/11 the administration, through programs developed and administered by the HHS Office of Public Health Preparedness and the CDC, has made significant progress building the programs necessary to strengthen the public health infrastructure for bioterrorism within this broader context. If these programs are carved out and moved into this new department, it will disconnect certain functions such as bioterrorism surveillance, laboratory networks and response, from other essential components of infectious disease response and control. It will thin out already limited expertise and enormously complicate the ability of our public health partners at the State and local level to work effectively. Rather than consolidating functions in a single agency, transferring the bioterrorism preparedness activities into this new department may actually require the creation of parallel and duplicative capabilities.

Therefore, HHS and CDC should continue to have direct responsibility for programs related to the public health infrastructure for infectious disease recognition, investigation and response, including bioterrorism.

However, we will need to integrate these activities into the framework for Homeland Security. One approach might be a coordination office placed within the new department working closely with CDC to achieve mutually agreed upon national security and public health priority for bioterrorism and response.

Similarly, future preparedness requires a comprehensive biodefense research agenda that links national security needs and research and development priorities and assures proper balance and integration of relevant research activities across many departments and agencies.

Coordination of such an agenda could be undertaken by the proposed department, engaging other departments like HHS, DOD, Commerce, DOE and USDA. However, the role of the Department of Homeland Security should be that of coordinator/facilitator only. The actual design implementation and oversight of the research agenda and its component programs must remain at the level of the mission agencies where the scientific and technical expertise resides.

For example, resources to develop and support the NIH biodefense research agenda should remain within that department.

Clearly, a new Department of Homeland Security will require significant expertise in public health infectious disease and biodefense. This must be seen as an important priority and reflected in the appropriate and high level appointments and in-house entities.

But in the final analysis, strengthening our homeland security programs will depend on achieving dramatically improved coordination and accountability. No matter where the lines are drawn in the new department critical activities will and should fall outside. So whatever the new department may look like, we must establish additional mechanisms to assure adequate oversight and coordination.

There are many more issues that will need to be raised and clarified before such important legislation is passed, but time does not allow all of that discussion now. I thank you for the interest of your committee, the holding of these hearings, and I stand ready to help you in any way that I can. I would be happy to answer questions.

[The prepared statement of Margaret A. Hamburg follows:]

Prepared Statement of Margaret A. Hamburg, Vice President of Biological Programs, Nuclear Threat Initiative

Mr. Chairman and members of the Committee: I appreciate your far-reaching interest in Homeland Security and particularly your attention to the public health and bioterrorism threats that are the focus of this hearing, and I thank you for the chance to participate in this hearing. My name is Margaret (Peggy) Hamburg. I am a physician and a public health professional, currently serving as Vice President for Biological Programs at NTI, a private foundation, co-chaired by Ted Turner and Sam Nunn, whose mission is to reduce the global threat from weapons of mass destruction. Previously, I have served as Assistant Secretary for Planning and Evaluation in the Department of Health and Human Services in the last Administration;

as New York City Health Commissioner for six years, under both Mayor Dinkins and Mayor Giuliani; and as Assistant Director of the National Institute of Allergy and Infectious Diseases, National Institutes of Health. I have spent much of my time over many years working on bioterrorism preparedness and response, and I welcome this opportunity to offer my views on the new Department of Homeland Security and improving US defenses against bioterrorism.

Events this past fall—including the attacks of September 11 and the dissemination of anthrax through the postal system—demonstrated our nation's vulnerability to terrorism, and underscored both the importance and complexity of homeland defense.

I applaud current efforts to give greater authority and accountability to our homeland security program, including the creation of a new federal Department of Homeland Security. There is a strong rationale for consolidating some of the many departments and agencies that share similar functions or provide various aspects of what is needed for comprehensive preparedness and response. Both the Administration's Bill to establish a Department of Homeland Security and S. 2452 to establish a Department of Homeland Security and a National Office for Combating Terrorism as introduced by Senator Lieberman and colleagues, offer important opportunities to strengthen leadership, focus and coordination of essential programs and policies. However, they also raise a number of concerns.

Preparing our nation against the threat of terrorist attack requires well-defined authority, accountability and coordination across an exceedingly broad array of agencies and activities. The existing Office of Homeland Security, despite the yeoman efforts of Governor Ridge and his staff, is clearly not structured for the task. A new cabinet-level Department of Homeland Security can potentially improve coordination of U.S. government activities such as border security, customs procedures and aspects of emergency response. But improving coordination of activities related to bioterrorism prevention, preparedness and response is a greater challenge.

In my testimony this morning, I want to briefly raise a number of issues that apply broadly to the creation of a new Department of Homeland Security, then focus my attention specifically on the biological threat.

DEPARTMENT OF HOMELAND SECURITY: SOME BROAD CONCERNS

The attacks of September 11 followed by the anthrax attacks have created great political pressure on the White House and Congress to take action to improve homeland security. Just as President Bush refused to be rushed with his post-September military response in Afghanistan and delayed the strikes until they could be timed for maximum effectiveness, so must Congress—in creating a Department of Homeland Security, act deliberately, with full analysis and without undue haste, before taking steps it will find hard to reverse. We need to move forward only after the most careful consideration of our goals and how best to achieve them. Several important concerns come to mind:

Need for a Strategic Framework

The creation of a new Department of Homeland Security represents an ambitious reorganization which will be difficult to implement and disruptive to many functions of government. Even under the best of circumstances, this restructuring will cost time and momentum in current programs. Thus the goals must be defined before legislation is passed, so the benefits of the new structure outweigh the costs of achieving that structure. We should be very clear about what we are doing and why—spelling out goals and objectives, as well as the related roles and responsibilities of the various partners.

Need for Balance

Current plans require that a great many agencies and agency components be pulled into one large Department focused primarily on terrorism preparedness and response. At the same time, this new Department of Homeland Security will still be responsible for dealing with a broad range of other activities. Many of these more routine activities will be important to the core Departmental mission because they will, on a regular basis, allow for the practice of systems that would be recruited into service in the event of an attack (e.g. disaster response and sheltering, FEMA). Similarly, routine non-terrorism activities might serve to identify unusual patterns or situations that might signal an impending terrorist event (e.g. monitoring shoreline for drug-runners or boating accident rescues, Coast Guard). However, there is serious concern that when you create a Department as diverse as this one would be, you will either lose focus on the organizing mission of countering terrorism or you will fail to effectively support those other routine functions. It is hard to imagine a Department remaining honed in on terrorism preparedness and response while responding to mudslides, hurricanes and fires, monitoring the fisheries, searching out drug traffickers, controlling hog cholera and investigating outbreaks of disease. It is also hard to imagine effective leadership for such a diverse array of tasks, requiring an equally diverse array of professional backgrounds and expertise.

Need to Address Existing Weaknesses (Not Just Move Pieces Around)

Given the above concerns about managing this complex and varied new Department, serious questions must be raised as to how the Department will remedy known weaknesses in certain of its component agencies and activities. Reorganizing defective components will not improve performance. Some of the problems may benefit from new leadership or enhanced attention and scrutiny, but without a clear game plan and focused strategy, others may continue to fester, or worse, their continuing dysfunction may be amplified in a new and confusing bureaucracy. A host of personnel, budgetary and jurisdictional issues may add to the difficulties of providing appropriate oversight, management and operational accountability.

Need to Maintain Program Connectivity / Coherence

In several domains, but particularly with respect to bioterrorism, the creation of a new and distinct Department may serve to disconnect certain functions such as bioterrorism surveillance, laboratory networks and response from the infrastructure needed to respond to routine, non-intentional public health issues. The response to a disease outbreak, whether naturally occurring or intentionally caused, will require the same critical components. Most likely, we will not initially know the cause of an emerging epidemic. What is more, our overall infrastructure for infectious disease recognition and response is far from robust. We must be careful not to further fragment our capacity, and inadvertently undermine our own best interests. We must also avoid the unnecessary development of duplicative systems at a time of limited resources.

HOMELAND SECURITY AND THE BIOLOGICAL THREAT

As our nation struggles to respond to the concerns posed by bioterrorism, both the nature of the threat and the role of public health, medicine and science continue to be poorly understood and underemphasized. The threat of bioterrorism is fundamentally different from other threats we face, such as "conventional" terrorism or attack with a chemical or nuclear weapon. By its very nature, the bioweapons threat—with its close links to naturally occurring infectious agents and disease—requires a different paradigm.

Designing that paradigm is proving to be a difficult challenge. Public health has never been traditionally viewed as an element of national security. Consequently, those who specialize in national security are largely unfamiliar with the public health system—what it is, how it works and why it is important to our overall mission of protecting the nation. It is not surprising that the various Commission Reports (e.g. Hart-Rudman) that have looked at national security/terrorism issues and current legislative proposals for the creation of a federal Department of Homeland Security have had trouble conceptualizing an appropriate organizational approach that includes bioterrorism preparedness and other biodefense ȧctivities. In fact, there is no clear and simple answer to the question of how best to organize the components of an effective bioterrorism prevention, preparedness and response program.

Critical Elements of a National Response

Certainly, before a major reorganization of the agencies and activities involved in biodefense, we must understand how these components need to mobilize and work together in every stage of defense—from prevention, to preparedness, to response. Accomplishment of this task would greatly benefit from a thorough and complete critical analysis of our response to the anthrax attacks.

It is stunning and disappointing that we haven't undertaken a systematic review of what happened. And I strongly recommend that an independent and comprehensive after-action review of the response to the anthrax letters be undertaken. It should be done in a rigorous fashion, looking within and across the relevant agencies of government, at all levels, and at the relationships with private sector organizations. We cannot afford to let these incidents go by without taking stock of what happened, what should have happened (but did not), and what needs to be done to improve response in the future. This must be more than a listing of lessons learned. It needs to be a well-researched report, with thoughtful and informed analysis, identification of gaps in preparedness and response, and realistic recommendations for improvement. To the best of my knowledge, no such exercise is currently underway in a crosscutting and systematic manner. Even as the aftermath and the investigation of the of the anthrax letters is still unfolding, there is still a real urgency to undertake such a process, before significant events fade from memory and before new events and priorities overwhelm us.

Such an analysis would give us indispensable insight into how we should structure our national response to bioterrorism, and how we should incorporate the following four essential elements.

(1) **Prevention.** Every effort must be made to reduce the likelihood that dangerous pathogens will be acquired or used by those that want to do harm. This must include improving intelligence, limiting inappropriate access to certain biological agents and establishing standards that will help prevent the development and spread of biological agents as weapons.

(2) **Strengthening public health.** Rapid detection and response will depend on enhanced disease surveillance and outbreak investigation from a well-trained cadre of public health professionals, educated and alert health care providers, upgraded laboratories to support diagnosis, and improved communications across all levels of government, across agencies and across the public and private sector.

(3) **Enhancing medical care capacity.** We must improve treatment for victims of an attack by enhancing local and federal emergency medical response teams, training health professionals to diagnose and treat these diseases, developing strategies to improve the ability of hospitals to rapidly increase emergency capacity, and providing necessary drugs or vaccines where they are needed through the National Pharmaceutical Stockpile.

(4) **Research.** A comprehensive research agenda will serve as the foundation of future preparedness. Perhaps most urgently, we need improved detectors/diagnostics, along with better vaccines and new medications.

Some of these activities are already underway, but need to be strengthened and extended. Other programs and policies still need to be developed and implemented. All are essential for homeland security. Yet it is important to note that while certain aspects of these activities are required to respond to the threat of bioterrorism specifically, these programs are just as important for the day-to-day, routine activities of public health and medical care.

Potential Benefits of Housing Biodefense Activities in a New Federal Department

There are certain real advantages to be gained from placing these programs within a new federal Department of Homeland Security. First and foremost, the biological threat, and the necessary programs to address it, is of profound importance to our national security. These activities require greatly enhanced priority and support. By residing within this new Department, they may be more likely to command that needed attention and support. Furthermore, experts in biological weapons threats, biodefense and public health preparedness must be full partners at the national security table, participating in strategic planning, policymaking and program design and implementation. Being part of the Department of Homeland Security might help to institutionalize this important participation.

In addition, legitimate concerns have been raised that if not housed within this new Department, crucial public health and bioterrorism programs may be neglected, and important operational public health and biomedical defense functions may not be integrated with national security objectives.[1]

Clearly, there is an urgent need for improved coordination and integration of bioterrorism programs and policies across agencies of government. The current patchwork—of programs that address bioterrorism prevention, preparedness and response, including research—is inadequate and unacceptable. These need to be brought together into a collective programmatic vision, and implemented in a manner that sets priorities, supports synergy, identifies gaps and avoids unnecessary overlap or duplication. To date, this has proved a difficult challenge. One might argue that the most effective way to address this concern is to pull these activities together under one roof.

There might be additional benefit of centralizing aspects of biodefense activities under one roof from the perspective of certain state and local government entities as well as private sector entities—including the medical care system and the pharmaceutical industry—all of whom are essential partners in combating bioterrorism and should also be integrated into an effective vision and framework for action. Looking at the federal government from the outside, it can be very confusing to discern where and how best to interact with the system. Again, the creation of a unified site within a Department of Homeland Security might reduce confusion,

[1] O'Toole, Tara. "Creating the Department of Homeland Security: Consideration of the Administration's Proposal." Testimony before the House Subcommittee on Oversight and Investigations, June 25, 2002

strengthen the ability to work across levels of government, and support the kinds of public-private partnerships that will prove essential to success.

Potential Disadvantages of Inclusion in a New Federal Department / Recommendations

While there clearly are benefits to be gained by moving certain aspects of bioterrorism and related public health issues into a consolidated new Department of Homeland Security, a serious cost/benefit analysis has to consider how best tc ensure that our overall governmental effort is maximally effective.

Organization of Bioterrorism Activities

As currently envisioned, the Administration's proposal for a Department of Homeland Security would seek to develop a single, government-wide, comprehensive and integrated research and preparedness plan to prevent chemical, biological, radiological and nuclear (CBRN) attacks, to reduce our nation's vulnerabilities to terrorism and to minimize damage and assure effective response should an attack occur.

This approach is intrinsically troubling—from the point of view of biodefense—because, as I noted earlier, the bioterrorism threat has some very distinctive features as compared to "conventional" terrorism or other weapons of mass destruction. Past experience tells us that many so-called bioterrorism programs failed to achieve their potential because they were addressed within the framework of CBRN or "Chem/Bio". There was an underlying assumption that these problems could be effectively approached with a "one size fits all" model, but in reality, such programs simply failed to address the biological component.

Meaningful progress against the bioterrorism threat depends on understanding it in the context of infectious and/or epidemic disease. It requires different investments and different partners. Until the distinctive nature of bioterrorism is fully taken into account at the level of policy, our nation's preparedness programs will continue to be inadequately designed: the wrong first responders will be trained and equipped; we will fail to fully build the critical infrastructure we need to detect and respond; the wrong research agendas will be developed; and we will never effectively grapple with the long-term consequence management needs that such an event would entail. We may also miss critical opportunities to prevent an attack from occurring in the first place.

Recommendations:

(1) Any new Department of Homeland Security must be staffed at the highest levels of leadership and decision-making with individuals who have significant expertise in public health, infectious disease and biodefense/bioterrorism.

(2) An Undersecretary for Biological Programs should be appointed to oversee and integrate the various activities going on within the Department of Homeland Security that relate to the biological threat. In addition, that individual should serve as liaison to the various other Departments with significant responsibilities and programs in the biological arena.

(3) An external advisory group for biological programs should be established with the responsibility of reviewing the appropriateness and comprehensiveness of biological threat related programs, policies and resource allocation / budget priorities.

Emergency Response / Role of Public Health Infrastructure

As noted earlier, a bioterrorism attack would differ in fundamental ways from other forms of terrorist assault. The requirements for effective bioterrorism preparedness and response are, for the most part, substantially different as well. Biological terrorism is not a "lights and sirens" kind of attack. Unless the release is announced or a fortuitous discovery occurs early on, there will be no discrete event to signal that an attack has happened, and no site you can cordon off while you take care of the casualties, search for clues and eventually clean up and repair the damage. Instead, a biological terrorism event would most likely unfold as a disease epidemic, spread out in time and place before authorities even recognize that an attack has occurred. We would see the first evidence of attack only after people begin appearing at their doctor's office or emergency rooms with unusual symptoms or an inexplicable disease. In fact, it may prove difficult to ever identify the perpetrators, the site of release, or even to determine whether the disease came from a natural outbreak or a terrorist attack.

Under most circumstances, the "first responders" to a bioterrorism event will be public health officials and health care workers. "Ground zero" will be in hospitals, health care facilities and laboratories. The "battlefield" response will come in the form of disease diagnosis, outbreak investigation, treatment of the sick and public health actions required to stop continuing contagion and stem disease. How swiftly we recognize and respond to a potential attack will dramatically influence our ability to reduce casualties and control disease. All of these recognition and response functions are more closely tied to public health and medical care activities than to the emergency response required for other types of catastrophic terrorism or even other kinds of natural disasters.

In the months since 9/11, the Bush administration—through programs developed and administered by the HHS Office of Public Health Preparedness (OPHP) and the Centers for Disease Control and Prevention (CDC)—has made some progress in building the programs necessary to strengthen public health infrastructure for bioterrorism within this broader context of infectious disease. If these programs are carved out of their current habitats and moved into this new Department, it will disconnect bioterrorism preparedness from other essential components of infectious disease response and control, thin out already limited expertise, and complicate the ability of our public health partners at the state and local level to work together effectively. If the nation develops two parallel systems for infectious disease surveillance and response—one (that for bioterrorism) of which is only really activated and practiced in a crisis—the likely outcome will be to weaken and fragment our nation's capacity to respond to infectious disease—whether occurring naturally or caused intentionally.

Recommendations:

(1) HHS and CDC should continue to have direct responsibility for programs related to the public health infrastructure for infectious disease recognition, investigation and response, including bioterrorism.

(2) A public health professional with appropriate background and experience could be placed within the Department of Homeland Security, perhaps with dual reporting to the DHS Secretary and the HHS Secretary. This individual could then work closely with the CDC Director to achieve mutually agreed upon public health priorities for bioterrorism preparedness and response

(3) The Department of Homeland Security should assure greater coordination, collaboration and program integration among the components of government doing infectious disease surveillance activities (e.g. DOD, USDA, Wildlife and Forestry).

Biodefense Research

Further investments must be made in biomedical research to develop new drugs, vaccines, rapid diagnostic tests and other medical weapons to add to the arsenal against bioterrorism. At the same time, it is also essential that we improve technologies to rapidly detect biological agents in environmental samples and develop other technologies to protect the health of the public. We must learn more about how these organisms cause disease and how the human immune system responds so that we can develop better treatments and disease containment strategies to protect us in the future. In addition, we must also devote more attention and resources to "systems research," in an effort to understand more about such issues as personal protective gear, environmental safety and decontamination.

Success will require collaboration among many agencies of government (HHS, DOD, DOE, USDA and others), academia and the private sector. Coordination of the development and budgetary support for such a comprehensive, integrated biodefense research agenda could certainly be offered under the auspices of the proposed Department of Homeland Security. This could help make sure that investment supports both national security needs and research and development priorities. It would also help integrate the bioterrorism-related research activities of the various mission agencies, including threats to humans, animals and crops. Hopefully, this would help foster proper recognition and support for elements of the research enterprise which are currently undervalued/under-resourced—such as the United States Army Medical Research Institute for Infectious Diseases (USAMRIID) and the Department of Agriculture's animal health research facility, Plum Island. It would also help identify program gaps, overlaps and opportunities for synergy.

At the same time, the role of the Department of Homeland Security should be that of coordinator/facilitator. The actual design and implementation of the research agenda and its component programs must remain at the level of the mission agencies, where the scientific and technical expertise resides. With a few possible exceptions, it would be unrealistic and inefficient to build the kind of sophisticated scientific expertise necessary to take on the direct conduct or management of research and development activities across a broad range of disciplines and technologies at the level of this new Department.

Recommendations:

(1) A research coordination office could be established within the Department of Homeland Security and charged with responsibility for assuring the development

and funding support for a comprehensive, integrated biodefense research agenda. This research coordination office could also help support the integration of threat and vulnerability analysis with the process of setting the research agenda. Such a research coordination office might also be effectively placed within a strong White House Office of Homeland Security, where it could work across the full set of cabinet agencies, including the Department of Homeland Security, to ensure a comprehensive, integrated and appropriately funded biodefense research agenda. An individual with appropriate scientific background and experience should head this office.

(2) Given the fact that HHS is the primary department with responsibility for biomedical research, and the unique role played by NIH, resources to support the NIH biodefense research agenda should remain within that Department.

(3) An external advisory mechanism should be established to encourage ongoing communication and collaboration with academic and industry partners. New mechanisms must be developed to engage participation from outstanding scientists from academe and industry, and to bring new young scientists into these endeavors.

(4) The highest level of government commitment is needed to address the national crisis in the development and production of new vaccines and antimicrobial drugs—a crisis that is growing in urgency in light of the bioterrorism threat. A new Department of Homeland Security, working closely with the appropriate agencies of government (e.g. FDA, NIH, DOD), industry and Congress, could lead such an effort, or it could be undertaken directly from the White House.

CONCLUDING REMARKS

Mr. Chairman and Members of the Committee: Government has no more important mission than protecting the lives of its citizens. A new Department of Homeland Security and a heightened defense against bioterror go directly to the heart of that mission. These tasks are as complicated as they are crucial. I thank you for the depth of the interest you've exhibited by holding this hearing. I stand ready to help in any way I can. And I would be happy to answer your questions—now or in the future.

Mr. GREENWOOD. Thank you. The Chair recognizes himself for 5 minutes for questions. This is a question I'd like to address to each of you. I think it's important to emphasize that the research programs targeted by the administration's bill are limited to only those dealing specifically with countermeasures to terrorist threats, such as smallpox and anthrax. Why shouldn't the new Secretary who will have access to a great deal of information about terrorist capabilities and interests have the authority to set the research priorities within this limited network?

Ms. HEINRICH. Our review of the proposed legislation states that the research would be broader than you suggest, Chairman. It says all biodefense research, which is—it is not only applied research and research that's focused onto particular pathogens, it's our understanding that it's an array of diseases. And what we have learned in discussions with experts is that there is a lot of interchange between those agents that could be used in naturally occurring infections, and in emerging infectious diseases.

Mr. GREENWOOD. If in fact, though, someone else's interpretation was more narrow than yours would you be happy with that?

Ms. HEINRICH. We still would have concerns because you're separating out the people that are responsible for actually conducting the research from the overall overarching authority and responsibility. It would seem to us that the role of Homeland Security can be that of coordinator, as Dr. Hamburg has suggested, that we have to have the strategic framework that we don't have and that your operating agencies that are actually conducting the work can be in a position to actually respond to areas that the Secretary of Homeland Security has said are areas of priorities. If in fact there are conflicts, there may be other mechanisms that can be used to resolve those. Certainly we've heard before this notion from the Office of the President and it also may be that Congress through the appropriations process would have a role to play in making sure that priorities were responded to.

Mr. GREENWOOD. Thank you. Ms. Cassell.

Ms. CASSELL. I would make the argument that having someone hold the purse string, so to speak, and being able to establish the priorities for research would be unwise if in fact that particular individual or agency, the head of that agency or department I should say, really does not have the scientific and medical infrastructure to wisely set the priorities. You stand a chance of not only disrupting research programs but, more importantly, I think what would happen is that you're not able to take full advantage of the scientific and medical infrastructure related to infectious disease research that already is ongoing and in place.

For example, I think it was only possible to establish the research agenda for biodefense, for developing countermeasures in a 3 to 4-month period, based on the wealth of basic knowledge and ongoing research that's already going on within NIAID. I think if you transfer the authority for establishment of priorities, whether it be for only one agent or two agents, you would miss those opportunities for the leveraging and the synergism, and that would be a major concern. I think for the next couple of years we're probably okay because the research agenda has been established.

However, giving the budget authority and the program authority to the Department of Homeland Security doesn't give me any assurance, in fact, that those efforts will have an opportunity to be materialized.

Mr. GREENWOOD. Isn't the concern on the other end of the spectrum that you could have research that is so academic and so far removed from the immediate threat of terrorism, that we are just not focused where we ought to be?

Dr. CASSELL. I can appreciate some of those concerns. However, I think that having something like the assistant secretary and an individual like the assistant secretary that will have dual reporting would, in fact, take care of that concern, because you would constantly have the input from that assistant secretary into the research agenda with respect to helping to set priorities, and also basically oversight in terms of meeting deadlines and time lines and research goals.

Mr. GREENWOOD. Ms. Hamburg?

Ms. HAMBURG. Yes. Well I think that clearly we need a stronger and more accountable system of coordination for a comprehensive integrated research agenda that engages the best talent within many parts of government and the private sector. But I think in terms of actually setting priorities and determining the elements of that research agenda, we will actually undermine our own best interests if we don't ensure that it actually is housed within agencies that are appropriately expert in the domains of research and connected to where the research is going.

Again, all of us have emphasized the close connection between the bioweapons threat and the threat of naturally occurring infectious disease. I think we also have to recognize that the bioweapons threat is evolving very quickly, because our capabilities in science and technology are evolving very swiftly. Some of the expertise in security only setting research agenda may actually miss emerging and important threats that are coming forward because of new capabilities in science, our ability to manipulate organisms, to understand what makes them infective and actually to manipulate them so they might be more infective, to actually manipulate them so they might be more lethal to create new organisms de novo, and I think looking forward, if we want to be prepared, if we want to be ahead of the curve in terms of evolving threats, then we really have to link this very, very closely to the scientists who know where science and technology is actually going, and where we can best target our resources and our capabilities to really have the kind of comprehensive short-term, long-term research agenda that we need for our protection.

Mr. GREENWOOD. Ms. Cassell, you wanted to add something?

Ms. CASSELL. Yes. Thank you. I think that we should be very careful to take advantage of lessons we can learn. We can learn by evaluating the defense research programs that have been in existence in terms of development countermeasures, very narrowly and with very narrow focus. We should look at the track record of those programs, I think, and again, lessons learned by being so narrowly focused, first having much broader focus in terms of taking advantage, as Dr. Hamburg has said, of other available knowledge.

Mr. GREENWOOD. Thank you. The Chair recognizes the gentlelady from Florida, but before doing so, would indicate for everyone's information—is this a series of votes? Never mind.

The gentleman from Florida for 5 minutes.

Mr. DEUTSCH. Thank you. Thank you, Mr. Chairman. The White House and Governor Ridge have told the committee that the new Department of Homeland Security would be quite capable prioritizing and managing the research and development programs and public health preparedness programs in the Department of Health and Human Services and contracting actual work back to HHS without any delay in those programs.

Is there any way that inserting another layer of decisionmakers over these programs would not close delay? If each of you can respond to that. Ms. Cassell?

Ms. CASSELL. I would say definitely not. I can't imagine that there won't be delay. First of all, you have to create appropriate scientific medical infrastructure within the new department in order to allow you to make those rational decisions that must be made.

Ms. HAMBURG. It's hard to imagine how that would increase efficiency and accountability. I think it will also require the addition of new layers of expertise within the Department of Homeland Security, and I think one needs to examine what are the benefits. Again, I come back to the crying need for better coordination, but that doesn't have to be achieved by creating a whole new systems of management.

Ms. HEINRICH. I'd like to suggest that we could learn from previous experience and actually look for places where we have successfully coordinated across Federal agencies in the private sector as well, especially in the area of R&D. I think there are examples of where agencies and programs have done that successfully.

Mr. DEUTSCH. This is really a follow-up. What type of expertise would homeland security have—or have to have in-house to prioritize and manage these programs?

Ms. Hamburg?

Ms. HAMBURG. Well, I think, you know, one contribution the new Department of Homeland Security could make would be to bring to bear the best possible information about the nature of threats and the credibility of emerging threats so that it could be integrated into preparedness and response programs. It also could help to ensure that the various elements that are being actually implemented by different parts of government are brought together into a more comprehensive picture so there aren't unintended gaps in programmatic activity or unnecessary duplication of effort.

And I think that it can offer an opportunity for individuals working outside of the Federal Government to have a place to go to in a coordinating way, to then find the services and programs that they need, get clarification of policies if you're at the State and local government, or if you're in the private sector, but not actually directly running those programs.

Mr. DEUTSCH. You know, one of the issues which really ties into this is really where would they get that expertise? These are human resources. That is really a question of trying to hire people. And one of the things that has impressed me incredibly, you know, from the jurisdiction of this subcommittee and the committee is, you know, CDC and HHS and NIH are, you know, on part with no part in terms of expertise. I mean, and there's a culture in each of those agencies that sort of breeds that. And I've never seen that created overnight. I mean, it seems impossible to create overnight, so I guess the real question is, if that's the level, the best of the best, the brightest of the brightest, the most creative of the most creative, how can you even expect that to happen in any short period of time in a new agency? I mean, I see you agree with me, so I guess, you know, Ms. Cassell in particular, if you want to respond.

Ms. CASSELL. I think you're right on target. In fact, I could not agree with you more.

I would just like to add to your comments in regards to what type of expertise would be required. One thing I think that is important if you're going to have the new department controlling, again, the research program, both from program development, setting of priorities and the budget, you need to have some expertise that's very knowledgeable with grant review and peer review process for ongoing research programs.

I can give you an example. I'm familiar with, based on participation in different reviews of biomedical research programs in this country, and that is when monies were awarded to the Department of Defense, for example, for breast cancer research, ultimately the authority for review of those programs and oversight of the programs actually was not transferred initially, but certainly the NIH ended up playing the lead role in terms of the administration of the program. Again, because that was—in terms of peer review research——

Mr. DEUTSCH. If I can make just one follow-up question, and that is really, you know—and this is just getting on the practical side of how you actually do this once—I mean, we are—once we create the agency. I mean, I have this real concern, and it's a practical concern, that there are people who are, you know, developers of—and really have the expertise, and it's not expertise you can just learn in school. It's expertise, that why would someone with that type of expertise in an environment that, you know, are in and they're happy, because otherwise these are people who clearly could leave and get jobs in different settings. Why would someone want to leave with a big question mark?

And it almost seems like the people that are going to—the incentive if you're an agency—CDC, whatever, you almost want to get rid of your sort of deadwood to a new agency. There would seem to be a sort of bureaucratic incentive at that point not to give up your best people, but your second best people. I mean, is that a concern, and how do we deal with that?

Ms. HAMBURG. Well, I think you clearly are experiencing the way of government and that's a valid concern and we've certainly seen it happen in many instances. I've worked in government in most of my career at the local and Federal level, and it's a concern that I immediately had when I heard this proposal. I think it's also the case that we have a limited supply of trained professionals in many of these critical areas, whether it's the bench researchers working on certain of these pathogens or the epidemiologists and infectious disease experts that we need to shape the research activities, the programs and the policies, and so we cannot afford to dislocate people from where they are currently working and functioning and working in a dual use, not exclusively a bioterrorism manner, and pull them into a new department that will not fully utilize their very important and limited talent in terms of our national personnel resource base.

Mr. DEUTSCH. Thank you.

Ms. Heinrich.

Ms. HEINRICH. I would just make a comment that I think that it's a challenge to draw experienced researchers and new researchers into the field of this biodefense work away from, you know, where their current focus is. So it's probably a problem that's more complex rather than easier.

Mr. DEUTSCH. Thank you.

Mr. BASS [presiding]. The Chair recognizes himself for 5 minutes. Just a process point here. I should note that Dr. Cassell and Dr. Hamburg are doctors, not—and should be addressed as such, even though they were mislabelled.

Dr. Hamburg, I just have a quick question for you. Nuclear threat initiative has been closely associated with Nunn-Lugar, the nuclear weapons initiative, and I'm wondering if you could give us some perspective which the other two witnesses might be able to comment upon about how a similar type program might be structured for the bioterrorist threat or the biological threat, because we don't have any such program currently to date.

Ms. HAMBURG. Well, Let me just give a little bit of background. The organization for which I work is cochaired by former Senator Nunn and Ted Turner who has funded it. It's a charitable organization whose focus is to reduce the threat of weapons of mass destruction, and on our board actually are many distinguished individuals, including Senators Lugar and Domenici who have been deeply involved with these activities for many years, along with Senator Nunn.

The Nunn-Lugar program really had focused predominantly on nuclear, but has had a biological component and it's been looking at how can we reduce the threat that exists from the weapons programs in the former Soviet Union that are now no longer active in many of the components, but facilities exist, people with expertise and know-how are now unemployed or underemployed. There are real security issues across many domains and concerns that important materials and capabilities could get in the hands of individuals who would use them to do harm.

On the biological side, I think there's enormous opportunity and opportunity that we can realize almost immediately by making greater commitment to that program. The former Soviet Union had a very large biological weapons program functioning in many different institutes with literally thousands of scientists and personnel working on different aspects of biological threats, animal, human and crop. We need to make sure that we, as a Nation, and in partnership with other nations, do everything that we can to ensure that both the material developed and the expertise developed can be redirected into many valuable prosocial research activities, both academic and entrepreneurial.

Mr. BASS. But Nunn-Lugar as it's currently constructed, can initiate and execute this kind of a program in biological containment as well as nuclear.

Ms. HAMBURG. It can, and there has been an element of it that has focused on the biological threat. It's been a small component, and it has, I think, been undersupported and undervalued in terms of the contribution that can be made. And I would be very eager to work with you if you'd like to explore opportunities in that realm to a greater degree.

Mr. BASS. Dr. Castle.

Ms. CASSELL. Yes. I'd just like to comment that you probably may be aware that there are small programs within HHS and DOD, some of which are administered by The National Academy of Sciences, to do exactly as Dr. Hamburg has described with respect to engaging former Russian bioweapons research scientists into meaningful infectious disease research. But again, it's a very small program and has been, I would argue, woefully underfunded for the last 6 years.

Mr. BASS. It's a huge problem. I yield back. I recognize the gentlelady from Colorado for 5 minutes.

Ms. DEGETTE. Thank you, Mr. Chairman. As I read the administration's proposed plan, it looks to me like the Department of Homeland Security would have the ability to shift research funds in both the NIH and in the CDC in any way they wanted. In other words, they could supersede decisions that those two agencies are currently making. Would that interpretation be correct, Dr. Cassell?

Ms. CASSELL. I'm going to agree with you 100 percent, and this is my major concern. I think it is the current of a lot of people. I think people may have trust that over the next 2 years because the agenda has already been established, this won't happen, but in years following the next two, I think that's a very real possibility that would occur——

Ms. DEGETTE. And looking at the legislation, I think people think it won't happen just because they think it won't be done, but it, in fact—the legislation gives the Department of Homeland Security to do exactly that. In other words, to say to the NIH, we think we need the resources you're using for other types of interdepartmental—any kind of research. We think, well, it might be important, but we think that this other thing is more important. So we're just superseding your decision, and we're redirecting it. That would be your understanding of the legislation as well?

Ms. HAMBURG. That is my impression that they have the final authority in terms of allocation of dollars and setting of priorities, and I think that is a real concern.

Ms. DEGETTE. Ms. Heinrich, do you agree that that's what the legislation says?

Ms. HEINRICH. The legislation gives—the proposed legislation gives the Department of Homeland Security the money and the authority to establish priorities. It also says that it should conduct the research through HHS predominantly and NIH. It also gives the President the prerogative to decide not to conduct and do research through this kind of arrangement, but it doesn't give us any indication of under what circumstances the President might use that prerogative.

Ms. DEGETTE. Okay. Thanks. See, here's the concern I have, and I think we're all agreeing. And by the way, Mr. Chairman, I think this is a wonderfully illuminating panel, and thank you very much for coming today.

The concern I have, Dr. Cassell, in listening to your testimony, there are a lot of infectious diseases that are killing millions of people every month, every year, and a great deal of money has been invested in trying to cure them. HIV is an example that I can think of, but yet we haven't done that, so the question is, if the Department of Homeland Security decides to shift the money to select agents, what happens to the research that's being done for these other diseases?

Ms. CASSELL. Well, I think through the regular appropriations process, NIH and—one might take confidence that these other programs would be protected, and I think that we have heard assurance from the doctor who directed the NIAID that the other research programs won't be compromised. However, I think that, you know, that is today as we've said, and what will happen in 2 years out, I think that might be another question.

I'd just like to add, if I might, to your concerns about some of the authorities that have been given, and it goes back to the oversight of select agents, and in fact, the way things are written now certainly, I think, gives a lot of room for going back and changing regulations and oversight of that program, and not that I want to change the direction you're going, but this does also potentially have the possibility of having a tremendous negative impact on the very research that we need to do in order to be able to get accounting measures.

Ms. DEGETTE. Right. You're not changing direction. That's exactly what I was trying to get at. And Mr. Hauer said, well, the problem is that we have limited resources, and we just have to recognize that, and so practically speaking, if you want to continue ongoing research and then have research into select agents, you're not going to be able to do both. You are going to have to shift resources away from some ongoing research, and I guess the question many of us are asking is who should be making those decisions, the scientists at CDC and NIH or somebody who is in this new department who's superseding their decisions. Correct? Dr. Hamburg, do you have——

Ms. HAMBURG. Well, I think—I think that, you know, clearly we live in a world, with limited resources and we can't do everything we might want to do in all areas of activity. I think that one of the great advantages of really housing our research activities, both the priority setting and implementation of the research at places like NIH is that you get synergy that you will lose if you try to carve it out into segments.

Fundamental understandings of how organisms cause disease, how the human immune system responds will have implications for both naturally occurring disease and intentionally caused disease. It will have implications for new drugs or vaccines we might develop against select agents that we're particularly concerned about as bioweapons threats, but also against organisms that might occur in nature. So I think you get more bang for your buck by having both biodefense-related research agenda but having it integrated with infectious disease research more broadly and understanding of immune response.

Ms. DEGETTE. Thank you. Mr. Chairman, let me just say in closing up here, I'm really concerned about what this bill—what the administration's proposal does for biological research within CDC and NIH, and I would hope that we could work in a bipartisan way to fix this, because some of the suggestions that this panel has had for having some coordination function but not a superseding function I think really make a lot of sense, and I yield back.

Mr. GREENWOOD. The Chair thanks the gentlelady.

I believe our questioning has been accomplished. So we thank the witnesses for your testimony in answering questions and excuse you and call forward our next panel, consisting of Dr. James McDonnell, the director of Energy Security and Assurance Program at the Department of Energy; Mr. John S. Tritak, director of Critical Infrastructure Assurance Office in the Department of Commerce; Mr. Robert Dacey, director of information security issues in the General Accounting Office; Dr. Samuel G. Varnado, director of the Infrastructure and Information Systems Center at Sandia National Laboratories; and Dr. Donald D. Cobb, associate director for threat reduction at Los Alamos National Laboratory.

Thank you. You understand that this subcommittee is holding an investigative hearing and in doing so it is our practice to take testimony under oath. Do any of you object to taking testimony under oath? Seeing no affirmative responses, the Chair would then inform you that pursuant to the rules of the committee and the House, you're entitled to be represented by counsel. Do any of you wish to be represented by counsel?

Seeing no affirmative responses, would you please stand and raise your right hand.

[Witnesses sworn.]

Mr. GREENWOOD. Thank you, you're under oath and Mr. Tritak, we'll begin with you. You're recognized for 5 minutes for your opening statement.

TESTIMONY OF JOHN S. TRITAK, DIRECTOR, CRITICAL INFRASTRUCTURE ASSURANCE OFFICE, DEPARTMENT OF COMMERCE; JAMES F. MCDONNELL, DIRECTOR, ENERGY SECURITY AND ASSURANCE PROGRAM, DEPARTMENT OF ENERGY; SAMUEL G. VARNADO, DIRECTOR, INFRASTRUCTURE AND INFORMATION SYSTEMS CENTER, SANDIA NATIONAL LABORATORIES; DONALD D. COBB, ASSOCIATE DIRECTOR FOR THREAT REDUCTION, LOS ALAMOS NATIONAL LABORATORY; AND ROBERT F. DACEY, DIRECTOR, INFORMATION SECURITY ISSUES, GENERAL ACCOUNTING OFFICE

Mr. TRITAK. Thank you, Mr. Chairman I'd like to have my written remarks in the record.

Mr. GREENWOOD. Without objection.

Mr. TRITAK. But I actually would like to touch on a couple of themes that I think are of interest to the committee and spend my 5 minutes on those and any follow-up questions we could take.

I think I'd like to start, Mr. Chairman, by trying to underscore how fundamentally different the homeland security mission is from what I would refer to as classic national security. When I got into this business back in the late 1980's, national security was something the government did. It was something the State Department did, the Defense Department did, the Justice Department did on behalf of the people. The role of the private industry really was a supplier of services and goods or as a taxpayer. But these were quintessentially government functions.

We're now entering a new age. Homeland security presents a national security problem that the government can't do alone. The target of terrorist activity, we know from statements made by bin Laden, is our economy, the pillars of economy specifically, which we take to mean the national infrastructure of the United States, and we also know that Osama bin Laden has urged his followers to exploit vulnerabilities, wherever they may be. On September 11, we saw how they were exploited in the physical sense, but we also have to take into account that the introduction of information systems and networks on a vast level create a veritable digitalness across the globe. It presents some new opportunities perhaps for exploitation.

Now, let's be clear what the goal of the terrorists is. It's to force us to turn inward and to disengage from our global responsibility, particularly the Middle East. They're going to fail in that mission. The notion is that by attacking the economy of the American people, we will fall to our knees. And whether or not they're going to succeed, which I know they are not, they're certainly going to try. So we have to recognize that homeland security is going to be a shared responsibility between the government and the owners and operators of our economy, the vast majority of which are private. And frankly, that is going to require redefining and clarification of the respective roles and responsibilities of the government industry on a level that we've never had to conceive of before.

The industry must be a full partner here. They bear responsibility to help secure our national infrastructure, and they need to work with government in a way that they are not used to. It's going to require a cultural investment on both sides. It's not easy for government to acknowledge that role—that change of role that government plays in this. It's not easy for industry either. But what I believe needs to take place and one of the most awesome tasks that the new Secretary is going to have to undertake is to create a culture of collaboration and partnership with industry that must permeate every level of organization in the new Department of Homeland Security.

We're not recreating a new Pentagon here. We're creating a new entity to achieve a common goal of protecting the American way of life within our borders against terrorism and to deal with episodic events where terrorism may find its mark.

And as I indicated, it will require a cultural adjustment, and that is not going to be easy and some need to be skeptical of whether or not that can take place. I happen to think it's inescapable and that a cultural collaboration brought on by a confrontation with the owners and operators is what is required. And so I want to underscore that whatever else is being discussed here today in terms of statutory changes in the bill or anything else is to recognize that I think this is a fundamental principle that is inescapable in this new age.

Now, going a little further, we also recognize that information sharing is an indispensable part of securing our infrastructure, indeed securing our homeland, and the administration's policy, and indeed the policy of the last administration, was to encourage information sharing. And information sharing has been taking place, the National Security Telecommunications Advisory Counsel, which you'll hear a little bit more about with Guy Copeland in the next panel.

But as much as information sharing is taking place, there is a reluctance to engage in the wholesale open exchange in a proactive manner, in a dynamic manner, because of concerns about existing laws and regulations. People can disagree over whether FOIA does or does not cover this sort of activity. My question is a little more basic. In the current statutory and regulatory environment, is it conducive to promoting or impeding voluntary information sharing?

And of course, resolving this is not going to be easy, because you may very well have two public goods that come into conflict, on the one hand, the need to encourage voluntary information sharing, and on the other, the demands of open government. Reconciling these two are inescapable, and frankly, they're going to fall on the shoulders of the Congress. I think it's important to recognize, however, that reconciliation and we need to address this issue.

Now, the administration has made it clear that a narrowly crafted FOIA exemption would help advance the cause of voluntary information. I know that there are people who look at section 204 of the present bill and have expressed some concern. And as I understand it, that section is in the process of being relooked at and revised in light of the dialog that has taken place between the Congress and the administration.

But what I'd like to be able to conclude is at least say something about the basic principles that the administration is trying to advance without going into details, which at this point, I'm not in a position to do. One, it's designed to be narrowly crafted, not overly broad. Two, it's only dealing within the zone of voluntary activity. There's no intention to role back mandatory disclosure requirements that exist in other parts of the law or regulation. And third, there's no intention to create safe havens for gross negligence or criminal activity. The idea here is to create an environment that is conducive and encourages this voluntary activity.

Now, I want to be very clear about something, because you probably will hear a little bit about this later. FOIA reform in this area is not a silver bullet. There is not going to be an avalanche of information pouring into the Federal Government the day after the passage of the bill, because one thing that information sharing requires, and it cannot be legislated, it cannot be regulated, is trust, and that trust evolves over time, and part of the experience with industry and government engaging on an ongoing basis on a constructive activity that advances the public interest.

What I referred to earlier is one example of a group that has been sharing information with the government for some time. They have already demonstrated the importance of it, but they too have indicated that more needs to be done in the way of FOIA.

Ultimately, Mr. Chairman, this process is going to be one of give and take. What this bill ultimately has to look like and what it will look like will be a consensus between the government, the Federal and executive branch and the Congress on homeland security. Recognizing that honest people can agree or disagree on any specific provision, let there be no doubt about what needs to happen, and I for one stand ready to support your efforts and yours of the committee in moving this along. I would also like to acknowledge one other thing, if I may, Mr. Chairman, and that is, both your leadership over the years on this issue and also the leadership of a very, very solid staff on both sides of the aisle. I've had the opportunity now to meet with a fair number of them, and they are a very tough bunch, but the one thing I will tell you is that their professionalism and their honesty and straightforwardness made it a pleasure to deal with them, even if we disagreed on some of the details of the policy. Thank you very much.

[The prepared statement of John S. Tritak follows:]

PREPARED STATEMENT OF JOHN S. TRITAK, DIRECTOR, CRITICAL INFRASTRUCTURE ASSURANCE OFFICE, BUREAU OF INDUSTRY AND SECURITY, UNITED STATES DEPARTMENT OF COMMERCE

I. INTRODUCTION

Mr. Chairman, members of the Committee, I am honored to appear before you today to discuss the critical infrastructure protection activities proposed for transfer to the new Department of Homeland Security. I look forward to discussing with you the important role that the Critical Infrastructure Assurance Office (CIAO) would play in this new Department.

It is very clear in this current environment that the country needs a single, unified homeland security structure that will improve protection against today's threats and be flexible enough to help meet the unknown threats of the future. The creation of the Department of Homeland Security is the most sweeping reorganization of our national security establishment in over 50 years. However, this decision was made on the basis of careful study and experience gained since September 11. The Administration considered a number of organizational approaches for the new Department proposed by various commissions, think tanks, and Members of Congress. The Secretary of Commerce, the Under Secretary and I—as well as all other senior management at the Commerce Department—fully support the President's plan and stand ready to undertake necessary efforts to facilitate the creation of the new Department as soon as possible.

The new Department of Homeland Security would be organized into four divisions: Border and Transportation Security; Emergency Preparedness and Response; Chemical, Biological, Radiological and Nuclear Countermeasures; and Information Analysis and Infrastructure Protection. The new department will be comprised mainly of existing organizational elements located in other Federal departments and agencies. For example, my office, the CIAO, now located in the Department of Commerce's Bureau of Industry and Security, will become part of the new Information Analysis and Infrastructure Protection Division.

I would like to take this opportunity to provide some background on the CIAO and to discuss briefly some of the specific activities and initiatives we are currently undertaking on cyber security and homeland security.

II. BACKGROUND ON THE CRITICAL INFRASTRUCTURE ASSURANCE OFFICE

The CIAO is not a new arrival to the homeland security effort: we have been working to realize the objective of critical infrastructure assurance for four years. The CIAO was created in May 1998 by Presidential Decision Directive 63 (PDD-63) to serve as an interagency office located at the Department of Commerce to coordinate the Federal Government's initiatives on critical infrastructure assurance. On October 18, 2001, Executive Order 13231 (the Order), was issued and entitled "Critical Infrastructure Protection in the Information Age," the CIAO began serving as a member of and an advisor to the newly created President's Critical Infrastructure Protection Board (the Board). The Board was created to coordinate Federal efforts and programs relating to the protection of information systems and networks essential to the operation of the nation's critical infrastructures. In carrying out its responsibilities, the Board fully coordinates its efforts and programs with the Assistant to the President for Homeland Security.

III. MAJOR CIAO ACTIVITIES AND INITIATIVES

CIAO's responsibilities for developing and coordinating national critical infrastructure policy focus on three key areas: (A) promoting national outreach and awareness campaigns both in the private sector and at the state and local government level; (B) assisting Federal agencies to analyze their own risk exposure and critical infrastructure dependencies; and (C) coordinating the preparation of an integrated national strategy for critical infrastructure assurance.

A. Outreach and Awareness

The Federal government acting alone cannot hope to secure our nation's critical infrastructures. The national policy of infrastructure assurance can only be achieved by a voluntary public-private partnership of unprecedented scope involving business and government at the Federal, State, and local levels. Forging a broad based partnership between industry and government lies at the heart of the CIAO's mission.

Private Sector Partnerships: CIAO has developed and implemented a nation-wide industry outreach program targeting senior corporate leadership responsible for setting company policy and allocating company resources. The challenge of such an effort is to present a compelling business case for corporate action. The primary focus of the CIAO's efforts continues to be on the critical infrastructure industries (i.e., information and communications, banking and finance, transportation, energy, and water supply). The basic thrust of these efforts is to communicate the message that critical infrastructure assurance is a matter of corporate governance and risk management. Senior management is responsible for securing corporate assets—including information and information systems. Corporate boards are accountable, as part of their fiduciary duty, to provide effective oversight of the development and implementation of appropriate infrastructure security policies and best practices.

In addition to infrastructure owners and operators, the CIAO's awareness and outreach efforts also target other influential stakeholders in the economy. The risk management community—including the audit and insurance professions—is particularly effective in raising matters of corporate governance and accountability with boards and senior management. In addition, the investment community is increasingly interested in how information security practices affect shareholder value—a concern of vital interest to corporate boards and management. In partnership with these communities, the CIAO has worked to translate potential threats to critical infrastructure into business case models that corporate boards and senior management can understand. Corporate leaders are beginning to understand that tools capable of disrupting their operations are readily available not merely to terrorists and hostile nation states but to a wide-range of potential "bad actors." As a consequence, they are beginning to grasp that the risks to their companies can and will affect operational survivability, shareholder value, customer relations, and public confidence. The CIAO has also worked actively to facilitate greater communication among the private infrastructure sectors themselves. As individual Federal lead agencies under PDD-63 formed partnerships with their respective critical infrastructure sectors, private industry representatives quickly identified a need for cross-industry dialogue and sharing of experience to improve the effectiveness and efficiency of individual sector assurance efforts. In response to that expressed need, the CIAO assisted its private sector partners in establishing the Partnership for Critical Infrastructure Security (PCIS). The PCIS provides a unique forum for government and private sector owners and operators of critical infrastructures to address issues of mutual interest and concern. It builds upon, without duplicating, the public-private efforts already being undertaken by the Federal Lead Agencies.

State and Local Government Partnerships: The CIAO has developed an outreach and awareness program for state and local governments to complement and support its outreach program to industry. State and local governments provide critical services that make them a critical infrastructure in themselves. They also play an important role as catalyst for public-private partnerships at the community level, particularly for emergency response planning and crisis management. The issue of securing the underlying information networks that support their critical services was a relatively new issue before September 11. State and local governments tend to be well organized as a sector, with multiple common interest groups. Similar to its program for industry, the CIAO has laid out a plan to implement outreach partnerships with respected and credible channels within state and local government. CIAO has also met with the National Governors Association and the National Association of State Chief Information Officers to encourage input into the National Strategy for Cyberspace Security. The front lines for the new types of threats facing our country, both physical and cyber, clearly are in our communities and in our individual institutions. Smaller communities and stakeholders have far fewer resources to collect information and analyze appropriate actions to take. Consequently, in February of this year, the CIAO began a series of four state conferences on Critical Infrastructures: Working Together in a New World, designed to collect lessons learned and applied from the events of September 11 from New York, Arlington, and communities across the United States. The intent of this conference series is to deliver a compendium of community best practices at the end of the first quarter of 2003. The first conference was held in Texas and the second in New Jersey. The last two will be held in the latter part of 2002 and the first quarter of 2003.

B. Support for Federal Government Infrastructure Activities

Homeland Security Information Integration Program: The Administration is proposing in the President's Fiscal Year 2003 budget request to establish an Information Integration Program Office (IIPO) within the CIAO to improve the coordination of information sharing essential to combating terrorism nationwide. The most important function of this office will be to design and help implement an interagency information architecture that will support efforts to find, track, and respond to terrorist threats within the United States and around the world, in a way that improves both the time of response and the quality of decisions. Together with the lead federal agencies, and guided strategically by the Office of Homeland Security, the IIPO will: (a) create an essential information inventory; (b) determine horizontal and vertical sharing requirements; (c) define a target architecture for information sharing; and (d) determine the personnel, software, hardware, and technical resources needed to implement the architecture. The foundation projects will produce roadmaps (migration strategies) that will be used by the agencies to move to the desired state.

Federal Asset Dependency Analysis—Project Matrix: The CIAO also is responsible for assisting civilian Federal departments and agencies in analyzing their dependencies on critical infrastructures to assure that the Federal government continues to be able to deliver services essential to the nation's security, economy, or the health and safety of its citizens, notwithstanding deliberate attempts by a variety of threats to disrupt such services through cyber or physical attacks.

To carry out this mission, the CIAO developed "Project Matrix," a program designed to identify and characterize accurately the assets and associated infrastructure dependencies and interdependencies that the U.S. Government requires to fulfill its most critical responsibilities to the nation. These are deemed "critical" because their incapacitation could jeopardize the nation's security, seriously disrupt the functioning of the national economy, or adversely affect the health or safety of large segments of the American public. Project Matrix involves a three-step process in which each civilian Federal department and agency identifies (i) its critical assets; (ii) other Federal government assets, systems, and networks on which those critical assets depend to operate; and (iii) all associated dependencies on privately owned and operated critical infrastructures. Early experience with the CIAO's Project Matrix process has demonstrated such significant utility that the Office of Management and Budget has recently issued a directive requiring all Federal civilian agencies under its authority to fund and perform the analysis.

C. Integrated National Strategy for Critical Infrastructure Assurance

Finally, the CIAO also plays a major role with respect to the development and drafting of the two national strategies relating to critical infrastructure protection—the National Strategy for Cyber Space Security and the National Strategy for Homeland Security. Specifically, the CIAO coordinates and facilitates input from private industry, as well as state and local government, to the national strategies. The Office of Homeland Security has enlisted the CIAO to provide coordination and support for its efforts to compile information and private sector input to its strategy to protect the physical facilities of critical infrastructure systems. The CIAO, working with its private sector partners, also has been instrumental in coordinating input from the private sector to the cyber space security strategy.

IV. CONCLUSION

The American economy is the most successful in the world. However, in the information age, the same technological capabilities that have enabled us to succeed can now also be turned against us. Powerful computing systems can be hijacked and used to launch attacks that can disrupt operations of critical services that support public safety and daily economic processes.

As the President and Governor Ridge have noted, today no Federal Agency has homeland security as its primary mission. Responsibilities for homeland security are dispersed throughout the Federal Government. The President's plan would combine key operating units that support homeland security so that the operations and activities of these units could be more closely directed and coordinated. This will serve to increase the efficiency and effectiveness of the Federal Government's critical infrastructure assurance and cyber security efforts.

The CIAO looks forward to continuing its role in advancing critical infrastructure protection policy in the new Department of Homeland Security. Thank you for the opportunity to appear before you today. I welcome any questions that you may have.

Mr. GREENWOOD. Thank you. Let me underscore your words about the staff. We couldn't do any of this without them.

Mr. McDonnell for 5 minutes.

TESTIMONY OF JAMES F. McDONNELL

Mr. MCDONNELL. Mr. Chairman and distinguished members of the committee, thank you for the opportunity to testify on the administration's proposal to create a Department of Homeland Security, and specifically, the critical infrastructure protection activities that will be assigned to the new department. I am James F. McDonnell, director of the Department of Energy, Office of Energy Assurance. I have been in this position since December of 2001, working with the Office of the Secretary to develop an integrated and streamlined management approach to protecting the national energy infrastructure. The Secretary of Energy has the responsibility as the lead Federal agency to coordinate protection activities in the energy sector.

Presidential decision directive 63 assigned this responsibility to DOE, and the Secretary expects the homeland security national strategy to continue that assignment of responsibility. The Office of Energy Assurance was established at the Department to better protect against severe energy disruptions in close collaboration with

State and local governments and the private sector, and where possible, to assist with emergency response efforts.

The Office provides technical expertise and management oversight to identify energy system critical components and interdependencies, identify threats to the system, recommend actions to correct or mitigate vulnerabilities, plan for response and recovery system disruption, and provide technical response support during energy emergencies. As originally conceived, the Office has four principle areas of management, which are energy reliability. The Office of Energy Assurance coordinates Department of Energy policy development and intergovernmental, interagency activities related to the protection and reliability of the national energy infrastructure.

The Office will utilize long-standing relationships with government appeared industry representatives to develop a national strategy for energy assurance and establish a national tracking and reporting process to assess the ongoing effectiveness of the national strategy, identifies shortfalls and develops corrective action plans; and coordinates efforts to expand cooperation on national energy infrastructure with friendly nations, international organizations and multinational corporations.

Energy emergencies: The Office of Energy Assurance ensures we are prepared to support States and industry efforts to plan for, respond to and mitigate actions that disrupt the Nation's energy supplies. This Office's primary missions are twofold. First is the identification of potential threats to the national energy infrastructure, including natural disasters and industrial accidents, and deliberate acts of terror, sabotage. The Office maintains an effective communications and liaison network with the energy sector to facilitate information flow during emergencies and communicate potential and actual threats to the appropriate authorities.

The second mission is to assist in the development of Federal energy emergency response plans. In carrying out this function, OEA will provide technical and professional assistance to States and industries for the development of local and regional response plans and conduct readiness exercises with States and industry to assist in identifying shortfalls prior to actual emergencies. Following such exercises, the Office will compile lessons learned during the conduct of emergencies and exercises for broad dissemination among relevant industries and facilities.

Energy infrastructure: The energy assurance team works with the companies whose resources comprise the Nation's energy sector to improve the protection of critical energy facilities. The infrastructure office works with the energy sector to introduce new security practices into the energy sector. The Office also interfaces with the DOE laboratory community to help identify and speed commercialization of new technologies designed to enhance the protection of sensitive facilities.

Infrastructure interdependencies: The Office of Energy Assurance had been designated to provide Federal oversight to the National Infrastructure Simulation and Analysis Center as a collaborative effort between the national laboratories, the Office of Energy Assurance, and other Federal agencies. The NISAC, once fully operational, will provide a fundamentally new technical planning and decision support environment for the analysis of critical infrastructures, their interdependencies, vulnerabilities and complexities for policy analysis and emergency planning. NISAC will use distributed information systems architectures to provide virtual analysis capabilities that will accommodate a large number of providers and a large number of users.

Tasking for the NISAC will be developed through an interagency planning process chaired by the Department's NISAC administrator, which includes representatives of the laboratories and industry and will ensure that the NISAC is truly a national asset to meet national strategy.

The Department of Homeland Security: The President's legislative proposal creating the Department of Homeland Security, includes moving the management of the National Infrastructure Simulation and Analysis Center, NISAC, and other functions of the Office of Energy Assurance from DOE to DHS.

The NISAC capability, once established, will provide a unique tool for planning and decisionmaking. The complexities of the physical and cyber interdependencies associated with the national energy infrastructure are vast by themselves. Once those complexities are overlaid with the other infrastructures, such as telecommunications, the interdependency complexities rise to a level that they become an issue that must be addressed at a national level. The transfer of the NISAC into the Department of Homeland Security will ensure that requirements develop and programmatic tasking for NISAC meet national priorities. DOE is planning to transfer funding and two staff members to DHS to provide program oversight for NISAC. DOE will continue to be a customer of NISAC, seeking to utilize this national capability to support energy sector analysis.

The transfer of the NISAC administrative functions with the Office of Energy Assurance into DHS will provide the new department with an integrated management structure to conduct activities associated with protecting the national energy infrastructure. The Office also manages a robust vulnerability assessment program that utilizes expertise from the private sector and the national laboratory complex, plans for and supports restoration and recovery efforts following natural disaster or acts of terrorism, assists States and industry in all aspects of energy emergency planning and supports the development of strategic energy policies.

The new Department of Homeland Security will thus have the ability to directly access the expertise located associated with the Office of Energy Assurance and the National Laboratories for Assessments of the energy sector. In addition, the new homeland security centers for excellence will provide the department with direct access to the capabilities currently resident in the national laboratories for research and analysis in other areas of the Nation's critical infrastructure.

Thank you, Mr. Chairman. I would be pleased to respond to any questions the committee may have.

[The prepared statement of James F. McDonnell follows:]

80-680 - 02 - 7

Prepared Statement of James F. McDonnell, Director, Office of Energy Assurance, U.S. Department of Energy

INTRODUCTION

Mr. Chairman and distinguished members of the committee, thank you for the opportunity to testify on the Administration's proposal to create a Department of Homeland Security, and specifically, the critical infrastructure protection activities that will be assigned to the new department. I am James F. McDonnell, Director of the Department of Energy Office of Energy Assurance. I have been in this position since December of 2001, working with the Office of the Secretary to develop an integrated and streamlined management approach to protecting the National Energy Infrastructure. The Secretary of Energy has the responsibility as the lead federal agency to coordinate protection activities in the Energy Sector. Presidential Decision Directive 63 assigned this responsibility to DOE and the Secretary expects the Homeland Security National Strategy to continue that assignment of responsibility. The Office of Energy Assurance was established at the Department to better protect against severe energy disruptions in close collaboration with State and local governments and the private sector and, where possible, to assist with emergency response efforts.

The Office provides technical expertise and management oversight to identify energy system critical components and interdependencies, identify threats to the system, recommend actions to correct or mitigate vulnerabilities, plan for response and recovery to system disruption, and provide technical response support during energy emergencies. As originally conceived, the Office has four principle areas of management, which are:

Energy Reliability

The Office of Energy Assurance coordinates Department of Energy policy development and intergovernmental, interagency activities related to the protection and reliability of the national energy infrastructure. The Office will utilize longstanding relationships with government and industry representatives to develop a national strategy for energy assurance and establish a national tracking and reporting process to assess the ongoing effectiveness of the national strategy, identifies shortfalls and develops corrective action plans; and coordinates efforts to expand cooperation on national energy infrastructure with friendly nations, international organizations and multinational corporations.

Energy Emergencies

The Office of Energy Assurance ensures we are prepared to support states and industry efforts to plan for, respond to and mitigate actions that disrupt the nation's energy supplies. This Office's primary missions are twofold; first is the identification of potential threats to the national energy infrastructure, including natural disasters and industrial accidents, and deliberate acts of terror, sabotage. The Office maintains an effective communications and liaison network with the energy sector to facilitate information flow during emergencies and communicate potential and actual threats to the appropriate authorities.

The second mission is to assist in the development of federal energy emergency response plans. In carrying out this function, OEA will provide technical and professional assistance to states and industries for the development of local and regional response plans and conduct readiness exercises with states and industry to assist in identifying shortfalls prior to actual emergencies. Following such exercises, the Office will compile lessons learned during the conduct of emergencies and exercises for broad dissemination among relevant industries and facilities.

Energy Infrastructure

The Energy Assurance Team works with the companies whose resources comprise the nation's energy sector to improve the protection of critical energy facilities. The Infrastructure Office works with the energy sector to introduce new security practices into the energy sector. The Office also interfaces with the DOE laboratory community to help identify and speed commercialization of new technologies designed to enhance the protection of sensitive facilities.

Infrastructure Interdependencies

The Office of Energy Assurance had been designated to provide federal oversight to the National Infrastructure Simulation and Analysis Center as a collaborative effort between the National Laboratories, the Office of Energy Assurance, and other federal agencies. The NISAC, once fully operational, will provide a fundamentally new technical planning and decision support environment for the analysis of critical infrastructures, their interdependencies, vulnerabilities, and complexities for policy analysis and emergency planning. NISAC will use distributed information systems architectures to provide virtual analysis capabilities that will accommodate a large number of providers and a large number of users. Tasking for the NISAC will be developed through an interagency planning process chaired by the Department's NISAC Administrator, which includes representatives of the laboratories and industry and will ensure the NISAC is truly a national asset meet national strategy.

The Department of Homeland Security

The President's legislative proposal creating the Department of Homeland Security includes moving the management of the National Infrastructure Simulation and Analysis Center (NISAC) and other functions of the Office of Energy Assurance from DOE to DHS.

The NISAC capability, once established, will provide a unique tool for planning and decision-making. The complexities of the physical and cyber interdependencies associated with the national energy infrastructure are vast by themselves. Once those complexities are overlaid with the other infrastructures, such as telecommunications, the interdependency complexities rise to a level that they become an issue that must be addressed at a national level. The transfer of the NISAC into the Department of Homeland Security will ensure that requirements development and programmatic tasking for NISAC meet national priorities. DOE is planning to transfer funding and two staff members to DHS to provide program oversight for NISAC. DOE will continue to be a customer of NISAC, seeking to utilize this national capability to support Energy Sector analysis.

The transfer of the NISAC administrative functions with the Office of Energy Assurance into DHS will provide the new Department with an integrated management structure to conduct activities associated with protecting the National Energy Infrastructure. The Office also manages a robust vulnerability assessment program that utilizes expertise from the private sector and the National Laboratory complex, plans for and supports restoration and recovery efforts following natural disaster or acts of terrorism, assists states and industry in all aspects of energy emergency planning and supports the development of strategic energy policies. The new Department of Homeland Security will thus have the ability to directly access the expertise located associated with the Office of Energy Assurance and the national laboratories for assessments of the energy sector. In addition, the new Homeland Security Centers for Excellence will provide the Department with direct access to the capabilities currently resident in the national laboratories for research and analysis in other areas of the nation's critical infrastructure.

Thank you, Mr. Chairman. I would be pleased to respond to any questions the Committee may have.

Mr. GREENWOOD. Thank you, Mr. McDonnell.

Mr. Varnado for 5 minutes.

TESTIMONY OF SAMUEL G. VARNADO

Mr. VARNADO. Mr. Chairman and distinguished members, thank you for this opportunity. I'm Stan Varnado, director of Sandia National Laboratories Programs Critical Infrastructure. The work you're doing here is very important and we're proud of being a part of it. My written statement has been entered into the record, and I'll just present a brief summary about of what is in that statement. I want to focus on two major problems in critical infrastructure protection. First is cyber security, and second is infrastructure interdependency.

In the cyber area, past research has shown that computer systems that control many of the Nation's infrastructures are highly vulnerable to cyber attack. These systems are called Supervisory Control and Data Acquisition, or SCADA systems. They are ubiquitous in the electric power, oil and gas, telecommunications and transportation industry. We are particularly worried about the SCADA systems for several reasons. First, many of the U.S. infrastructure elements depend upon their reliable operations. Second, the systems in which the electric power industry are used—are using are being used in ways to which they were never designed because of the way the grid is being operated under the restructuring environment. Third, the consequences of attacks on the SCADA systems can be significant including loss of life, burnout of equipment that is difficult to replace, environmental impacts and others.

Fourth, the industry is coming to us now and asking for help. Fifth, according to an article in the June 27th addition of The Washington Post, the Al Qaeda terrorist network is looking for information on the SCADA system to maybe plan an attack. In our research, we found many vulnerabilities in the SCADA systems and these will increase as the industry moves toward Internet-based control systems. Some of these vulnerabilities are listed in my written statement. We believe that the security of these systems must be up there. DHS should make the cyber security issue a prominent one.

In the case of SCADA systems, DHS may want to work with the program that DOE has already staffed. They can supply requirements to DOE and could support DOE's request for resources. The second major area of concern is that of infrastructure interdependency. U.S. now depends upon an increasingly interdependent staff critical infrastructure elements that include electric power, oil and gas, transportation, water, communication, banking and finance and others. These systems depend upon each other for reliable operations. For example, banking and finance depend upon telecommunications which is dependent upon electricity, which is dependent upon coal, oil, nuclear and gas, which is dependent upon water and so on. The list is endless.

Currently no two exists that allow an adequate understanding of the operation of such a complex system. The system interdependencies make it hard to identify critical nodes that must be protected, to define the consequences of outages and to optimize mitigation strategies.

The National Infrastructure Simulation and Analysis Center, NISAC, which is proposed to now become a part of the new Department of Homeland Security, was established to address this problem. We use Sandia and Los Alamos National Laboratory's extensive computing and simulation capabilities to allow comprehensive assessments of the vulnerabilities of the Nation's infrastructure to allow identification of critical nodes and to develop and optimize mitigation strategy. I will provide some background on NISAC for you, and my colleague will provide additional information on NISAC capabilities.

NISAC was formally established last year in the USA PATRIOT Act. The current status is that it's funded at $20 million in fiscal year 2002. The money this year came through DOD's DTRA's organization. In fiscal year 2003, the President's budget calls for the money to come through DOE. We have established a NISAC joint program office to represent both laboratories that are involved. We have selected a NISAC joint program director to manage the program. We are currently developing strategies and partnerships with public entities, private industry and universities who will also participate in this activity as technology suppliers. We are already developing models of the electric power grid, oil and gas distribution systems, telecommunications networks and economic models of the consequences. For example, in models of interdependencies of the electricity and telecommunications infrastructure in California has been developed. We are currently studying their interdependency and the consequences.

The proposal to place NISAC in the new Department of Homeland Security is very sound. We agree with it. The reason is that it allows the NISAC to address national needs rather than just the need of any simulation. To achieve this, however, we need a national level multiagency committee that represents the interests of all the agencies, and that should be established in order to set requirements for NISAC. So our concern is we offer the following recommendations.

The homeland security actions requires DHS to establish a national level multiagency process to solicit needs from all concerned agencies and to define requirements for NISAC. The Act should give DHS the authority to pass the DOE NSA laboratories directly, just as the nonNSA portions of DOE do now. This would eliminate bureaucratic red tape and additional costs and delay associated with the process. We further recommend the legislation specify that NISAC be managed for DHS by the existing NISAC joint program office in order to take advantage of the significant amount of research that has already been done.

We believe that the inclusion of these recommendations in the Homeland Security Act will provide the greatest utilization advances to important issues facing critical infrastructure protection. Thank you, Mr. Chairman, and members of the committee for this opportunity.

[The prepared statement of Samuel G. Varnado follows:]

Prepared Statement of Samuel G. Varnado,, Sandia National Laboratories

INTRODUCTION

Mr. Chairman and distinguished members of the committee, thank you for the opportunity to testify on the Administration's proposal to create a Department of Homeland Security, and specifically, the critical infrastructure protection activities that will be assigned to the new department. I am Dr. Samuel G. Varnado, Director of Sandia National Laboratories' Infrastructure and Information Systems Center. I have more than thirty-eight years' experience in energy, information, and infrastructure systems development. I currently coordinate the Laboratories' activities in critical infrastructure protection.

Sandia National Laboratories is managed and operated for the National Nuclear Security Administration (NNSA) of the U.S. Department of Energy (DOE) by Sandia Corporation, a subsidiary of the Lockheed Martin Corporation. Sandia's unique role in the nation's nuclear weapons program is the design, development, qualification, and certification of nearly all of the nonnuclear subsystems of nuclear warheads. We perform substantial work in programs closely related to nuclear weapons, including intelligence, non-proliferation, and treaty verification technologies. As a multiprogram national laboratory, Sandia also conducts research and development for other Federal agencies when our special capabilities can make significant contributions.

At Sandia National Laboratories, we perform scientific and engineering work with a mission in mind—never solely for its own sake. Even the fundamental scientific work that we do (and we do a great deal of it) is strategic for the mission needs of our sponsors. Sandia's management philosophy has always stressed the ultimate linkage of research to application. When someone refers to Sandia as "the nation's premier engineering laboratory," that statement does not tell the whole story: We are a science and engineering laboratory with a focus on developing technical solutions to the most challenging problems that threaten peace and freedom.

My statement, which amplifies my colleague David Nokes' testimony to this committee on June 25, 2002, will describe some of the key problems posed in protecting the nation's critical infrastructure and Sandia National Laboratories' contributions and capabilities in that area. I will also comment on the proposed relationship of that work to the Department of Homeland Security.

SANDIA'S CONTRIBUTIONS TO CRITICAL INFRASTRUCTURE PROTECTION

Like most Americans, the people of Sandia National Laboratories responded to the atrocities of September 11, 2001, with newfound resolve on both a personal and professional level. As a result of our own strategic planning, our LDRD investments, and the foresight of sponsors to invest resources toward critical infrastructure protection, Sandia was in a position to immediately address some urgent needs.

For example, we quickly completed vulnerability assessments of a number of dams in the Western U.S. and worked with the electricity sector to improve the robustness of their supervisory control and data acquisition (SCADA) systems to cyber attacks. These and other contributions to critical infrastructure protection are possible because of strategic planning we had conducted years ago and early investment in the capabilities that were needed to respond. The outstanding technology base supported by NNSA for its core missions is the primary source of this capability. We also made strategic decisions to invest laboratory-directed research and development funds (LDRD) in the very things that we knew were urgent needs: physical security technology, modeling and simulation of infrastructure elements, and cyber security. We were heavily involved in supporting the President's Critical Infrastructure Protection Committee during the Clinton administration, and that activity provided impetus for our current activities. In recent months, requests for Sandia's services from federal agencies other than DOE for work in emerging areas of need have increased. Approximately twenty-eight percent of our total laboratory-operating budget is now provided by federal agencies other than DOE.

SANDIA CAPABILITIES FOR CRITICAL INFRASTRUCTURE PROTECTION

Sandia National Laboratories and the other nuclear weapon laboratories constitute a broad, multidisciplinary technology base in nearly all the physical sciences and engineering disciplines. We leverage those capabilities to support other national security needs germane to our missions, including homeland security, when our capabilities can make significant contributions.

Physical Security

For over 25 years, Sandia has been the lead laboratory for the DOE in safeguards and security. During this time, we have developed risk assessment methodology and used it to design the security approaches for storage and shipment of nuclear weapons and special nuclear material. We have developed vulnerability assessment capabilities and models to optimize mitigation strategies. These models were used in the early days to design protection systems for nuclear power plants as well as for our traditional missions. Recently, the same technology has been used to assess the vulnerabilities and improve the robustness of dams, chemical plants, water systems, conventional electric power plants, and pipelines.

We have developed numerous airport security sensors and systems, including design of secure portals and explosives detectors. Today, a commercially produced, walk-through portal for detecting trace amounts of explosive compounds on a person is available for purchase and installation at airports and other public facilities. The technology for this device was developed, prototyped, and demonstrated by Sandia National Laboratories over a period of several years and licensed to Barringer Instruments of Warren, New Jersey, for commercialization and manufacture. The instrument is so sensitive that microscopic quantities of explosive compounds are detected in a few seconds.

Using similar technology, we have developed and successfully tested a prototype vehicle portal that detects minute amounts of common explosives in cars and trucks. Detecting explosives in vehicles is a major concern at airports, military bases, government facilities, and border crossings. The system uses Sandia's patented sample collection and preconcentrator technology that has previously been licensed to Barringer for use in screening airline passengers. The same technology has been incorporated into Sandia's line of "Hound™" portable and hand-held sensors, capable of detecting parts-per-trillion explosives and other compounds. These devices can be of great value to customs and border agents at ports of entry.

Sandia pioneered a tool called Probabilistic Risk Assessment (PRA) to evaluate the risks in high-consequence systems such as nuclear weapons and nuclear power generation plants. We use this tool to assess the risks in critical infrastructure systems such as dams, water utilities, chemical plants, and power plants. Combined with our expertise in security systems for nuclear facilities, we have helped utilities and industrial associations create security assessment methodologies that help owners and operators determine vulnerabilities and identify mitigation options. Methodologies have been developed for water utilities, chemical storage facilities, dams, power plants, and electrical power transmission systems.

Cyber Security

Sandia has significant ongoing work in the technology areas intended to protect cyber and network resources and the information that resides on such systems. Programs that assess the vulnerabilities associated with these systems are in place for our own resources as well as for those at other federal government agencies. We conduct red-teaming to challenge information systems and identify and remove vulnerabilities. Our objectives are to enhance the robustness of critical information systems and develop solutions for survivability and response options for systems under attack. Sandia operates a supervisory control and data acquisition (SCADA) laboratory to study the real-time control systems that are used to control the power grid, the pipelines, transportation systems, and water systems. Sandia's capabilities in cyber security arise from our nuclear weapons mission, in which we design the cryptographic systems needed for secure command and control systems for the nuclear stockpile. Sandia is the only DOE laboratory that is approved by NSA to conduct cryptographic research. We have helped many infrastructure owners perform vulnerability assessments and develop risk mitigation strategies.

Modeling and Simulation

National security and the quality of life in the United States rely on the continuous, reliable operation of a complex set of interdependent infrastructures: electric power, oil and gas, transportation, water, communications, banking and finance, emergency services, law enforcement, government continuity, agriculture, health services, and others. Today, these systems depend heavily on one another; that interdependency is increasing. Disruptions in any one of them could jeopardize the continued operation of the entire infrastructure system. Many of these systems are known to be vulnerable to physical and cyber threats and to failures induced by system complexity.

In the past, the nation's critical infrastructures operated fairly independently. Today, however, they are increasingly linked, automated, and interdependent. What previously would have been an isolated failure could cascade into a widespread, crippling, multi-infrastructure disruption today. Currently, there are no tools that allow understanding of the operation of this complex, interdependent system. This makes it difficult to identify critical nodes, determine the consequences of outages, and develop optimized mitigation strategies.

The National Infrastructure Simulation and Analysis Center (NISAC) concept, which would be transferred to the Department of Homeland Security under the Administration's bill, is also an example of our experience with critical infrastructures and will be described and discussed later in this statement.

CRITICAL INFRASTRUCTURE PROTECTION PROBLEMS

The U.S. infrastructure is difficult to protect because of its size and complexity. There are many avenues for possible exploitation by an adversary. In this statement, I will address two of the problems we consider to be the most serious.

Cyber Security

Computerized supervisory control and data acquisition (SCADA) systems control the operations of critical infrastructures such as power utilities, distribution networks, and municipal water supplies. These systems have generally been designed and installed with little attention to security. They are highly vulnerable to cyber attack. In fact, it has been claimed that it is possible to turn the lights off in many major cities with a cyber attack. An article in the June 27, 2002, edition of the Washington Post adds credence to this claim, and states that these systems have been the targets of probing by Al Qaeda terrorists. Some government experts conclude that the terrorists plan to use the internet as an instrument of bloodshed by attacking the juncture of cyber systems and the physical systems they control. The article further postulates that combined cyber and physical attacks could generate nightmare consequences.

Sandia has been investigating vulnerabilities in SCADA systems for five years. During this time, many have been found. Our assessments show that security implementations are, in many cases, non-existent or based on false premises. Some of the vulnerabilities in legacy SCADA systems include inadequate password policies and security administration, no data protection mechanisms, and information links that are prone to snooping, interruption, and interception. When firewalls are used, they are sometimes not adequately configured, and there is often "back-door" access because of connections to contractors and maintenance staff. We have found many cases in which there is unprotected remote access that circumvents the firewall. From a security perspective, it should be noted that most of the SCADA manufacturers are foreign-owned. In summary, it is possible to covertly and easily take over control of one of these systems and cause disruptions with significant consequences. Recognition of that fact led numerous federal agencies and municipal water and transportation systems to request Sandia help following September 11.

Of even more concern is the fact that the control systems are now evolving to the use of the internet as the control backbone. The electric power grid is now, under restructuring, being operated in a way for which it was never designed. More access to control systems is being granted to more users; there is more demand for real time control; and business and control systems are being connected. Typically, these new systems are not designed with security in mind. More vulnerabilities are being found, and consequences of disruptions are increasing rapidly. Industry is now asking for our help in understanding vulnerabilities, consequences, and mitigation strategies. After September 11, Sandia also received requests for help from private companies and professional societies.

Interdependencies

The U.S. infrastructure is becoming increasingly interdependent. For example, the banking and finance sector depends upon telecommunications, which depends on electricity, which depends on coal, gas, oil, nuclear sources, water, and transportation. These interdependencies create the potential for high consequence, cascading failures in which a failure in one element of the infrastructure leads to failures in others. Further, interdependencies make it difficult to identify critical nodes, vulnerabilities, and optimized mitigation strategies. We have studied one case, for example, in which the best way to assure operation of the electric power grid is to protect the gas pipeline that feeds the generation stations in that area. The bottom line is that interdependencies cause the infrastructure to behave as a complex system whose behavior is difficult to predict.

Most of the current federal critical infrastructure protection activities are directed at individual infrastructure elements. This stovepiped approach was reinforced by PDD-63, in which various agencies were assigned responsibility for protecting specific infrastructure elements (e.g., DOE was assigned electricity and oil and gas, DOT was assigned transportation, etc.). While it is necessary to understand these individual elements, the more compelling problem is to address the interdependent nature of the behavior of the infrastructure in order to prevent more severe consequences. We believe that this modeling and simulation effort is essential and will lead to the ability to define the critical nodes at the system level, identify consequences of outages, and define optimized protection strategies.

Possible Solutions to Critical Infrastructure Problems

It is unreasonable to expect that every part of the infrastructure can be completely protected. Rather, a risk management strategy must be used to decide where to invest limited protection resources. Three steps are needed:

- Define the infrastructure elements that are truly critical. Criteria must be established that define "critical". These could include, for example, loss of life, economic impact, time to rebuild, cost to rebuild, potential for loss of confidence in the government, etc.
- Perform vulnerability assessments for these critical elements.
- Develop optimized prevention and mitigation strategies.

It will be necessary to work closely with private industry in all these steps, since they own 85% of the US infrastructure. They must see a business case, based on risk analysis, before they are willing to invest in protection. Vulnerability assessment methodology is well known to Sandia, other DOE labs, and certain private companies. They can play important roles in all three steps, but especially in identifying, from a systems perspective, the critical nodes and in evaluating the consequences of disruptions so that business cases can be developed. The methodology for conducting the required analysis is known. What is needed from a technology development perspective is additional research in cyber security techniques and development of additional simulation and modeling capability, since modeling of the behavior of complex systems will require high performance computing. Additionally, help is needed in working with private industry. Many of the private owners of the infrastructure feel that identification of critical nodes and vulnerabilities is sensitive information, and they are reluctant to share it with the government. Government action is needed to create a process under which sensitive information can be shared among those in government and industry with a need-to-know.

Congressional support is needed to help implement the following steps that will lead to a more robust national infrastructure:

- Establish a new category of sensitive, restricted information for Critical Infrastructure Protection applications. Procedures for protecting the information and processes for granting access to both industry and government personnel are needed.
- Provide training in vulnerability and risk assessment methodology to private industry.
- Support additional research into cyber security issues, including cryptographic methods such as authentication, low power encryption methods, and standards. The establishment of test beds to allow evaluation of competing technologies should be encouraged.
- Support development of tools needed for identifying critical nodes, consequences of outages, and optimized mitigation strategies.

NATIONAL INFRASTRUCTURE SIMULATION AND ANALYSIS CENTER (NISAC)

The President's bill to establish a Department of Homeland Security provides for an Under Secretary for Information Analysis and Infrastructure Protection. It further proposes, under Title II, to transfer the responsibility for NISAC to the Department of Homeland Security. NISAC was formally chartered by the USA Patriot Act of 2001 (Oct 26, 2001) to serve as "a source of national competence to address critical infrastructure protection and continuity through support for activities related to counter terrorism, threat assessment, and risk mitigation." (Section 1016 of Public Law 107-56, the USA Patriot Act, 10/26/2001). NISAC, a partnership of Sandia and Los Alamos national laboratories, is leveraging current modeling, simulation, and analysis expertise to develop higher fidelity simulations crucial to the success of the Nation's critical infrastructure protection program. These labs were chosen to manage NISAC because of their considerable investment in infrastructure and interdependencies modeling over the last decade, the availability of high performance computers at the labs, and their modeling and simulation capabilities.

Status

The President's FY03 budget request called for the FY03 NISAC activities to be funded through the Department of Energy. NISAC, with Sandia and Los Alamos national laboratories as core partners, has devoted considerable effort to expanding the critical infrastructure modeling, simulation, and analysis capabilities of the two laboratories. A Joint Program Director has been selected to manage the NISAC program on behalf of both labs. NISAC has built consensus in the government and private sector on the importance of infrastructure interdependency analysis to the nation's critical infrastructure protection program. The NISAC Joint Program Office is developing strategic plans and associated research and development programs to meet its national charter. These plans include the identification of key strategic partners from other labs, universities, and private industry who will serve as technical collaborators in the performance of the tasks assigned to NISAC. Further, NISAC has proposed a senior-level, national, interagency process, including DHS, to generate, prioritize, and set national-level requirements for its modeling and simulation activities.

Observations

The proposal to move NISAC to the Department of Homeland Security is sound. It will allow NISAC to serve as a national resource that can address critical infrastructures and, most importantly, their *interdependencies* across the entire range of infrastructure elements—energy, telecommunications, transportation, banking and finance, water, etc. It will allow the NISAC work to be prioritized by national needs, rather than the by the interests of a single agency. Further, it will be possible to implement a national level requirements-setting process for NISAC activities, which fulfills the intent of the Patriot Act.

It is important that the existing NISAC Joint Program Office continue to serve as the managing entity for NISAC, serving under the oversight of the new DHS, in order to capitalize on the previous decade's investment in the technology base. An added benefit to the proposed organizational structure within DHS is that it would place NISAC and the National Communications System (NCS) under the same Under Secretary. NCS has significant capability in modeling the telecommunications infrastructure, while Sandia and Los Alamos have similar capabilities in modeling the energy infrastructure, chem./bio problems, and infrastructure interdependencies. This concentration of technical capability in one organization will provide a demonstrated competence that should lead to early and useful results.

Recommendations

- The legislation that establishes the Department of Homeland Security should clearly state that NISAC will be managed by the NISAC Joint Program Office for the Department of Homeland Security.
- The legislation should state that DHS will assume both funding and oversight responsibilities for NISAC as soon as DHS is established. A NISAC program manager within DHS should be named.
- The Homeland Security Act should give the Department of Homeland Security the power to task the NNSA laboratories directly, just as do the Science, Energy, Environmental, and other non-NNSA offices of DOE. That authority would eliminate the bureaucratic red tape and additional costs associated with the Work-for-Others (WFO) process.
- The legislation should require that DHS establish a national level, multi-agency process to solicit needs and define requirements for NISAC. Participating agencies could include DOE, DOT, DOC, OSTP, DOS, Treasury, and others. Final approval for all NISAC activities should reside with a senior DHS official.

SUMMARY AND CONCLUSION

Sandia National Laboratories and the other NNSA laboratories constitute a broad, multidisciplinary technology base in nearly all of the physical sciences and engineering disciplines. We are eager to leverage those capabilities to support the science and technology needs of the Department of Homeland Security when our capabilities can make significant contributions.

Sandia possesses strong competencies in physical and cyber security and in modeling and simulation. Most of this technology is suitable for transfer to industry and deployment in homeland security applications. We have been proactive in addressing the challenges of infrastructure protection. We have a track record of anticipating emerging homeland security threats and investing in technology development to counter them through our Laboratory-Directed Research and Development program and sponsor-directed programs. We are one of the premier laboratories for working with industry to transform laboratory technologies into deployable commercial applications. Bureaucratic and regulatory roadblocks exist that limit access to the DOE/NNSA national laboratories by other federal agencies, and those obstacles should be removed by the homeland security legislation in order to facilitate direct access to those resources.

On behalf of the dedicated and talented people who constitute Sandia National Laboratories, I want to emphasize our commitment to strengthening United States security and combating the threat to our nation's critical infrastructures. It is our highest goal to be a national laboratory that delivers technology solutions to the most challenging problems that threaten peace and freedom. Thank you, Mr. Chairman. I would be pleased to respond to any questions you may have.

Mr. GREENWOOD. Thank you, Dr. Varnado.

Dr. Cobb for 5 minutes.

TESTIMONY OF DONALD D. COBB

Mr. COBB. Thank you, Mr. Chairman, and distinguished members of the committee for inviting Sandia and Los Alamos here today to discuss the issue of critical infrastructure protection, and in particular, the national infrastructure simulation and analysis center, or NISAC.

This morning I'd like to discuss with you the efforts to protect the Nation's critical infrastructure in the form of this joint, and I think, unique Los Alamos and Sandia partnership. NISAC brings to bear on the problem of protecting the Nation's critical infrastructure, some of the most sophisticated modeling simulation technology to be found anywhere. This technology is based on a decade-long, $150 million investment by the Federal Government in work at both laboratories. The work is to do complex modeling and simulation of some of the most complex systems, namely our infrastructure.

It also is supported by two of the largest secure computing environments. I think that is an important point in that we have the experience to use massive computing as tools, and also the environment to protect the information in the appropriate fashion. NISAC, when it's fully operational, is envisioned to provide the type, scale, comprehensive level of information that will enable the Nation's senior leadership, our decisionmakers at the highest levels, to proactively work to deny terrorist attacks against high targets, key nodes and our critical infrastructure.

For the first time, we'll be able to simulate the operations of and the interdependencies among the elements of our infrastructure, including telecommunications, electricity, oil and gas, transportation, public health and so forth.

We will have confidence that these results can be used by decisionmakers to identify key gaps and vulnerabilities, and thereby set the priorities for investment in protection measures. Today NISAC is already providing important information to the Office of Homeland Security and other government agencies. Permit me to just describe one example.

Recently we were asked to complete a short fuse study for the Office of Homeland Security looking at various scenarios for distributing vaccine. This study used a new simulation tool called EpiSims, which stands for epidemiological simulation. EpiSims, in turn, builds on a decade of transportation modeling simulation that was carried out for the Department of Transportation. This latter capability called TranSims literally reproduces the complex nonlinear pattern of traffic in major urban areas on a minute-by-minute basis.

How are these two things connected? Basically, the methodology in TranSims, in order to replicate how dynamic interactions occur among members of a diverse population such as a major city in the United States and that synthetic urban population which is derived from demographic information are the tools that we need to do many of this type of model and simulation. So EpiSims used that basic methodology and those synthetic populations in the studies that we did.

Along with input from some of the experts that you heard earlier on public health interactions so we could have the lead people in the area of public health allegation provide input and then looking at our results to confirm that they do, in fact, match their experience.

So in recognition of this type of capability that has been developed over the years and building on this and leveraging it, Congress chartered NISAC under the U.S. PATRIOT Act of 2001 to, quote, serve as a national source of competence to address critical infrastructure protection and continuity through support for activities related to counterterrorism, threat assessment and risk mitigation.

As was stated earlier, the President's homeland security legislation calls for the transfer of NISAC to the new Department of Homeland Security. Because the purpose of NISAC and its true realization is the responsibility—has the responsibility across all the infrastructure sectors that are interdependencies, it seems to us that we concur that this is the appropriate place for NISAC to be.

In other words, it should report directly to the agency that will inherit this responsibility for protecting our infrastructure.

So in closing, let me say that through the NISAC collaboration, Sandia and Los Alamos look forward to continuing support the new Department of Homeland Security and in protecting our Nation's critical infrastructure and I thank you for the opportunity, and we will be happy to answer your questions later.

[The prepared statement of Donald D. Cobb follows:]

PREPARED STATEMENT OF DON COBB, ASSOCIATE DIRECTOR, THREAT REDUCTION, LOS ALAMOS NATIONAL LABORATORY

Thank you Mr. Chairman and distinguished members of the House Energy and Commerce Subcommittee on Oversight and Investigations for inviting me here today to discuss the administration's proposal for creating the Department of Homeland Security. I am Don Cobb, Associate Director for Threat Reduction at Los Alamos National Laboratory. At Los Alamos, I am responsible for all programs directed at reducing threats associated with weapons of mass destruction. I personally have more than 30 years experience working to reduce these threats. Los Alamos is operated by the University of California for the DOE/NNSA and is one of three NNSA laboratories, along with Lawrence Livermore National Laboratory and Sandia National Laboratories, responsible for maintaining the nation's nuclear stockpile. In addition to our stockpile responsibilities, the three NNSA laboratories have been involved for decades in technology development and problem solving in the realm of arms control and nonproliferation. Through our work in these areas, Los Alamos has developed a skill and technology base that enabled us to respond immediately following the September 11 attacks, to calls for assistance in counter terrorism and homeland security. With the President's call for a new Department of Homeland Security, Los Alamos stands ready to focus its capabilities in support of this new department.

Today, I would like to discuss with you the broad set of capabilities that Los Alamos brings to U.S. efforts to protect our homeland from future terrorist attacks. While my testimony is Los Alamos centric, progress in science and technology depends on collaboration among the national laboratories, government, industry and academia.

Los Alamos National Laboratory firmly supports the creation of a Department of Homeland Security (DHS). Consolidation of federal homeland security agencies has the potential to protect the nation against terrorism.

The President's proposal would give the Department four divisions: Information Analysis and Infrastructure Protection; Chemical, Biological, Radiological, and Nuclear Countermeasures; Border and Transportation Security; and Emergency Preparedness and Response. Each of these mission areas will require focused research and development (R&D). My statement will describe some of the key contributions Los Alamos and the other national laboratories can make to homeland security in each of these areas.

ENGAGING THE SCIENCE AND TECHNOLOGY (S&T) COMMUNITY

"The government will need mechanisms to engage the technical capabilities of the government and the nation's scientific, engineering, and medical communities in pursuit of homeland security goals," says a new National Academies report.[1] Every division of the DHS will require research, development, testing, and evaluation (RDT&E) to solve the technical challenges it will face.

At Los Alamos, we have asked the question, "How can a newly formed DHS best engage with the S&T community, including the national laboratories, universities and industry?" I believe that in order to succeed, DHS requires a single, focused S&T office that serves as the central R&D organization for the Department. As suggested by the House and Senate bills, this office could be placed under a separate Director of Science and Technology. The best and brightest human resources, including federal staff augmented by scientists and engineers assigned from national laboratories, industry and academia, must staff this S&T office. Boundaries with other organizations must be "permeable," enabling people to move back and forth easily.

[1] National Research Council Committee on Science and Technology for Countering Terrorism, *Making the Nation Safer: The Role of Science and Technology in Countering Terrorism* (Washington, DC: National Academy Press, June 2002).

The S&T office would be responsible for the planning and oversight of focused RDT&E, including both rapid technology acquisition and long-term, high-risk, high-payoff research. Functional responsibilities for the agency would therefore include:

- Threat and vulnerability assessment;
- Identification of needs through interactions with other agencies, and with state and local governments;
- Strategic planning and prioritization for RDT&E investments;
- Program planning, budgeting, funding and oversight;
- Systems architectures;
- Science and technology acquisition from universities, industry and national laboratories;
- Technology integration;
- Evaluation of technologies and systems effectiveness; and
- Close coordination with end-users during initial system deployments.

The office should be established quickly, in place and functioning concurrently with the establishment of the DHS—we want to maintain, and even accelerate, the momentum which has built since September 11. I now will describe some of the key contributions Los Alamos is making to homeland security.

INFORMATION ANALYSIS AND INFRASTRUCTURE PROTECTION

National Infrastructure Simulation and Analysis Center (NISAC). Los Alamos is partnering with Sandia National Laboratories to establish NISAC. NISAC is intended to provide improved technical planning, simulation, and decision support for the analysis of critical infrastructures, their interdependencies, and vulnerabilities for policy analysis and emergency planning. This technology is based on a decade long, $150M investment in basic research and software development, supported by the world's largest secure, scientific computing environment. NISAC will provide the type, scale, and comprehensive level of information that will enable the nation's senior leadership proactively to deny terrorist attack options against potentially high-value targets, instead of simply reacting to the latest threat scenarios. NISAC will provide essential analytic support for discovering and overcoming gaps in our homeland security.

NISAC was created as part of the U.S.A. Patriot Act of 2001 (P.L. 107-56). The President's proposal calls for the transfer of NISAC to the DHS. Because NISAC has responsibility across all infrastructure sectors, it is appropriate that NISAC should directly support the agency charged with cross-infrastructure responsibilities. NISAC is part of a broader portfolio of infrastructure modeling and simulation work at the two laboratories. This is significant. The technical and programmatic synergies that accrue to NISAC as a result of this association allow for immediate application of the R&D efforts to real problems today. From vulnerability assessments of actual infrastructures to "what if" simulations of biological event scenarios, NISAC is providing insights and information to senior decision makers now. As this capability matures, we will do more.

National Transportation Modeling and Analysis Program (NATMAP). NATMAP, currently being developed for the Department of Transportation, builds on Los Alamos' transportation modeling technology developed over the past decade. NATMAP simulates individual carriers—trucks, trains, planes, and waterborne vessels—and the transportation infrastructures used by these carriers to simulate freight commodity shipments of the U.S. transportation network. It moves individual freight shipments from production areas, through intermodal transfer facilities and distribution centers, to points of consumption. The advantage of the NATMAP is that the nation's system can be represented at any level of detail—from trucks and goods moving among counties and within regions, to national multi-modal traffic flows including cross border trade with Mexico and Canada. This strength can be exploited for transportation policy, security and infrastructure investment purposes.

Vulnerability/Threat Assessments: Nuclear Facilities. Over the last 20 years, Los Alamos and Sandia have analyzed physical security and identified vulnerabilities at numerous nuclear facilities throughout DOE, DoD, and U.S. Nuclear Regulatory Commission (NRC) facilities. These facilities include nuclear reactors, plutonium-handling facilities, nuclear weapons storage facilities, commercial nuclear power plants, and spent nuclear fuel facilities. We routinely train external agencies on developing protection strategies for low-probability/high-consequence scenarios, such as aircraft crash, sabotage, and fire. Fundamental to these activities are the unique facilities and capabilities that Los Alamos brings to these analyses. We are the only site where highly radioactive materials can be studied experimentally for their response to postulated threat scenarios. Such an understanding is essential for analyzing threats and their potential consequences.

Threat Analysis and Warning. Following the September 11 attacks, we established a multidisciplinary team of analysts searching for evidence of terrorist activity. Such analysis requires the latest information management technologies, advanced computational methods, and automated pattern identification to search enormous amounts of electronic information. This tremendous task is complicated by the fact that the vast majority of data represents completely innocent activity. Under the new Department, a major effort will be needed to develop the tools that will provide the ability to accurately synthesize information from intelligence, law enforcement, and open sources. Using our experience in solving related problems over the years, for example in identifying activities indicating WMD proliferation, Los Alamos will continue to provide analytic capability in this area.

Immigration and Naturalization Service: Entry/Exit System. The Immigration and Naturalization Service Data Management Improvement Act (DMIA) of 2000 (P.L 106-215) created a Task Force to evaluate how the flow of traffic at United States ports of entry can be improved while enhancing security and implementing systems for data collection and data sharing. The Task Force is advisory in nature, and as such, will develop recommendations regarding the development and deployment of an integrated, automated entry/exit system. A team of experts from Los Alamos is working with the Task Force to provide advice and objective recommendations regarding the design and development of the system.

GENetic Imagery Exploitation (GENIE). Los Alamos has developed a sophisticated image analysis technology called GENIE to create high-resolution maps. Current sensor platforms collect a flood of high-quality imagery. Automatic feature extraction is key to enabling human analysts to keep up with the flow. Machine learning tools, such as the genetic algorithm-based GENIE, have been successfully used in military and intelligence applications of broad area search and object detection, evaluation of environmental disasters, space imaging, and diagnosis from medical imagery. GENIE has been quickly deployed on a wide range of processing systems across the nation, and was recently recognized with an R&D 100 award.

Gigabit Computer Network Traffic Monitoring. Los Alamos has recently developed technology that can monitor computer network traffic at gigabit/gigabyte rates, which could be applied to the problem of terrorist activity detection. By being able to scan network traffic at gigabit rates, both for trends as well as between specific sources and destinations, our tools can be used to provide indicators or early warning of suspicious communications. While many of these traffic analysis techniques are well known, they have been limited until now by the inability to collect and process data at gigabit rates.

Geographic Information Systems (GIS). Los Alamos has high-end computer systems capable of assembling, storing, manipulating, and displaying geographically referenced information. Our GIS make it possible to link, or integrate, information that is difficult to associate through any other means. For example, a GIS might allow emergency planners to easily calculate emergency response times in the event of a disaster; we can predict water quality, air quality, contaminant transport, wildfires and other natural hazards based on defined threat scenarios. A critical component of Los Alamos' GIS is our 3D modeling and visualization capability. We can produce wall maps and other graphics, allowing the viewer to visualize and thereby understand the results of analyses or simulations of potential events.

CHEMICAL, BIOLOGICAL, RADIOLOGICAL, AND NUCLEAR COUNTERMEASURES

The response to chemical, biological, radiological and nuclear threats necessarily take very different approaches. The dual-use nature of chemical and biological materials makes them easily accessible. For instance, feritlizer can be used to help plants grow, but the same chemicals can also be used in the construction of a bomb. In addition, hazardous microorganisms can be grown from very small starting samples. Given the prevalence of these materials, the primary focus in countering chemical and biological threats is on early detection of attack, early warning to authorities and first responders, and rapid characterization of the agent to guide response. Radiological and nuclear materials, on the other hand, have a much longer history of being regulated and safeguarded at their source. Consequently, the best way to respond to this variety of threat is to prevent terrorists from ever acquiring the necessary materials, protecting them at their source. Thus, we have an opportunity for a layered protection strategy to counter nuclear terrorism.

Chemical and Biological Countermeasures

Los Alamos has a long history of working in the biological sciences, born out of initial work done on the effects of radiation on humans. Over the years, this has developed into a significant expertise, including leadership in the international Human Genome Project and the development of now widely used biomedical technologies, based on our expertise in lasers and isotope chemistry. For example, Los Alamos created the field of flow cytometry, which allows researchers to flow objects past a laser that can rapidly answer questions about individual cells or molecules, like DNA. Thanks to this strong foundation in the biosciences, Los Alamos was able to make contributions during the recent anthrax attacks, as well as in the broader area of biothreat reduction, primarily through our work for NNSA's Chemical and Biological National Security Program (CBNP).

Field Detection and Early Identification of Pathogens

The Biological Aerosol Sentry and Information System (BASIS), a joint Los Alamos-Livermore project, provides early warning of airborne biological weapons attacks for special events such as the Olympics. Planned for use in civilian settings, BASIS can detect a biological attack within a few hours, early enough to treat exposed victims and limit casualties significantly. It was deployed at the 2002 Winter Olympics in Salt Lake City. The BASIS system incorporates distributed sampling units (sensors), a re-locatable field laboratory, and an operations center that employs a secure web-based communications system.

Advanced BASIS technology is currently being integrated into the Biosurveillance Defense Initiative. The Initiative, which is sponsored by the Defense Threat Reduction Agency of the Department of Defense and the NNSA, is a joint Los Alamos, Livermore, and Sandia program. The tri-lab effort will establish an urban test bed for biosurveillance in a U.S. metropolitan area this fall. Technologies developed by the three NNSA laboratories for early detection of biological incidents, as well as Department of Defense systems, will be included in the test bed.

Pathogen Characterization for Forensics, Attribution and Response

Once an attack has occurred, it is up to the biological science and medical communities to respond to the aftermath. These communities, Los Alamos included, responded to the challenge posed by the fall 2001 anthrax attacks. Los Alamos assisted the federal response to the attacks from the beginning, providing DNA forensics expertise to the investigation, determining what strain of anthrax was used, as well as other characteristics important for response (e.g., antibiotic resistance or genetic manipulation).

Los Alamos was able to respond to the attacks as we did because we have been working for the past ten years on analyzing the DNA of anthrax and building a comprehensive database of strains from around the world. Beyond just anthrax, the Laboratory is working on a variety of pathogen strain analysis approaches for detection, characterization and attribution of threat pathogens. This work, along with that of our colleagues at Livermore and Northern Arizona University, has provided the assays being used in BASIS. Sophisticated analysis capability resides at Los Alamos for more comprehensive pathogen characterization and, importantly, for the identification of unknown microbes.

Los Alamos works with a broad range of characterization and identification technologies. For instance, Los Alamos has established a DNA fingerprinting method for rapidly identifying the "genetic barcode" for each threat agent species. We have established an archive of such "barcodes" so that, when we conduct an analysis on a new sample, we can rapidly compare its signature to all those in the database. Additionally, if a threat pathogen is known, Los Alamos can use our DNA analysis methods to detect a broad range of agent properties that are important for understanding the attack and guiding prophylaxis and treatment; including evidence of genetic manipulation and antibiotic resistance. We can also differentiate strains of the known threat agents and can, for some species, determine their original geographic origin.

Biological Demonstration and Application Program. The forensic technologies described above, as well as routine analytical techniques, are being evaluated and standardized in the Biological Demonstration and Application Program (BDAP). BDAP is a collaborative NNSA-sponsored effort between Los Alamos, Livermore and the Northern Arizona University. The BDAP will facilitate rapid transition of NNSA-developed forensic technology into use by the public health, law enforcement and intelligence communities.

Biological Toxin Detection. We have developed a prototype of a simple, compact sensor system for detection of biological toxins, viruses, and bacteria. The prototype has been sent to a customer for use and evaluation. Our initial efforts have been focused on the development of a single-channel, hand-held, battery operated instrument for detection of cholera and ricin toxins within environmental samples. This sensor approach offers high sensitivity and specificity, simplicity of use, and rapid response time (5-10 minutes).

Chemical Detection. Los Alamos has also developed sensors for detecting chemical threats. For instance, the Swept Frequency Acoustic Interferometer (SFAI) can be used to determine the composition of suspected chemical weapons without opening up the weapon or disturbing it. These devices are hand-carried and have been tested extensively. The technology is so sensitive that it can easily distinguish between the contents of cans of Coke® and Diet Coke®. Research is also moving forward employing fuel cell technology for development of an inexpensive, small and highly sensitive chemical agent vapor detector.

Nuclear and Radiological Countermeasures

As described earlier, the radiological and nuclear threat must be dealt with in marked contrast to how the chemical and biological threat is managed. For example, if you wait to detect the use of a radiological or nuclear device, in most cases, it's too late. Instead, what is critical in this area is making every effort possible to secure materials at their source and ensure that terrorists cannot access them.

Securing Materials at their Source

The DOE/NNSA Materials Protection, Control and Accounting (MPC&A) program is the first line of defense against nuclear terrorism. With the dissolution of the Soviet Union, NNSA/DOE estimates that Russia inherited approximately 850 tons of highly enriched uranium (HEU) and plutonium. Considering the International Atomic Energy Agency definition of significant quantities, this is enough material to make more than 50,000 nuclear explosive devices. MPC&A security upgrades are complete for about ⅓ of the fissile material identified as being at risk of theft or diversion in Russia. Rapid progress is being made to increase the security of the remaining materials, but completing the effort will take several more years of intensive work.

Whereas in the past nonproliferation efforts were focused on weapons-usable materials, today there is a recognition that other radiological materials (used for industrial, medical and research purposes) pose a threat in the form of radiological dispersal devices (RDDs), or "dirty bombs." Los Alamos is actively working with DOE/NNSA and counterparts in Russia to develop strategies to secure radiological sources that pose a threat in the form of a dirty bomb.

Thousands of radiological sources are used in the U.S. for research, medical and industrial applications. The Nuclear Regulatory Commission plans to strengthen control of the sources it licenses for these uses. The DOE and its predecessor agencies originally produced radiological—sources for a variety of defense and civilian applications. These so-called "orphan sources" are being recovered by Los Alamos and repackaged as transuranic waste. More than 3,000 sources have been recovered to date. The pace of this recovery effort will likely increase to cover the more than 5,000 sources remaining.**Second Line of Defense**

The Second Line of Defense (SLD) program has the mission to detect and recover any nuclear material that may slip through the first line of defense described above. The program works to strengthen Russia's overall capability to prevent the illegal transfer of nuclear materials, equipment, and technology to would-be proliferators. The immediate goal of the program is to equip Russia's most vulnerable border sites with nuclear detection equipment. A future goal is to establish a sustainable counter-nuclear smuggling capability in Russia. SLD provides training programs for front-line inspectors, and purchases detection equipment that can "sniff" out nuclear materials.

Protecting U.S. Borders, Bases and Cities

This area, in effect the third line of defense, strives to detect radiological or nuclear materials at U.S. ports of entry. For several federal agencies, including the U.S. Coast Guard and the U.S. Customs Service, we are providing information on handheld radiation detectors and isotope identifiers. We are providing advice on what instruments to buy, and instructing operators in their use. Los Alamos is actively involved in a maritime surveillance study that analyzes potential vulnerabilities of commercial shipping.

Los Alamos is also playing a role in helping to protect U.S. military bases. One example of this is a joint NNSA and Defense Threat Reduction Agency effort. Its goal is to improve the Department of Defense's ability to detect, identify, respond, and prevent unconventional nuclear attacks by national, sub-national, or terrorist entities. The project combines technology and resources from both agencies to develop, deploy, test and demonstrate nuclear protection systems and networks at four different U.S. military installations. This effort is currently underway and involves Los Alamos and several other NNSA and DOE laboratories. If successful, the systems will be applicable to civilian urban areas.

Radiation Sensors and Detection Systems

Handheld Search Instruments. Handheld instruments are those that a police officer, customs inspector, or similar official can use to search for radioactive material on a person or in a suspicious package. They can identify the isotope emitting the radiation—an enhancement that allows a user to distinguish between benign radiation emitters such as radiopharmaceuticals or smoke alarms, and the weapons-usable material that we want to interdict. Los Alamos has developed a new handheld instrument with a Palm™ interface that enables users to distinguish between radiation sources within seconds. The Palm™ unit can provide data about the nature of the nuclear source at hand and the isotopes present. Los Alamos is exploring commercial licensing and production for this handheld search instrument. Earlier versions, the so-called GN (gamma-neutron) series of handheld instruments have already been commercialized.

Package Monitor. The Laboratory has developed systems to detect nuclear materials, particularly hard-to-detect ones such as uranium-235, which might be missed by regular search instruments. An example of this is a newly developed package monitor that detects nuclear material in parcels via neutron interrogation. A prototype of the package monitor is currently being field-tested at a U.S. Customs facility.

Portal Monitors. Portal monitors are specialized radiation sensors in physical packages that are optimized for detecting radiation from nuclear materials as a pedestrian or vehicle passes through a choke point. Los Alamos is the DOE repository of portal-monitoring expertise and has helped developed the technical standards for portal monitor performance. LANL has placed portal monitors around the world in support of the nuclear Second Line of Defense program as well as domestic and international safeguards programs. Currently, LANL is involved in the technical evaluation of portal monitors from all U.S. vendors against the technical standards.

Active Interrogation of Cargo Containers. Los Alamos is working with Idaho National Engineering and Environmental Laboratory and commercial partner ARACOR to develop and test a system that actively interrogates large cargo containers (air, sea, rail, and road) to determine if there is any nuclear material present. The system, a large U-shaped structure with a linear accelerator on one side and x-ray detectors on the other, can be driven over a cargo container to produce an x-ray image. The image shows neutron emissions, which are a signature of nuclear material.

Long-Range Alpha Detector. The LRAD is potentially valuable for sampling volumes of air or extensive surfaces where an alpha emitter may have been dispersed, and thus might be used in response to radiation-dispersal attacks. LRADs have been used for environmental monitoring at places where dispersed uranium is a problem. An LRAD implementation for radon monitoring has been commercialized by Eberline and could be rapidly adapted to the contamination-monitoring role.

EMERGENCY PREPAREDNESS AND RESPONSE

Los Alamos plays an important role within the area of nuclear emergency response. The largest and the most well-known team in this area is the DOE-managed NEST team. NEST was created in 1975 in response to concerns over nuclear terrorism activity. Its effectiveness is due to well-established interagency relationships including significant Department of Defense and FBI collaboration. NEST is focused on responding to a threatened act involving radiological or nuclear materials or devices. Among the range of potential terrorist threats involving weapons of mass destruction, the nuclear response infrastructure and capabilities are the most mature and capable of addressing the threat. NEST includes the capabilities to search for, diagnose, and disable an improvised nuclear device.

NEST depends on a team of highly dedicated individuals at the national laboratories and facilities throughout the DOE-complex who volunteer their expertise to this program. Los Alamos' NEST and related activities are funded at approximately $10 million in fiscal year 2002. More than 100 Los Alamos scientists and engineers are involved in various aspects of the NEST program. Nearly all are involved in other parts of the Laboratory's research in nuclear weapons or threat reduction. Many of the employees who work part-time on NEST are involved with more than one team within the NEST program.

It is important to note that NEST is more than a group of scientists who stand at the ready with pagers on their belts, waiting to be contacted to respond to a crisis. NEST team members at the DOE and NNSA laboratories, including Los Alamos, are involved in a wide range of related activities including research and development into diagnostic tools, disablement techniques, and computer simulations and modeling; working with the intelligence and law enforcement communities on the analysis of threats and the development of analytical tools; training of employees from other government agencies in environments that allow hands-on work with the actual nuclear materials that they might encounter in the field; and providing subject-matter experts when required. Los Alamos has the lead within NEST for development of nuclear diagnostic tools to help determine the nature of the suspected threat device and for maintenance of what is called the "home team," a group of experts parallel to those that would be deployed in the field who can provide analysis, advice and technical support.

Los Alamos is involved to varying degrees in all aspects of the national NEST program. The activities of the national team, and Los Alamos' role, are as follows.

Search Activities. Los Alamos is primarily involved in research and evaluation of detectors used for search.

Joint Tactical Operations Team (JTOT). JTOT is a partnering of DOE and DoD expertise that provides advice or direct assistance to render safe a suspect malevolent employment of a nuclear device by terrorists or others and to perform a nuclear safety assessment for the eventual safe disposition of the device. Los Alamos plays a major role in the JTOT mission and is involved in maintaining management oversight, render-safe capability, diagnostics capability, emergency response home-team capability, a watchbill (a group of experts who are on call 24 hours-a-day, seven days-a-week, year-round), communications support and deployable equipment, and contingency planning.

Real Time Radiography. This system uses a portable source of x-rays to look at a suspect object in real time, without moving or disturbing the object. Using this technique, we can identify electronic components within the object, yielding important data for action decisions. Just as a dentist uses an x-ray to locate a cavity, we can use this system to locate where to drill a suspect object, disrupting its electronics and disabling other components. This system was adapted from commercially available equipment and enhances what is available to most emergency responder units.

Accident Response Group (ARG). ARG is responsible for dealing with incidents involving a U.S. weapon, commonly referred to as a "Broken Arrow." Los Alamos has experts on the ARG roster that may be called upon if their particular set of knowledge is necessary to deal with the given situation.

Disposition. These assets support both the JTOT and the ARG team, making decisions about the ultimate disassembly and disposition of a device after it has been made safe to move and ship to a remote location.

Consequence Management. Following an incident, this team is involved in the immediate monitoring of any potential radiological dispersal and in monitoring and forecasting that can advise responders on issues of evacuation and treatment.

Attribution. This area involves drawing upon capabilities from the U.S. weapons testing program to analyze samples and draw forensic inferences about a threat device. In the case of a nuclear detonation or seizure of a weapon (or precursor material) it will be necessary to attribute quickly and accurately the material/item/incident to the perpetrators through an understanding of the materials used, type of device, yield produced or anticipated, the source of the technology and the pathway(s) that lead to the event. This requires an integrated national security program that draws on the broad based technical expertise available in NNSA as well as key NNSA facilities and analytical capabilities.

Radiological Assistance Program (RAP). Related to but separate from NEST, DOE and Los Alamos maintain response plans and resources to provide radiological assistance to other federal agencies; state local, and tribal governments; and private groups requesting such assistance in the event of a real or potential radiological emergency. The Los Alamos RAP organization provides trained personnel and equipment to evaluate, assess, advise, and assist in the mitigation of actual or perceived radiological hazards or risks to workers, the public, and the environment. This Los Alamos capability supports associated activities throughout RAP Region Four: Kansas, Oklahoma, Texas, Arizona, and New Mexico.

CONCLUSION

Los Alamos is a national laboratory with a broad set of capabilities in the area of homeland security and a long history of serving the nation in this area. As President Bush stated in his June 6, 2002, address to the nation, "In the war against terrorism, America's vast science and technology base provides us with a key advantage." Our capabilities will continue to be at the service of the nation.

Mr. GREENWOOD. Thank you very much, Dr. Cobb.

Mr. Dacey for 5 minutes.

TESTIMONY OF ROBERT F. DACEY

Mr. DACEY. Mr. Chairman and members of the subcommittee, I'm pleased to be here today to discuss the potential benefits and the challenges in implementing the information analysis and infrastructure protection division in the proposed Department of Homeland Security. As you requested, I will briefly summarize my written statement. As proposed by the President, the division's functions would include (1) receiving and analyzing law enforcement, intelligence, and other information to detect and identify potential threats of terrorism to the United States, (2) assessing the vulnerabilities of the key resources and infrastructures and developing a comprehensive national plan to secure them, and (3) taking necessary measures to protect these resources in coordination with executive agencies and in cooperation with State and local government personnel, the private sector and other entities.

It is important to note, as has been said earlier today, that non-Federal entities control most of our Nation's critical infrastructures. The consolidation of these critical infrastructure protection functions and organizations may, if properly organized and implemented, lead over time to a more efficient, effective and coordinated program. Combining related efforts such as incidents reporting, warning, and analysis could not only eliminate possibly duplicative efforts, but might also result in stronger and more coordinated capabilities. Other potential benefits include better control of funding and the consolidation of points for Federal contact in coordinating CIP activities.

Also, the division will face tremendous human capital, information management and technology, and other challenges, not the least of which will include integrating the diverse communications and information systems in the programs and agencies being brought together and securing the sensitive information that these networks and systems will likely process.

Further, through our past work, we have identified other significant challenges for many of the aspects and the functions that are to be transferred to the new Department. For each of these challenges, significant improvements have been made and numerous continuing efforts are in progress. However, much more needs to be done to address them.

These challenges which face the new Department include No. 1, developing a national CIP strategy. A more complete strategy is needed that addresses specific CIP roles and responsibilities, both within the department and the many agencies that will remain outside of the Department. Also the strategy needs to clearly identify interim milestones and objectives and set timeframes for achieving them, establish performance measures and clarify how CIP entities will coordinate their activities with each other. A national strategy that covers both cyber and physical CIP is expected to be issued within the next several months.

The second challenge is improving analysis and warning capabilities. More robust analysis and warning capabilities are still needed to identify threats and provide timely warnings. Such capabilities need to include both cyber threats which has been the historical focus of many of our national CIP efforts, as well as physical threats. The third area is improving information sharing on threats and vulnerabilities which needs to be improved both within the Federal Government and between the Federal Government and the private sector and State and local governments.

The fourth and last area is addressing pervasive weaknesses in Federal information security. One of the principle tenets of PDD 63 was that the Federal Government would serve as a model for information security. At this point, a comprehensive strategy for information security is needed, again, in which roles and responsibilities for Federal systems will be delineated, appropriate guidance is given, regular monitoring is undertaken, and security information and expertise are shared to maximize its value. Resolving these significant challenges will be critical to the success of the new Department.

Mr. Chairman, that concludes my statement. I'd be happy to answer any questions you or other members of the subcommittee may have.

[The prepared statement of Robert F. Dacey appears at the end of the hearing.]

Mr. GREENWOOD. Thank you, Mr. Dacey.

The Chair recognizes himself for 5 minutes and I'll address the question first to you, Mr. Tritak. The last time you testified before this subcommittee was last year in April. I asked you to provide us with a worst-case scenario for a major terrorist attack using computer systems on a critical infrastructure, and we had a substantial discussion about those threats. You indicated then that your primary mission was awareness, essentially to work with the Federal civilian agencies and private sector companies to make sure that they were aware of the risks in planning for such terrorist attacks.

Well, since then we've unfortunately had catastrophic and tragic terrorist attacks on the United States, albeit not primarily computer-based. Can you tell us how your role has changed and what you see it being or becoming in the new Department of Homeland Security?

Mr. TRITAK. Yes, I will, Mr. Chairman. First, obviously 9/11, for many of us, was a shock and a surprise. We had—the kinds of things we saw, were things that kept us up at night for quite sometime.

In turning to the worst-case, I tend to try to avoid characterizing it that way, but I think it is important because it builds on some of the things you have heard in the opening remarks, is that a terrorist is not going to limit himself to one means of disruption, and in fact, one could envision—in fact, if you will recall the article that came out in The Washington Post recently, a combined cyber physical attack may very well be contemplated by terrorists. If, God forbid, there should be another attack of this sort using terribly destructive physical means and then through cyber means disrupting communications, emergency communications and the like, that would create panic, but would also preclude our emergency people from getting to where they need when they needed to get there.

I think where I view the role moving in this new organization is that the need to engage industry much more so even than we were able to before, but quite frankly before 9/11, while people intellectually accepted the challenges, it seemed so farfetched. Now nothing seems other-worldly anymore. We need to figure out exactly how far they can go in dealing with this on a voluntary basis under—by incentivizing market forces as much as possible.

Recognizing that we may have to bring in other measures to make—bridge the gap between where the business case can go and what homeland security demands. I think it requires collaboration, because what we want is we want to have an economy that remains robust and effective. We don't want to throw a bone to bin Laden in this regard.

I think what we're seeing though, since 9/11 is I've been hearing from companies. They want to do the right thing. One great benefit of the new Department is I think there's going to be a clear message about what is needed. There's going to be maybe not one voice, but one message as to what needs to be done, and we're going to—I like to think that the work I do now is complementary to those things being done at other organizations, but I'm not going to kid you. There will be some overlap, and where efficiencies can be gained and an effective approach by industry will not be obtained by consolidating some of these networks.

Mr. GREENWOOD. Thank you. Let me address the question to you, Mr. McDonnell. The NISAC, when fully staffed and operational, is intended to serve as a premiere operation for conducting complex interdependency analyses of the Nation's critical infrastructures, many of which are privately owned. How will these analyses be done, and would you be relying on data and modeling generated primarily from the DOE? And last, will you be working with the private sector infrastructures to jointly model the interdependencies? And following that, who can and will be the primary clients for these new capabilities?

Mr. MCDONNELL. Sir, I can address the programmatic oversight of that; and they may choose to jump in here as well, addressing the technical aspects. It is a vision that NISAC will sort of use all source data which will use information from the industry, from different levels of government, and use different community capabilities throughout universities that are in collaboration with the national agencies.

The development of a collaborative effort that is geared on sharing information is the common understanding of the provision. As Dr. Cobb mentioned, there is utilization of simulation capabilities that were designed for different functions some years ago. The Argonne National Laboratory has this dependence capability that's also being put into that.

The envisioned principal customer prior to the announcement of the homeland security is the national government, the Federal Government, in a national program and that the priorities for the requirements for NISAC will be driven based on the collaborative process where private industry can come in, academia could go through the laboratory complex or through the member universities working at the laboratories and the laboratories can come in with science-based proposals to push the technology forward. Those requirements, those initial requests would be vetted in the emergency process.

It's already been established this is an interdependency community. That community then kicks into the Executive Office of the

President for a decision, and it will now move to the Department of Homeland Security to establish what the requirements, the priorities are going to be for the NISAC for the fiscal year. That way the States, academia, pretty much anyone who has an interest in this will have an opportunity to get their concerns on the table.

That will be made in a collaborative effort. There won't be one agency that will be saying "this is mine." it truly will be a national team effort to determine what should be done and how it should be done. The deliverables will be to myself as an emergency response planner to be able to take this data back from NISAC instead of waiting for disaster. I would sit down with other agencies and say let's game this out, let's model an infrastructure with, for example, a terrorist attack on a region with the intent of the destruction of the economy, as opposed to a specific site, and think through where the vulnerabilities are, what actions we want to make, make sure you've thought through these things.

Mr. GREENWOOD. Let me interrupt you there. In your opening statement, your testimony, you indicated that knowing where the greatest nodes of vulnerability are is a difficult thing to accomplish and you've set about to do that. What occurred to me when you said that was, well, if you don't—if it's difficult for you to know right now, it's probably pretty darn difficult for the bad guys to know right now. It's going to be very difficult for them to assemble the information that you're going to be able to assemble. So when you complete your job you'll know something that the other guys don't know. They'll want to know very much, and you'll be sitting on some pretty critical information.

What can you tell me—it's your sort of putting a genie in the bottle here—whatever the right words are—but you're going to create some information that doesn't exist. You're going to identify some vulnerabilities. By the very act of identifying those vulnerabilities you're going to create a vulnerability in the publication of information that's very dangerous. Can you tell me a little bit about how much thought has gone into how to protect that information?

Mr. MCDONNELL. Yes. It's sort of a two-part question.

First, I would submit to you that there probably are people overseas looking at our infrastructure with the intent of doing what I was discussing. Our military and our strategic planning has done that as part of our national war planning computers. So it's not so much that we're starting something new in looking at our infrastructure as an effort to attack the United States, it's getting the collaborative team of industry, State and Federal folks doing it together.

The second part of the question is the protection of the infrastructure. As Dr. Cobb mentioned, part of the reason for the national laboratories wanting control and development of this information is they have for years done this and have the protocols in place to ensure that this information is protected. The Defense Department has been a partner in discussions on how we protect the community. The actual controlling and protecting the information has to be done as national security information, classified appropriate as national security information and protected.

Mr. GREENWOOD. My time has expired.

The Chair recognizes the gentleman from Florida.

Mr. DEUTSCH. Mr. Dacey, in your testimony you detail the shortcomings of Federal agencies regarding the implementation of Presidential Directive 63 some 4 years after the Presidential Directive speech. My question is very simple: What does it take for agencies to take these Presidential Directives seriously?

Mr. DACEY. As you pointed out, we did comment on some of the shortcomings in the implementation of PD 63. In that particular case, it had to do with Federal agencies. I think we have for quite a number of years indicated there are significant challenges in getting Federal agencies to adequately secure their systems, not just with respect to CIP but the broader issues we bring out in our testimony later on. I think there are a variety of issues that need to be considered, and I think some are being considered currently under legislation that would extend the existing GISRA requirements. That would be a requirement for regular reporting performance measurement by the agencies as well as independent analysis by the inspector reports going forward to OMB and to the Congress.

Part of that process and structure would really require regular oversight, too, by agencies as well as by Congress to ensure that actions are taking place. There have been a number of improvements, but there are substantial challenges that go forward.

I think, given initial implementation, GISRA, which we testified on earlier this year, it's clear agency heads are now starting to become aware of it and actions are taking place, but we're not close to having a secure system in the Federal Government.

Mr. DEUTSCH. Has Congress provided the resources?

Mr. DACEY. One of the challenges that agencies offered to us was one of having adequate technical resources to do the work. I know that in the fiscal year 2003 budget there is a substantial increase in computer security requested budget funding to help fund some of those requirements. So I think that will go a long way toward providing some of those resources. Whether that's enough or not I can't say. Because one of our criticisms at the time was this is the first time in the GISRA reporting that agencies really tried to assess what their actual costs were, and OMB came out in their report and indicated they didn't find a correlation between the amounts expended and the security of those systems.

So I think it's going to take a little bit of time so that we have systems that measure those benefits of those costs to see if we're spending money appropriately and what additional funding will be necessary, if any.

Mr. DEUTSCH. Have the agencies whose operations have been transferred to the new Department successfully implemented the Presidential Directive?

Mr. DACEY. A lot of the efforts have been at the agency level. I'm not sure of the specific components, most of which are subcomponents of the larger agencies, how they fit specifically into that process. I can get back to you if we have any information. I'm just not sure we do at this point——

Mr. DEUTSCH. If you can, I would appreciate it.

[The information referred to follows:]

As we stated in our written testimony, both GAO and the inspectors general have issued reports highlighting concerns about PDD 63 implementation in federal agencies, such as development of critical infrastructure protection plans, identification of mission-essential infrastructure assets, and performance of vulnerability assessments and preparation of related remediation plans. PDD 63 required agencies to appoint a critical infrastructure assurance officer and specified reporting at the agency level. Consequently, we do not have information at the agency subcomponent level that would correspond to almost all of the functions proposed to be transferred to the new Department of Homeland Security.

Mr. DEUTSCH. Would the transfers to the new Department make it more or less likely that PD 63 will be implemented promptly?

Mr. DACEY. I think there are certainly some tremendous benefits in putting into one central place some of these functions that are directly related to critical infrastructure protection. The main focus of the Department is really in terms of gathering a lot of information and trying to assess what is the nature of threats and what is the nature of vulnerabilities in our current critical infrastructure, and to begin identifying them and coming up with a national strategy.

So there are a lot of tasks there. So that will lead to some improvements in the coordination of some of the functions that are currently carried out by the separate entities, as we pointed out in our testimony.

The key part of it is, though, there are a significant number of entities outside of the new Department that are involved in CIP, particularly cyber CIP. It's going to be important, as we point out in the testimony, that these functions be appropriately coordinated, whether it be by the Critical Infrastructure Protection Board or otherwise. We're looking forward to a strategic plan that would include some discussion of how those entities will work together.

We're talking in terms of entities working together in CIP, cyber CIP and those need to be coordinated with the ones that are being transferred to the new department.

Mr. DEUTSCH. Has the Critical Infrastructure Assurance Office assigned tasks adequately?

Mr. DACEY. In terms of that, we have not done an analysis of the functions being performed by the Critical Infrastructure Assurance Office. We have been aware that they have been doing outreach. Certainly that's one of their functions.

Second, in terms of the national strategy, we understand we're about to see that. We're waiting and have made comments on the types of information that we would expect to be in the national strategy. I understand it will be issued in the some time in the next few months, probably September.

The other area is in the project matrix, certainly they've been making significant efforts to work with the Federal Government. In fact, OMB has now required other agencies to undergo a project matrix which is really to see what are the critical assets that exist in these agencies and what are the critical infrastructure that they rely on and how do we protect them. I think that's an important project because I think some of the initial plans and programs submitted under PDD 63 which have been criticized didn't really get about fully identifying what those critical assets are in trying to determine what needs to be done to protect them. So I think project matrix, if it continues to be properly implemented, will be an important program to try to get at some of the initial objectives of PDD 63.

Mr. DEUTSCH. If we were to move the Critical Infrastructure Assurance Office to the new Department, would the performance—would you predict the performance would be better at that point?

Mr. DACEY. I think some of the issues that we pointed out. They do have some similar types of activities to some of the other entities being sent over. There's a lot of outreach going on by the Critical Infrastructure Assurance Office. Similarly, the NIPC is doing outreach through InfraGuard and other programs to the private sector and State and local governments.

To some extent those activities, not that they're not coordinated now, but certainly bringing those together could provide an opportunity to consolidate the outreach efforts as well as to consolidate the points of contact in the Federal Government. As we said earlier, I think it helps to have some place where everybody knows here's where we need to call to deal with these issues. I think that's an important point.

The other areas have to do with identifying critical assets. As I said the project matrix, the NIPC, has had a key asset initiative under way to try to identify key critical infrastructure assets. I think those two programs could be better coordinated under one program. I assume some of these would be combined in this new Department into one or a few programs versus the many that now exist.

Mr. WHITFIELD [presiding]. Mr. Tritak, since that is one of your areas of responsibility, would you like to comment on Mr. Dacey's remarks?

Mr. TRITAK. I actually agree with everything he's just said. CIAO is a creation of the 1998 governmental organization. We operate on a fairly modest budget. I would like to think we made some headway in terms of reaching out to industry and the project matrix program which was an attempt and is an attempt to focus efforts—if you have scarce resources and you have to allocate them, be sure you allocate them against those functions which we can't afford to lose even for brief periods of time.

I also agree with Mr. Dacey by creating this new Department we can make better use of not only the assets of CIAO but also the other agencies brought in under it. When you hear the word "outreach," there's a tendency to think of one model approach. There are different parts of audiences we're trying to reach. We spend much of our time trying to focus on corporate leadership. They're the ones that make the investment choices, and they set the policy. Once you get that kind of buy-in, the other is information sharing across agencies. Lots of the good work that's been done by NIPC and others is engaging in that level. So I think, however, when we—they're able to leverage this out much more effectively now.

Bob Dacey and I talk at least once a week, but it's easier if I'm turning around a corner and talking to him.

Mr. WHITFIELD. Thank you very much.

There's been a lot of discussion about the public-private partnership. I would like to make a couple of comments, then ask each of you to comment on it, if you would.

There's been some argument made that building this public-private partnership is too much carrot and not enough stick and that a much more regulatory approach is needed. I'd like to know what your view is on that. Do you believe efforts to regulate security across these private sectors is warranted or even remotely likely to be effective?

Mr. TRITAK. The policy of both the past administration and this is to try and encourage, incentivize the market and to look to owners and operators to manage this risk in an effective manner. Clearly, we haven't begun I think to explore all the ways in which that can be done; and we think information sharing is one piece that can actually produce good results. In order for these modeling simulations to work, there has to be a collaboration between the owners and operators. By doing that, it creates all forms of possibilities in terms of work. So I think that it can work.

I think that, more importantly, if we want this homeland security function to work properly we have to create a proposition in the mission statement that calls for collaboration and partnering. I think, as I indicated before, it's going to take some adjustment of government and industry. Both have to realize we're in a different ball game here. It's not just industry. It's on our side. They shouldn't just wait for government to come to tell them what needs to be done.

Mr. WHITFIELD. Mr. McDonnell.

Mr. MCDONNELL. First off, I wouldn't believe at this point that we have determined that regulation is warranted. There has been a lot of discussion about regulating security, but I've had teams go out with 25 key energy assets since the first of January and, in general, I truly believe that security managers and the executives out there are trying to do the right thing. They're looking to us for our leadership, to provide them the information they need to do the right thing. That is going to take a lot of work. That's the outreach efforts that I do think are complementary, where John's staff and mine tend to bump into each other and step on each other a little bit. It's a good, positive tension. We're working extremely hard to work with industry to give them the information they need to make intelligent decisions.

I do not think that security should be regulated because I believe that we need to protect security information as national security information. We need to protect vulnerabilities from public disclosure. As we start regulating those then the nature of the information stops being a collaborative effort to protect the asset itself and starts being an effort to formulate a requirement of the Federal Government which then requires some level of disclosure.

When the industry operators—when I send a team to the vulnerability sector, they know they're not going to see that information turn around back to them as regulation, as an increased burden that they have to deal with. They can share that information among staff and with the national laboratory folks that we that do these assessments and get real, honest observations and advice without worrying about having have us come back.

That being said, if as we go forward with vulnerability assessments—let me back up real quickly. We've developed with industry voluntary security guidelines that have just been published with all of the energy sector. If we go forward with our vulnerability sectors and industry is not taking care of the assets, then maybe we need to revisit what regulation is required. At this point, we don't have any justification.

Mr. WHITFIELD. Dr. Varnado.

Mr. VARNADO. I believe we're going to have sell industry a business case for why they should invest. Right now, we don't have tools to show them what are the consequences. So one of the things that NISAC is trying to do is to present to industry a cost-benefit program saying, if you protect against this particular threat, you will avoid this kind of consequence. But until this point we haven't really had a way to show them what the results will be of all of that. So one of the thrusts of our work is to do the economic modeling, and that's going on reasonably well at this point. At some point we'll be able to say to industry, this is your risk problem to manage.

Industry is excellent in managing risk. That's what they do. But they need the tools to understand how to understand the risk. So at some point we will understand that. We will talk like insurance salesmen at that point. The government down the road will say, the risk that the industry is willing to accept is not acceptable to us. I think we need to work this business case risk management problem first and see how the industry responds.

Mr. WHITFIELD. Dr. Cobb.

Mr. COBB. Just a short comment.

I think in the balance between regulation and trying to move forward jointly there's a premium, at least from our perspective, on trust building; and trust building is somewhat difficult if it's totally governed by rule.

I think—as Mr. McDonnell said, I think our experience in protecting proprietary and very sensitive information from major sectors of the American industry, for example, which is privately run, transportation information that affects a local urban area and so forth, that we have the experience—I know at Sandia and Los Alamos we have the experience of protecting this kind of vulnerability in securing information and working with the people who have those responsibilities to do the protection to their benefit. I would hope you would be able to maintain that.

Mr. WHITFIELD. Mr. Dacey.

Mr. DACEY. In terms of this area, we certainly have provided a lot of commentary in this testimony and the prior reports in determining some of the challenges that are faced by implementing the PDD 63. At the same time, I think it's been said already, that we're dealing with a little different situation.

I think what we're looking at here is a need to have a relatively free flow of information taking place between private sector local government and the Federal Government and not so much in terms of periodic reports and things of that nature but regular free flow of information in both directions, both from those entities to the government and back. I think that makes it a little bit different of a dynamic in terms of the cooperation.

We have done studies on information sharing and have pointed out a number of the key areas that are important to that process. At the same time, I think there are some things that maybe not have been explored as thoroughly as possible in terms of providing appropriate incentives for industry and the State and local governments to deal with it. For example, when we did our report information sharing and other prior work, it's been consistently pointed out that there are concerns with both FOIA, antitrust and civil liabilities issues with respect to the willingness of some of those entities to provide information voluntarily to the Federal Government.

We also point out that, in terms of Y2K, we had similar concerns that were expressed in sharing information, very sensitive information, often times and now we're getting into some incident information too, specific breakins and attacks that have occurred that are fairly proprietary. Trying to get that kind of information, we may need consider some of these types of things—I know there's current draft legislation out on that issue exactly and some provisions certainly in the bill with respect to FOIA.

In terms of other areas, there have been some bright spots. I think it's been consistently shown where we had a long-term relationship, and we pointed this out last April, with the electric power industry, the success of those efforts have been much greater than the other ISACs because the Federal Government has had a longstanding relationship with that industry. There is more a degree of trust that exists between them, as we pointed out. That could perhaps be a model. I'm not sure we're there in the other ISACs to quite to that level as we are with the electric power industry.

Second, certain sectors—I believe it was the American Chemical Council is now requiring their members to undergo vulnerability assessments as a condition for membership in that organization. So there are some efforts taking place.

The real question is, can the government get the information it needs on a voluntary basis? We haven't done a thorough analysis to assess that. But other witnesses can provide some insights about whether or not they are willing to provide this information.

Again, it has to do with an incident that might signal an attack or scanning activity that might indicate that someone is trying to get some information. We know there's been activity in scanning and that area. So it's that vulnerability analysis. So I think there are a lot of things that need to be explored. The question for the Congress and the new Department is whether or not those have been adequately explored to make a change in the current direction.

Mr. WHITFIELD. Well, thank you for your comments.

I'll recognize the gentleman from Michigan, Mr. Stupak.

Mr. STUPAK. Thank you.

Let me pick up where the Chair has just left off. Mr. Dacey, in your testimony, though, you conclude that—I'm quoting—the new Department will face tremendous information management and technology challenges, not the least of which will be integrating the diverse communications and information systems of the programs and agencies that need to be brought together.

So if we have this tremendous challenge in trying to adjust management information and the information you get from other Federal agencies and private sector, there have been experiences you can point to within the Federal Government where there've been integrated Federal programs that suggest that this can't even be accomplished. Are other departments and agencies willing to give up their turf or their information, if you will?

Mr. DACEY. I think one of the critical elements, success factors, if you will, of the new Department is to make sure all the relevant information is getting focused in this new sector. We now have a fairly active process taking place currently at NIPC in coordination with a lot of other entities, some of which will be transferred to the new Department in trying to deal real time with some these issues, trying to identify some of the concerns that exist out there. I think it's going to be important to try to deal with that.

I think that's an issue that's going to have to be faced, regardless of the movement of this new Department. I think we really need to have a comprehensive way to bring this information together to properly analyze and see what the issues are. As we pointed out in numerous reports after September 11, the government had a lot of information relative to the attack. We need to figure out how to get that information together. I think that's going to be a challenge again, regardless of what gets moved to the new Department. The Federal Government needs to look at ways to get that information together.

In terms of successes in putting together similar types of information, I'd have to get back to you on that and see if we have any examples that we can provide.

[The information referred to follows:]

The size of the integration of communication and information systems for the proposed Department of Homeland Security is unprecedented in the federal government. Although the federal government has made improvements in its IT management, our work shows that agencies continue to face challenges in (1) information technology investment selection and management control processes, (2) enterprise architecture, (3) software development, cost estimating, and systems acquisition practices, (4) effective chief information officer leadership and organizations, and (5) information security. The department will need to overcome challenges such as these to develop effective systems. We understand that the administration is working on the development of an enterprise architecture for the new department. An effective enterprise architecture is a key element of an such effort.

Mr. STUPAK. It seems to me if you have face-to-face contact with these departments and agencies and discuss it—because I can't think of—I've been here 10 years now. I remember going all the Y2K hearings. None of the computers seem to be compatible with the Federal Government. Every time you spent a—whether it's HHS, it doesn't work with this one, doesn't work with this, we have to start all over again.

I think most of this information is stored within the computers. The computer systems aren't compatible. What happens to cyber security when these software programs are learning to talk with each other? GAO even said that there's pervasive weakness in the Federal information security. How do we accomplish that with this new Department when the computers won't talk to each other and cyber security is still a real threat out there?

Mr. DACEY. I think the cyber security issue needs to be dealt with. What we talk about in our testimony and what we understand is currently under way is to look at standardized enterprise architecture for the new systems and the new Department. I think that's going to be critical. Otherwise, you will end up with stovepipe systems that won't talk to each other. You need a model. Here is where we are now, and here's where we want to be in the future. I think that's critical that be done as part of the process.

Mr. STUPAK. How long do you think it will take to us get there?

Mr. DACEY. I don't know how long it will take. Certainly there have been challenges for the Federal Government in developing enterprise architecture in other settings, as we point out in your testimony as well.

Mr. STUPAK. None of them have worked.

Mr. DACEY. I wouldn't say none, but certainly we have lots of examples.

Mr. STUPAK. Can you tell me of one that worked?

Mr. DACEY. I'll get back to you on that.

[The information referred to follows:]

Federal agencies are at different stages of maturity in their development and use of enterprise architectures, with most agencies now establishing the management foundation needed to successfully develop and use them. But few agencies have actually developed and are currently using them. The Customs Service and the Internal Revenue Service, both within the Department of Treasury, are examples of agencies that have and are using enterprise architecture.

Mr. STUPAK. I'm not making a point. I just sat through so many of these hearings. It seems like nothing had compatibility and doesn't seem to work so well.

You know, in your testimony you do speak a lot of the need to share information on the principle of vulnerabilities, including the private sector. There's been an interesting discussion going on there, but what really are the barriers to the sharing of sensitive information with private firms that may clearly have a need to know that the information comes from a law enforcement office, classified intelligence information? I mean, what are the barriers we're facing here?

Mr. DACEY. Well, certainly one of the challenges which we pointed out last April in our report is the need to try to sanitize the information to protect from an intelligence standpoint sources and methods, from a law enforcement standpoint information that would be critical for potential prosecution. The key area there, it just takes a little time. I think we have seen evidence that those processes have worked.

In looking at NIPC and some of the examples they could provide initially, there was a lot of information that was withheld pending trying to figure out what is the law enforcement value and protecting that. I think they've worked out mechanisms now to better allow them to disseminate the information and still have the law enforcement process ongoing. So I think there are definitely challenges. I think they can be accommodated.

One of the other issues that's currently taking place is getting security clearances for certain people so more secure information can be shared with those people who do have clearances.

Mr. STUPAK. It would seem that—not only trying to figure out if there's a law enforcement angle to the information, it seemed like there was more of a turf war. We don't trust this with this information and this is our information and it's not going further or second guessing of the person who wrote the memos, wherever they may be. So I don't think it's all just computer-related problems or security-related problems but really leadership problems in trying to trust the information we have and not be afraid to share it with someone else who may get credit before. Having spent quite a bit of time in law enforcement, I certainly witnessed firsthand many times how that happens.

What can we do—I know I'm over, Mr. Chairman, but let me ask this question—what can we do about the problem some people would say crying wolf with too much unspecific information regarding a possible terrorist attack or threat that is released to the public? How do we deal with that?

Mr. DACEY. That is a challenge.

One of the issues that we raised in our report last year was to make sure that we aren't too extensive in the number of reports that go out for that very reason. I think those that—we've gotten a lot of more sensitive certainly since September 11. There have been a lot of issues that have been out there, if you follow the NIPC. But a lot of the other sites, a lot more activity seems to be taking place right now in identification of potential threats that need to be communicated. I think there will be a continuing challenge.

With respect to your other comment in terms of sharing, I think one of the issues that NIPC set up is a model—Ron Dick had indicated the critical success factor in his mind was really representation from the different sectors that contribute information. He has folks there from the CIA, from NSA, and from DOD intelligence agencies that are serving very key roles in the staff at NIPC. In fact, we criticize that in our April report. Since then, actually, I've—they've had consistent representation.

So I think one of the keys is really to make sure we have a really multi-agency representation and capability. Otherwise, you're going to certainly have the potential for people not willing to share. If we have people there that are part of those organizations, hopefully they can help facilitate better communications.

Mr. WHITFIELD. The gentleman's time has expired.

I want to thank those members of Panel 3 for their testimony today. We appreciate your being here. Unless there's additional questions, we will dismiss this panel.

The Chair calls forward Panel 4. The Chair welcomes those of you on Panel 4. We appreciate you being here today. We look forward to your testimony.

We have as witnesses today: Mr. William Smith, Executive Vice President, Network Operations, BellSouth; Mr. Guy Copeland, Vice President, Federal Sector, Computer Sciences Corporation. We have Ms. Lynn Costantini, Director of Online Services at the North American Electric Reliability Council. We have Mr. John Sullivan, President and Chief Engineer of the Boston Water and Sewer Commission. We have Mr. Kenneth Watson, who's the President of the Partnership for Critical Infrastructure Security at Cisco Systems. We have Mr. David Sobel, who's General Counsel of the Electronic Privacy Information Center; and we have Mr. Jeremiah Baumann, Environmental Health Advocate with the U.S. Public Interest Research Group.

You are aware that the committee is holding an investigative hearing and when doing so has had the practice of taking testimony under oath. Do you have any objection to testifying under oath today?

The Chair advises you under the rules of the House and the committee you are entitled to be advised by counsel. Do you desire to be advised by counsel during your testimony today?

In that case, if you would please rise and raise your right hand, I will swear you in.

[Witnesses sworn.]

Mr. WHITFIELD. You are now under oath, and you may proceed with your 5-minute summary of your written statement.

Mr. Smith, the Chair will recognize you to start.

TESTIMONY OF WILLIAM SMITH, EXECUTIVE VICE PRESIDENT, NETWORK OPERATIONS, BELLSOUTH; GUY COPELAND, VICE PRESIDENT, INFORMATION INFRASTRUCTURE ADVISORY PROGRAMS, FEDERAL SECTOR, COMPUTER SCIENCES CORPORATION; LYNN P. COSTANTINI, DIRECTOR—ONLINE SERVICES, NORTH AMERICAN ELECTRIC RELIABILITY COUNCIL; JOHN P. SULLIVAN, JR., PRESIDENT AND CHIEF ENGINEER, BOSTON WATER AND SEWER COMMISSION; KENNETH C. WATSON, PRESIDENT, PARTNERSHIP FOR CRITICAL INFRASTRUCTURE SECURITY, CISCO SYSTEMS, INC.; JEREMIAH D. BAUMANN, ENVIRONMENTAL HEALTH ADVOCATE, U.S. PUBLIC INTEREST RESEARCH GROUP; AND DAVID L. SOBEL, GENERAL COUNSEL, ELECTRONIC PRIVACY INFORMATION CENTER

Mr. SMITH. Mr. Chairman and members of the subcommittee, good afternoon. My name is Bill Smith. I'm Chief Technology Officer for BellSouth Corporation. I appreciate the opportunity to be here with you today to discuss a vital national security issue and that is information sharing between government and the private sector and the role of the proposed Department of Homeland Security.

As a major telecommunications network operator, the challenge we face is reliability, security, and robustness of the critical national and international infrastructures. Furthermore, we need a comprehensive strategy that's flexible enough to prepare for and respond to an evolving spectrum of threats. Such a strategy should both increase the protection of vital industry assets and ensure public safety.

The cause of the increased reliance and interdependency and the potential for infrastructure disruption may come from multiple sources, including rapid growth, regulation, deregulation, terrorism, natural disturbances such as hurricanes and earthquakes. Telecommunications systems constitute a fundamental infrastructure of modern society, and a successful terrorist attempt to disrupt them could have devastating effects on national security, the economy and every citizen's life.

It is clear to all that the telecommunications industry is facing some of the greatest challenges in our economy today. There is fierce competition and eroding market shares compromising the environment in which we operate.

Despite these challenges, BellSouth continues to support numerous infrastructure protection initiatives formed pursuant to PDD

63, but like others in our industry we find that there are many duplicative efforts under way, all competing for the same scarce resources. In the wake of the September 11 terrorist attacks, our industry as well as those supporting other infrastructures has seen dramatic increases in the number of requests to participate in these efforts.

In addition, we've received numerous requests for sensitive information such as lists of critical facilities and Federal, State and local authorities. From the perspective of a corporation such as BellSouth these requests are troubling, because if such a list were released publicly, whether through accidental disclosure, it would provide terrorists with essentially a road map directing them to our most critical locations. Therefore, we would support efforts of the Department of Homeland Security to, among other things, serve as a focal point to coordinate these efforts and to allow us to make best use of our expertise and resources such as the National Coordinating Center, or NCC, for Telecommunications.

I have included a number of our concerns in the written testimony. However, with FOIA our concerns are largest. With respect to FOIA, many companies are hesitant to voluntarily share information with the government because of the possible release of this information to the public.

BellSouth currently shares cyber-related institutions information with the Telecom Sharing and Analysis Center or the telecom ISAC located in NC. However, whatever the cause of the concerns, information sharing is done on a limited basis within entrusted circles and strictly within a fashion that will eliminate any liability of harm from FOIA requests from BellSouth.

This is not to say that ISACs do not provide value. BellSouth and the other ISAC participants have benefited from the advanced warning of viruses. Our first notification of the Nimba worm enabled us to successfully defend our networks. In turn, BellSouth was the first company notified in telephone ISACs the problems associated with simple network management protocol.

Eventually, all the Nation's critical infrastructures will have ISACs, and their level of success will depend on several factors. First, information must be shared voluntarily in a trusted form. PDD 63 and the national plan clearly say that critical infrastructure protection must be in a public-private partnership. Legislating information sharing will not foster the type of cooperation that's needed to address these critical issues.

Second but of equal concern is the need to improve information sharing within and amongst government agencies.

In closing, I would like to reaffirm BellSouth's commitment to protecting our Nation's critical infrastructure. Thank you for the opportunity to appear here today. I look forward to answering any questions you may have.

[The prepared statement of William Smith follows:]

PREPARED STATEMENT OF BILL SMITH, CHIEF TECHNOLOGY OFFICER, BELLSOUTH CORPORATION

Mr. Chairman and Members of the Subcommittee. [Good morning/afternoon.] My name is Bill Smith and I am Chief Technology Officer for BellSouth Corporation. I appreciate the opportunity to appear before you today to discuss a vital national

80-680 - 02 - 8

security issue—information sharing between the government and the private sector and the role of the proposed Department of Homeland Security.

Virtually every crucial economic and social function in our society depends on the secure and reliable operation of infrastructures. Indeed, they have enabled our country to achieve levels of productivity and a standard of living that is the benchmark for the rest of the world. However, these benefits have come at the cost of increased complexity, interdependency and risk. Critical infrastructures such as energy, banking and finance and transportation depend on the robustness of our telecommunications networks, while the explosive growth of the Internet's ability to interconnect computer networks, and our digital economy have increased the demand for reliable and disturbance-free communications.

As a major telecommunications network operator, the challenge we face is maintaining the reliability, security and robustness of critical national and international infrastructures. And, we need a comprehensive strategy flexible enough to prepare for, and respond to, an evolving spectrum of threats. Such a strategy should both increase protection of vital industry assets and ensure public safety.

Because of increased reliance and interdependency, the potential for infrastructure disruption may come from multiple sources, including system complexity, rapid growth, regulation, deregulation, terrorism, and natural disturbances such as hurricanes and earthquakes. Telecommunications systems constitute a fundamental infrastructure of modern society, and a successful terrorist attempt to disrupt them could have devastating effects on national security, the economy, and every citizen's life. At BellSouth, we continue to improve the security of our telecommunications systems, but our widely dispersed physical assets, unfortunately, can never be defended absolutely against a determined attack.

It is clear to all that the telecommunications industry is facing some of the greatest challenges in our economy today. Fierce competition, eroding market shares and tenuous market conditions compromise the environment in which we operate.

Despite these challenges, BellSouth continues to support the numerous infrastructure protection initiatives formed pursuant to Presidential Decision Directive 63 (PDD 63), but like others in our industry, find that there are many duplicative efforts underway, all competing for the same scarce resources.

In the wake of the September 11th terrorist attacks, our industry, as well as those supporting other infrastructures, have seen dramatic increases in the number of requests to participate in these efforts. In addition, we have received numerous requests for sensitive information—such as lists of critical facilities—from federal, state and local authorities. From the perspective of a corporation such as BellSouth, these requests are troubling because if such a list were released publicly, whether through a FOIA request or through accidental disclosure, it could provide terrorists with a road map directing them to our most critical locations.

Therefore, we would support efforts of a Department of Homeland Security to, among things, serve as a focal point to coordinate these efforts, and allow us to make the best use of our expertise and resources such as in the National Coordinating Center (NCC) for Telecommunications.

In the current environment, we have the following concerns about information sharing:

- liability under the Freedom of Information Act
- third-party liability (e.g., sharing suspected problems about a piece of equipment before thoroughly tested and verified)
- the lack of a defined antitrust exemption for appropriate information sharing concerning infrastructure vulnerabilities
- possible disclosure of information under state sunshine laws
- disclosure of sensitive corporate information to competitors
- declassification of threat/intelligence information to a level that can be acted upon by company personnel and,
- the natural inclination of law enforcement, DoD, and intelligence agencies to dissuade the sharing of information related to criminal investigations.

With respect to FOIA, many companies are hesitant to voluntarily share sensitive information with the government because of the possible release of this information to the public. BellSouth currently shares cyber-related intrusion information with the Telecom Information Sharing and Analysis Center—the Telecom ISAC—located within the NCC. However, because of the concerns just noted, the information sharing is done on a limited basis, within trusted circles, and strictly within a fashion that will eliminate any liability or harm from FOIA requests for BellSouth information. This is neither maximally efficient nor effective.

This is not to say that the ISACs do not provide value. BellSouth and the other ISAC participants have benefited from advance warnings of worms and viruses. For example, the ISAC provided us our first notification of the NIMDA worm in a clear and timely manner that enabled us to successfully defend our networks. In turn, BellSouth was the first company to notify the Telecom ISAC of problems associated with the simple network management protocol (SNMP).

Eventually, all of the Nation's critical infrastructures will have ISACs, and their level of success will depend on several factors. First, information must be shared voluntarily in a trusted forum. PDD-63 and the National Plan (Version 1.0 for Information Systems Protection) clearly state that critical infrastructure protection must be a public/private partnership. Legislating or regulating information sharing will not foster the type of cooperation that is needed to address these critical issues. Second, but of equal concern, is the need to improve information sharing and communication within and amongst governmental agencies.

As an owner and operator of a significant portion of the Nation's critical infrastructure, BellSouth assumes a proactive stance regarding critical infrastructure protection. For this reason, we routinely monitor legislation addressing these issues. Although the House recently passed H.R. 4598, the "Homeland Security Information Sharing Act," BellSouth hopes it is refined further as it moves through the legislative process. Specifically, it is not enough to share classified or sensitive information with select individuals as cited in the legislation. What is important is that that information be "actionable"—that is, recipients of such information must have the flexibility to act on that information by passing it on to other appropriate parties. With respect to H.R. 5005, the "Homeland Security Act of 2002," we support this legislation and believe that Section 201(5) will best be implemented through a public-private sector partnership, rather than through an expansion of regulatory authority and the imposition of new regulation. We also support Section 204 which provides an important FOIA exemption for information regarding infrastructure and other vulnerabilities that is provided voluntarily. Finally, we support the FOIA and antitrust protections embodied in H.R. 2435, the "Cyber Security Information Act."

In closing, I would like to reaffirm BellSouth's commitment to protecting our Nation's critical infrastructures. Thank you for the opportunity to appear here today. And I look forward to answering any questions you may have.

Mr. WHITFIELD. Mr. Copeland, you're recognized for 5 minutes.

TESTIMONY OF GUY COPELAND

Mr. COPELAND. Thank you, Mr. Chairman and members of the subcommittee. I am honored to be here today testifying on behalf of the Information Technology Association of America, known generally as ITAA, where I serve as Co-Chair of the Information Security Committee and Vice Chair of the Homeland Defense Task Group. I'm also a board member of the Information Sharing and Analysis Center for the Information Technology industry sector of the IT ISAC in which my company is a founding member and which ITAA was instrumental in forming.

ITAA represents a broad spectrum of information technology and communications companies and is a strong public advocate for the very important goals of homeland security, including cyber security.

A recent Washington Post article quoted the Chief of Staff of the President's Critical Infrastructure Protection Board. He said we were underestimating the amount of attention al Qaeda was paying to the Internet. Now we know that they see it as a potential attack vehicle. Al Qaeda spends more time mapping our vulnerabilities in cyberspace than we previously thought. An attack is a question of when, not if.

A study just released by Internet security firm Riptech Incorporated found that Internet attacks against public and private organizations around the world leapt 28 percent in the last 6 months, with most targeting technology, financial services and power companies.

Government and industry can work together to address this threat. That is why, for example, ITAA helped found the IT ISAC.

It is the reason that ITAA has worked to help get the information and communications sector input into the President's developing a national strategy for critical infrastructure and cyberspace security. In turn, this Critical Infrastructure Assurance Office and the Critical Information Protection Board under Dick Clarke are working closely with industry.

Corporations own and operate the majority of systems that make up and protect our country's critical infrastructure. ITAA joins with other sectors in believing that effective information sharing between government and private sector ultimately is critical to address insider threats.

Sharing information about information security experiences is difficult. No company wants to risk information that they have volunteered in good faith and confidence may be misused or misinterpreted to their detriment, and certainly no company wants information to surface to any terrorists or criminals.

Government agencies seek detailed data about infrastructure and computer attacks. The private sector wants consistently to provide comprehensive and detailed information to government on a voluntary basis but only with the guaranty that it be protected. Today, however, corporate counsels frequently raise the unacceptable risk that such information could be ultimately be divulged through the Freedom of Information Act. If the government wants to include the way America responds to the threat of critical infrastructure attacks, government needs to give CEOs the certainty that voluntarily provided sensitive information would be protected.

As Mr. Dacey noted in the previous panel, various legislative proposals address this. Among them, ITAA has endorsed H.R. 2435, the Cyber Security Information Act, cosponsored by Representatives Tom Davis and Jim Moran and S. 1456, the Critical Infrastructure Information Security Act, cosponsored by Senators Bob Bennett and John Kyl.

Today, we would like to express our support for a proposed amendment to title II of H.R. 5005 being offered by Congressman Tom Davis, establishing relationships of trust and confidence for information sharing which are still all too rare.

An excellent example cited by John Tritak is the President's Advisory Committee and related bodies such as the National Coordinating Center for Telecommunications.

Dating to September 1982, the NSTAC is perhaps the oldest and most successful industry and government partnership to address telecommunications and information systems issues impacting national security and emergency preparedness. I suggest you examine the NSTAC and its partnering government organization, the National Communications System, and their ongoing joint government and industry processes as a successful foundation on which to build.

Despite their past experience with sharing of operational information and in light of the need for even more sensitive sharing to address tomorrow's threats, the NSTAC is on record as twice endorsing the need for FOIA protection for voluntarily shared critical infrastructure information.

Attached to my testimony is a list of several reasons why current FOIA interpretation may not be sufficient. Ambiguity and discretion remain the order of the day when it comes to agency decision-making and remains the top concerns. That is why there is clear unity in the private sector in favor of removing disincentives to information sharing, and that is why we support legislation in the U.S. House of Representatives and U.S. Senate and specifically we recommend adopting Tom Davis' amendment to H.R. 5005, the Homeland Security Act. We call on this committee and Members of the U.S. Congress that have not already indicated their support for this legislation to do so today.

Also, Mr. Chairman, I would like to mention the proposal from the ITAA president to Congressmen Tauzin and Dingell last week.

As the committee reviews and considers possible changes to the Homeland Security Act for 2002, ITAA encourages you and the administration to ensure proper priority for information security in the new Department. Toward this end, ITAA recommends creating a Bureau of Cyber Security headed by the Senate confirmed Assistant Secretary for Cyber Security. This proposal would have the Assistant Secretary for Cyber Security reporting to the Under Secretary for Information Analysis and Infrastructure Protection. We believe that such a structure will provide appropriate focus of resources and management visibility to address all cyber threats, including physical attacks on cyber assets and lead to better security in cyberspace.

Thank you, Mr. Chairman. I would be pleased to answer questions.

[The prepared statement of Guy Copeland follows:]

PREPARED STATEMENT OF GUY COPELAND, VICE PRESIDENT, INFORMATION INFRASTRUCTURE ADVISORY PROGRAMS, COMPUTER SCIENCES CORPORATION ON BEHALF OF THE INFORMATION TECHNOLOGY ASSOCIATION OF AMERICA

Good morning Mr. Chairman and distinguished Members of the Subcommittee. I am honored to be here today. I am testifying on behalf of the Information Technology Association of America—known as ITAA—(http://www.itaa.org/infosec) where I serve as Co-Chair of the Information Security Committee and Vice Chair of the Homeland Defense Task Group. I also am a Board Member of the Information Sharing and Analysis Center for the Information Technology industry sector—the IT ISAC (http://www.it-isac.org)—in which my company is a founding member and which ITAA was instrumental in forming. And I represent Mr. Van Honeycutt, the CEO of my company, Computer Sciences Corporation, in the President's National Security Telecommunications Advisory Committee—more easily pronounced as the acronym NSTAC—(http://www.ncs.gov/nstac/nstac.htm), a body that provides the President of the United States with industry advice regarding critical, information and telecommunications services supporting our national economy and other critical functions of society. Mr. Honeycutt chaired the NSTAC from September 1998, to September 2000. During that period I served as the chair of the working body of the NSTAC, the Industry Executive Subcommittee Working Session. Many of the companies represented in the NSTAC membership are also members of ITAA.

ITAA represents a broad spectrum of information technology and communications companies, and supports the very important goal of increasing information sharing 1.) within the private sector and 2.) between industry and government in order to better protect our nation's critical infrastructure and to promote and sustain global physical and economic security.

Also, Mr. Chairman, I would like to reference a proposal that ITAA noted in letters to Commerce Committee Chairman Tauzin and Ranking Member Dingell last week. As this Subcommittee and the full Committee review the Homeland Security Act of 2002 or H.R. 5005 and considers possible changes to the bill, ITAA encourages you and your colleagues to work with the Bush Administration to highlight information security in the new Department.

Towards this end, ITAA recommends creating a Bureau of Cyber Security headed by an Assistant Secretary for Cyber Security. Under the current proposal, the components that would be merged into the Department of Homeland Security from other departments and agencies that focus on cyber security (e.g. NIPC, NCS, CIAO, and Cybercorps) would be included with those that focus on physical security in the new Information Analysis and Infrastructure Protection division. This melding would be a mistake. The challenges in the cyber world are sufficiently different from those in the physical world to merit a Bureau that focuses on Cyber Security and that is headed by a Senate-confirmed public official.

This proposal would have the Assistant Secretary for Cyber Security reporting to the Under Secretary for Information Analysis and Infrastructure Protection. There would be three bureaus in this new bureau under the revised structure: 1. Bureau of Analysis and Warning, which would analyze all source intelligence, 2. Bureau of Critical Infrastructure, which would develop protection for physical assets, and 3. Bureau of Cyber Security, which would conduct programs within the USG and with the private sector to protect communications, the Internet, computer systems, and IT networks.

We believe that such a structure would enhance the internal cohesion of U.S. cyber terrorism fighting efforts, provide appropriate focus of resources and management visibility, and lead to better homeland security in cyberspace. This only addresses one piece of the equation, however. Just as the Internet interconnects a vast array of public institutions and private entities, so too must the security policies and practices of public and private domains be linked to bolster the safety of all concerned.

As you may know, Mr. Chairman, ITAA has endorsed H.R. 2435, the Cyber Security Information Act co-sponsored by U.S. Representatives Tom Davis and Jim Moran, and S. 1456, The Critical Infrastructure Information Security Act, co-sponsored by Senators Bob Bennett and John Kyl. Today, we would like to express our support for a proposed amendment to Title II of H.R. 5005 by Congressman Tom Davis. We call on this Committee and Members of U.S. Congress that have not already indicated their support for this legislation to do so today. For reasons I will outline below, the certainty and trust these bills engender are key to preventing or at least minimizing future threats to critical infrastructures.

You may have heard the numbers before. According to the 2002 FBI / Computer Security Institute Survey:

- 90% of large corporations and government agencies responding detected computer security breaches within the last twelve months.
- 80% acknowledged financial losses due to computer breaches.
- 44% were willing and/or able to quantify their financial losses. These 223 respondents reported $455,848,000 in financial losses.
- 34% reported the intrusions to law enforcement.

A December 2001 ITAA / Tumbleweed Communications survey found:

- 70% of Americans concerned about Internet and computer security.
- 74% expressed fears that their personal information on the Internet could be stolen or used for malicious purposes.
- 74% said they are concerned that cyber-attacks could target critical infrastructure assets like telephone networks or power plants.

A study released yesterday by Internet security firm Riptech, Inc. found that "... Internet attacks against public and private organizations around the world leapt 28 percent in the past six months, with most targeting technology, financial services and power companies." [1]

While these numbers show the magnitude of the economic impact and also the concerns of the American people about cyber attacks on our critical infrastructure, let me read a passage from an article in late June 2002 from the Washington Post to emphasize the sheer magnitude of the threat in this age of terrorism that we are living in:

"Unsettling signs of al Qaeda's aims and skills in cyberspace have led some government experts to conclude that terrorists are at the threshold of using the Internet as a direct instrument of bloodshed. The new threat bears little resemblance to familiar disruptions by hackers responsible for viruses and worms. It comes instead at the meeting points of computers and the physical structures they control." [2]

Sobering, isn't it? But, government and industry can work together to address this threat, reduce the economic impact of cyber attacks, and help reduce Americans' very understandable and justified concern about the possibility of cyber attacks on our nation's critical infrastructure. Information sharing between government and the private sector is a very important part of detecting and mitigating cyber threats.

---

[1] "Internet Attacks on Companies Up 28 Percent, Report Says," by Michael Barbaro, *Washington Post*, July 8, 2002.

[2] "Cyber-Attacks by Al Qaeda Feared," by Barton Gellman, *Washington Post*, June 27, 2002

As the U.S. General Accounting Office (GAO) stated in an October 15, 2001 report entitled "Information Sharing: Practices That Can Benefit Critical Infrastructure Protection," information sharing and coordination "are key elements in developing comprehensive and practical approaches to defending against computer-based, or cyber, attacks which could threaten the national welfare."

"...The importance of sharing information and coordinating the response to cyber threats among various stakeholders has increased as our government and our nation have become ever more reliant on interconnected computer systems to support critical operations and infrastructures, such as telecommunications, power distribution, financial services, national defense, and critical government operations. Information on threats and incidents experienced by others can help stakeholders identify trends, better understand the risks they face, and determine what preventative measures should be implemented." [3]

Many of the same concerns regarding information sharing existed in the period leading up to the Year 2000 date rollover, and resulted in an unprecedented effort between industry, government and the public interest sectors to support the drafting and passage of Federal legislation to remove legal obstacles—FOIA, antitrust, and civil liability—from "Y2K readiness disclosures" that were an essential element of our successful addressing of the date change challenge. Indeed, many of the same elements in the Year 2000 Information and Readiness Disclosure Act of 1998 are found in the Davis-Moran and Bennett-Kyl bills. This is not surprising, given that many of the same individuals who labored to assure our successful meeting of the Y2K challenge occurred have been in leading roles among critical infrastructure providers to assure that terrorism does not succeed where father time did not, for example, by helping to draft this legislation.

In short, ITAA joins with our critical infrastructure providers in believing that effective information sharing can: 1) reduce the harm and impact of attacks on critical infrastructures; 2) help the owners and operators of critical infrastructure systems in multiple sectors to determine the nature of an attack; 3) provide timely warnings; 4) provide analysis to both industry and government to prevent future attacks; 5) mitigate attacks in real-time; and 6) assist in re-constitution and recovery efforts.

As I stated at the outset, ITAA supports the very important goal of information sharing. Strong and unwavering support of that goal is why ITAA and its members are cooperating with several other sectors and a variety of government partners in the National Cyber Security Alliance (http://www.staysafeonline.info), the Partnership for Critical Infrastructure Security (http://www.pcis.org), and the Cyber Citizen Partnership (http://www.cybercitizenship.org).

Support of that goal is why ITAA helped found the IT Information Sharing and Analysis Center (http://www.it-isac.org) and is the reason that ITAA has worked to help develop and facilitate private sector input for the Information & Communications Sector into the President's National Strategy for Critical Infrastructure and Cyberspace Security, a plan that Presidential Advisor Dick Clarke calls "a living document" that will change as the threats change.

Support of that goal is why ITAA and its sister associations from around the world have prioritized e-security and critical infrastructure assurance as public policy priorities in the 46-country World Information Technology and Services Alliance or WITSA (http://www.witsa.org), and is why ITAA and WITSA sponsored the first Global InfoSec Summit now nearly two years ago.

Support of that goal is why ITAA continues to raise awareness of critical infrastructure assurance and e-security challenges as a business continuity issue, if not a business survivability issue at the CXO (CFO, CTO, etc.) and Board level among our member companies and throughout the private sector.

Support of that goal is why ITAA and its members are so committed to building trust-based relationships with law enforcement officials and agencies at every level of government and internationally.

Support of that goal is why ITAA and many of its sister associations—which represent millions of small and medium business as well as large corporations—have been in strong support of the bi-partisan legislation that I referenced earlier. H.R. 2435 and S. 1456 were introduced in both the U.S. House of Representatives and U.S. Senate last year to remove narrowly defined legal barriers to information sharing within the private sector and between the private sector and government.

Better information sharing is a necessary step to leveling the playing field in the critical infrastructure assurance world. How so? "Bad actors" have great advantages when it comes to pooling what they know about hacking tools, malicious code, net-

---

[3] Report to Senator Robert F. Bennett, Ranking Minority Member, Joint Economic Committee, Congress of the United States by the U.S. General Accounting Office, October 15, 2001, page 1.

work vulnerabilities and the like. One of the ironies of the Internet is that it can serve as a school for scoundrels, fostering hacker communities, serving as a classroom for future attacks and helping cyber-psychos communicate their exploits.

Meanwhile, sharing information about corporate information security practices is inherently difficult. Companies are understandably reluctant to share sensitive proprietary information about prevention practices, intrusions, and actual crimes with either government agencies or competitors. Information sharing is a risky proposition with less than clear benefits. No company wants information to surface that they have given in confidence, and that may jeopardize—through misunderstanding or misperception—their market position, strategies, customer base, investor confidence or capital investments, and certainly no company wants information to surface that could aide terrorists or criminals.

Government agencies seek detailed data about computer attacks for the purposes of better law enforcement, earlier detection, and the promotion of best practices in government and industry. Today, however, corporate counsels advise their clients not to share voluntarily the details of computer attacks with government agencies because the risk that such data could ultimately be divulged through the Freedom of Information Act (FOIA)—even over the agency's objections—is unacceptably high.

The bottom line? Uncertainty. Uncertainty about whether existing law may expose companies and industries that voluntarily share sensitive information with the federal government to unintended and potentially harmful consequences. This uncertainty has a chilling effect on the growth of all information sharing organizations and the quality and quantity of information that they are able to gather and share with the federal government. We are not talking about a Harvard moot court debate. If we want to improve the way corporate America responds to the threat of critical infrastructure attacks, government needs to give CEOs and their corporate counsels the certainty that this legislation would provide.

I would like to report on steps industry has already taken to promote information sharing and how this process can be improved; I would also like to emphasize two points about the proposed legislation:

1. Government partners have come to the private sector to ask for information concerning current and potential vulnerabilities in various sectors of our national critical infrastructure. The private sector wants consistently to provide comprehensive and detailed information to government on a voluntary basis, but in order to do so have asked that that information be protected.
2. The private sector AND the Federal Government both have agreed for years that it is important to develop and strengthen information sharing processes and organizations within the private sector since we own and operate the majority of systems that make up and protect our country's critical infrastructure.

The IT industry is one of several industries to adopt a formal approach to the information sharing challenge. In January 2001, nineteen of the nation's leading high tech companies announced the formation of a new Information Technology Information Sharing and Analysis Center (IT-ISAC) to cooperate on cyber security issues. The objective of the IT-ISAC is to enhance the availability, confidentiality, and integrity of networked information systems. The organization is a not-for-profit corporation that allows the information technology industry to report and exchange information concerning electronic incidents, threats, attacks, vulnerabilities, solutions and countermeasures, best security practices and other protective measures. I am proud to be a Founding Board Member of that organization.

On the telecommunications side of the Information and Communications—or "I&C" -Sector, an ISAC has been formed by the National Coordinating Center for Telecommunications (NCC). Building on NCC's traditional role as the operational focal point for the coordination, restoration, and reconstitution of national security and emergency preparedness—or "NS/EP"—Telecommunications and facilities, the NCC-ISAC facilitates voluntary collaboration and information sharing among government and industry participants. The NCC-ISAC gathers information about network vulnerabilities, threats, intrusions, and anomalies from various sources, including the telecommunications industry and the U.S. government. That information is then analyzed with the goal of averting or mitigating the effects of computer intrusions on the telecommunications infrastructure.

The value of the ISAC approach is found in the ability to acquire and share information with the group in a way that individual group members cannot accomplish. This process often involves the rapid assessment and conversion of information that individual ISAC members had held as proprietary and confidential into a form that can be shared both with ISAC members and with other affected or interested parties. ISACs are exchanging some "sanitized" information between sectors and at times, on a very limited basis, with the National Infrastructure Protection Center or NIPC. The ISAC information product commonly deals with the provision of early warnings of impending attacks, and the establishment of trends in types and severity of attacks. If more legal protections were in place, there could be more sharing of Internet threat and solution information among the ISAC membership and other appropriate organizations, including the Federal Government. ISACs operate successfully because they are a closed community, founded on mutual trust, and focused on prevention before a large attack occurs. They differ markedly from other open communities whose duties are to alert the more general networked public after a breach has occurred.

As the world economy continues to become more international in nature, ISACs will provide a rich source of useful, validated security threat information, for those enterprises that do not, or are not able to, participate in the information security structure. It is by sharing security data that the nation and the world will be able to respond effectively to the continuing and growing threat, both internally and externally, against critical infrastructures.

Two additional points need to be made: First, this entire process is just getting underway. While there are a few examples of the most competitive companies sharing information within a few ISACs, more time is needed before we will be able to measure real success. Relationships of trust and confidence need to be built. That is why the government, through legislation, has a critical role to play NOW, in the formation of the process, and its encouragement.

Second, many in the business community believe that their efforts are hampered by the government's apparent desire for a limited, one-way form of information sharing. The government seems to conduct much of its internal conversations about critical infrastructure on the basis of classified information—the kind that can only be shared in very restricted ways—and yet it expects the business community to share its own sensitive information without any ironclad assurances of confidentiality, certainly nothing like the treatment accorded classified information. We are not seeking that level of protection, but as we encourage greater sharing we must likewise promote the notion that the communication must flow in both directions.

A lack of certainty is also a decided impediment to sharing critical infrastructure information with government. That kind of information is not "ordinary" and should be entitled to the extraordinary treatment of a complete ban on FOIA disclosure. Legislative proposals address this defect by taking the subject information out of the realm of agency discretion to disclose. We need to close the gate firmly when this information is shared with government.

Concerns about inappropriate release of sensitive infrastructure information via FOIA have impeded current sharing with government. Dating to September 1982, the NSTAC is perhaps the oldest and most successful industry and government partnership to address telecommunications and information systems issues impacting national security and emergency preparedness (NS/EP).

NSTAC activities are the genesis for technical reports, recommendations to the President, and NS/EP operational programs. Showing how industry and government partnership is an integral part of the success of the NSTAC, the primary working body of the NSTAC, the Industry Executive Subcommittee (IES) is chaired by a government executive, the Deputy Manager, National Communications System. The IES consists of executive representatives appointed by each NSTAC Principal. The IES holds regular Working Sessions to consider issues, analyses, or recommendations for presentation to the NSTAC members for their approval. When an issue requires research or other examination, the IES forms a task force to address it. For example, the National Coordinating Center for Telecommunications (NCC), an industry/Government coordination center for day-to-day operational support to NS/EP telecommunications, began in 1984 from an NSTAC recommendation. More recently, the NCC has established an Information Sharing and Analysis Center (ISAC) function as part of its NS/EP telecommunications mission. The Telecommunications Service Priority (TSP) System, once an NSTAC issue, is also now an operational program. TSP is the regulatory, administrative, and operational authority that enables priority provisioning and restoration of telecommunications services for Federal, State, and local government users, as well as nongovernmental users. Also originating from NSTAC activities, an industry-based Network Security Information Exchange (NSIE) was created and meets regularly with a Government NSIE in a classified forum to address the threat posed to the public network as a result of actual or possible electronic exploitation of system vulnerabilities.

Despite this track record of success, their past experience with sharing of operational information, and in light of the need for even more sensitive sharing to address tomorrow's threats, the NSTAC is on record as twice endorsing the need for FOIA protection for voluntarily shared, critical infrastructure information.

Antitrust concerns are another potential legal hurdle to information sharing. We understand that the Department of Justice has offered assurances that its program of business review letters would be forthcoming for information sharing and analysis centers constituted under the Administration's policies. Yet the issuance of even a set of such letters would prove inadequate, for at least three reasons. First, such ISACs would have to be constituted with a view toward satisfying the Department, as opposed to maximally fulfilling their primary mission. Second, there is the unavoidable negative implication for numerous other affected parties not in possession of a business review letter. Third, the ISACs are not the only organizations that have been constituted to share cyber threat information among industry sector members or with Federal agencies.

Beyond federal FOIA and antitrust—and let me emphasize the ITAA believes that addressing the FOIA issue is the heart of the proposed legislation—the current bills go on to clarify that critical infrastructure threat data shared voluntarily with the government would not be disclosed either under the Federal Advisory Committee Act (FACA) or under state FOIA laws. We do recognize the federalism question that the second provision raises. At the same time, homeland defense is creating a need for federal, state, and local bodies to work jointly to a previously unprecedented degree. In some instances, first responders will not be from federal agencies. Information sharing ought not to dead-end at the federal level but should flow all the way down to the first responders. Without the same protection at the state level as at the federal, state agencies will face the same lack of revealing detail that federal agencies are experiencing today. In this regard, language in § 3(e) of H.R. 4598 recently passed by the House dealing with the sharing among law enforcement agencies of homeland security information may provide a model for treatment of FOIA-excluded critical infrastructure threat information moving to the states and local governments.

Finally, the bills also call for limited use protection—not immunity—so that critical infrastructure information disclosed to the government cannot subsequently be used against the person submitting the information. Opponents of this legislation state that the provision is a smokescreen for promising unlimited liability to the corporate community. Nothing could be further from the truth. Once again, it bears repeating: the subject of this legislation is information that the government has requested informally from the business community. There is ample reason to grant limited use protection in return for full disclosure of this information intended to help the government accomplish its mission.

A comparison with the legislative, public policy and marketplace purposes behind this legislation and that underlying the Y2K legislation may be instructive. In 1998, as today, many of the leading proponents of that legislation were uncertain about the extent of the need to alter FOIA's exemptions, in order to assure that information would flow from the private sector custodians to the government and beyond. But, lacking the luxury of time to wait for a court test case, consensus in Washington was that a Congressional imprimatur of approval of limited FOIA, antitrust and civil liability exposure (later provided in the "Y2K Act of 1999") was appropriate, indeed, critical, in view of the scope of risk, and extreme reticence of many corporate holders of information to share that.

A very similar situation exists today with regard to custodians of critical infrastructure threat and risk information. Whatever position a legal scholar may take on the extent of FOIA's present shield, an affirmative statement of Congressional approval of ISACs and other information sharing organizations is essential to our meeting the challenge of the terrorist threat.

Attached to my testimony is a list of several reasons why current FOIA language may not be sufficient to protect critical infrastructure information from disclosure. Ambiguity and discretion remain the order of the day when it comes to agency decisions about disclosure of any kind of business confidential data, despite its importance and despite good precedents in some of the Federal Courts. The lack of certainty is of course acceptable in the ordinary course of business; it simply reflects the bias of FOIA in favor of disclosure, a bias with which we do not quarrel. However, critical infrastructure assurance cannot be considered business as usual.

With the appropriate protections in place, legitimate businesses, law enforcement agencies, intelligence agencies, and the Homeland Security organization—in whatever form it may take—can share the information needed to ward off attacks and track down attackers.

There has been, in ITAA's view—and this view has also been expressed by other associations such as the Edison Electric Institute, the U.S. Chamber of Commerce, the National Association of Manufacturers, the Financial Services Roundtable, Americans for Computer Privacy, and the American Chemistry Council —a misunderstanding of the legislation by some critics. Again, we are not calling into question the existing FOIA case law, which taken together suggests that a federal agency would win a test case. Rather, we are saying only that the risk of a loss of such a test case—as viewed by the parties bearing the risk—remains unacceptably high. More importantly, corporations should not be required to accept such risks, or the cost of litigation, when reporting significant cyber events in an attempt to protect the public interest. Second, the proposed legislation has only to do with disclosure of computer attack data and critical infrastructure protection. Normal regulatory information gathering will proceed unimpeded, as it should.

In closing, I would like to cite another passage from the Washington Post article that I referred to earlier in my testimony: "We were underestimating the amount of attention [al Qaeda was] paying to the Internet," said Roger Cressey, a longtime counterterrorism official who became chief of staff of the President's Critical Infrastructure Protection Board in October. "Now we know they see it as a potential attack vehicle. Al Qaeda spent more time mapping our vulnerabilities in cyberspace than we previously though. An attack is a question of when, not if."[4]

The threats are out there. Our critical infrastructure is vulnerable. The private sector and public sector must work together to understand, respond to, and prevent these threats. That is why there is clear unity in the private sector in favor of removing disincentives to information sharing and that is why we support legislation in the U.S. House of Representatives and U.S. Senate—and specifically, we recommend adopting Tom Davis' amendment to H.R. 5005, the Homeland Security Act of 2002. We call on this Committee and Members of U.S. Congress that have not already indicated their support for this legislation to do so today.

Thank you, Mr. Chairman. I would be pleased to answer any questions that you and/or Members of this Committee may have at this time.

Appendix 1:

FOCUS ON THE FREEDOM OF INFORMATION ACT

reasons current law fails to adequately protect critical infrastructure threat information

The Freedom of Information Act (FOIA, 5 USC 552) expresses the policy of the United States in favor of disclosure of information in the government's possession, to the greatest possible extent. No one argues with this basic premise of government in America. Transparency and open government are important parts of the foundation of our democracy.

At the same time, no one disputes that when the government engages in strategic planning and discussions about the national security and national defense in the emerging and dangerous world spawned by the resurgence of terrorism and the necessity of making war on it, the sensitive information generated should be exempt from disclosure on grounds of overriding national defense and foreign policy considerations.

In addition, no one disputes that the "Critical Infrastructure" of the United States—from pipelines and electric utilities to information networks and telecommunications, transportation systems for goods and people and more—is at risk of attack both prior to, and now, during the war on terrorism.

The bulk of this critical infrastructure, however, is under the ownership and control of America's private sector, not the national security umbrella of government. It is time to recognize the important role in national security and foreign policy that America's critical infrastructure plays, and treat information related to "any threat to the security of critical infrastructure" just as any other information exempt from disclosure as a matter of national security.

That is not the case today. Information generated by the government and properly classified under "criteria established by an Executive order to be kept secret in the interest of national security or foreign policy" is exempt from disclosure. Period. 5 USC 552 (b)(1)(A)(B). Information generated by the private sector owners and operators of the nation's critical infrastructure and voluntarily shared with a government agency may be treated as "confidential business information"[1], but only if the agency makes a number of determinations in its discretion, and it does not exercise its discretion to change its mind in the future. Such information may also fit within the FOIA exclusion for "law enforcement information" when disclosure "could reasonably be expected to endanger the life or physical safety of any individual" (5 USC 552(b)(7)(F)), but the same reservations about agency discretion apply here as well. Treatment of critical infrastructure threat information should be "upgraded" by pro-

[4] "Cyber-Attacks by Al Qaeda Feared," by Barton Gellman, Washington Post, June 27, 2002

[1] The statutory phrase is "trade secrets and commercial or financial information obtained from a person and privileged or confidential." 5 USC 552 (b)(4).

viding that it is specifically exempted from disclosure by statute (5 USC 552(b)(3)), removing the extra burden of discretionary treatment.

The change will not open the floodgates to a host of other exemptions from disclosure. This change would respond to a limited need for specific relief in the case of information that rises to the level of a national security concern, but resides outside the national security umbrella. It does not seem likely that other requests for new exemption could meet this test.

It should be the case that upgrading this specific type of information is in the interest not just of the business community, but also of the government itself and the citizenry in general. It is in everyone's interest to take the steps reasonably necessary to protect critical infrastructure from attack, and learn from incidents and recoveries that have taken place in the past.

What is clear is that current FOIA treatment of critical infrastructure threat information makes the private sector reluctant to engage in the full and frank disclosure of information to government that should be taking place right now. Why is the current FOIA treatment of critical infrastructure threat information less than adequate? There are a number of reasons. Here are several:

1. Under current rules the submitter of information does not know whether it will be treated as confidential by the agency, and the agency will not make a commitment at the time of submission. This lack of certainty alone prevents many disclosures.
2. Current policy requires that agencies not exercise their discretionary authority unless and until a disclosure request under FOIA is received. When a request is received, agencies have discretion to inform the submitter of the need to defend the confidentiality of their information. The agencies can decide they have enough information to make the decision without informing the submitter.
3. Recent precedents (the *Critical Mass* case and its progeny) suggest that "voluntarily" submitted "trade secret, commercial or financial information" may be protected from disclosure if not "customarily" disclosed by the submitter. Nevertheless, every word in quotes represents a different discretionary determination that must be made by the agency at the time of a FOIA request. Submitters have their arguments to make, but no assurance that those arguments will be accepted.
4. Recent precedents are not necessarily accepted throughout the United States in every judicial circuit. Submission of critical infrastructure threat information should not be expected to be limited to agencies in Washington, D.C.
5. Information disclosed to competitors in an ISAC under the terms of binding nondisclosure agreements (NDA) conditioning ISAC membership may qualify for confidential treatment under the *Critical Mass* case, but absent strict compliance with such formal requirements—as could happen in the case of an incident recovery crisis or other emergency—disclosure by the submitter could lead to a finding that *Critical Mass* protections do not apply.
6. Agencies always have discretion to decide that, despite a submitter's claim of confidentiality and the reasons for it, the submitter's claim in light of the passage of time or other considerations cannot be valid and the policy interests expressed by FOIA are stronger and enough to justify disclosure. That is a risk the business community has come to accept in its ongoing dialogue with government. It is not a risk that should have to be assumed for the treatment of critical infrastructure threat information.
7. Some confidential business information turns stale with the passage of time, justifying the exercise of agency discretion. Critical infrastructure threat information does not. That alone should be reason enough to upgrade its treatment under FOIA.

In sum, it is essential to eliminate discretionary treatment for this limited class of information. The owners and operators of the nation's critical infrastructures should be able to have confidence that the information they share with government will not be made public at a later date. Today they do not have that confidence.

Mr. WHITFIELD. Thank you.

Ms. Costantini, you're recognized for 5 minutes.

TESTIMONY OF LYNN P. COSTANTINI

Ms. COSTANTINI. Chairman Greenwood, ladies and gentlemen of the subcommittee, thank you for the opportunity to testify on behalf of the North American Electricity Reliability Council. We are in support of the President's proposal for a Department of Homeland Security. NERC is a not-for-profit organization formed in 1965 to promote the reliability of electric systems that serve North America. It accounts for all the electricity supplied in the United States, Canada and a portion of Baja California, Mexico.

In addition to its job of "keeping the lights on," NERC services the electric industry's contact and coordinator in the United States and Canada for bulk electric systems security matters and it operates the Electricity Sector's Information Sharing and Analysis Center.

As the director of Information Technology, it is my responsibility to ensure NERC's information assets and the environment in which they operate are secure. I serve on the Critical Infrastructure Protection Advisory Group, and I am a member of the ES-ISAC team.

Generally, NERC supports the administration's Department of Homeland Security and appreciates the recognition in this proposal of the role of the private sector in protecting critical infrastructure. More than 80 percent of assets that drive our economy are privately held. Without the assistance of the U.S. Government to help the owners of these assets understand the threat environment and warn them when they are hauled out as targets, these assets may be vulnerable.

The public-private partnership is crucial to helping us understand such complicated potential vulnerabilities as the independencies between and among different infrastructures, such as telecommunications, electricity, transportation and natural gas. NERC believes it's imperative to national security to refine and strengthen that public-private partnership. Organizing the authority and responsibilities for critical infrastructure protection under the Department of Homeland Security supports that goal.

We recognize, however, that there exists barriers which prevent a flow of information between and among the public and private sectors. Except in special circumstances, information provided to the government is subject to disclosure to the citizenry and others via FOIA. Information sharing among members of private industry is subject to antitrust regulation, and trust is as much a concern as antitrust.

The effect of these concerns is that some valuable information necessary to fully analyze risks to critical infrastructure interests is not being employed.

These concerns are more than theoretical. For instance, the United States Department of Energy, working with the Office of Homeland Security, has asked the electric utility industry to provide the government with a list of nationally critical facilities. While we understand how this information can be useful, NERC and its members are unwilling to prepare a target list without adequate assurance that such information will receive appropriate protection. FOIA exemptions do not provide that level of assurance.

Furthermore, in response to September 11, entire industries must now decide whether and how to share spare parts or other finite resources. The issue of sharing also involves potential allocations of scarce supplies. Entire industries may need to determine security-related requirements to ask of their suppliers. At the very least, entire industries want to discuss the security-related shortcomings of existing product supply industries. Each of these actions is ripe for antitrust allegation.

NERC does believe these barriers to public-private partnership are surmountable. We will overcome them by clarifying the Freedom of Information Act exemption to provide indisputable, consistent rules for the nondisclosure of critical infrastructure protection information. Alternatively, create new statutes stipulating nondisclosure of specific, sensitive data provided to the U.S. Government for the purposes of critical infrastructure protection; grant security clearances for personnel in critical infrastructure industries so that the flow of information between the public and private sectors can remain intact and secure; provide limited antitrust exemptions such as those that enabled cross-sector coordination during the year 2000 rollover; continuing to build trust.

NERC believes that centralizing leadership authority and responsibility under the Department of Homeland Security is a step toward this building trust.

Recognizing the voluntary system of information sharing between the public and private sector as an effective means of promoting critical infrastructure assurance is vital. Helping the private sector overcome barriers to participation and providing antitrust protection will allow the trust relationship to grow and be fruitful.

On behalf of NERC, I thank you for your time.

[The prepared statement of Lynn P. Costantini follows:]

PREPARED STATEMENT OF LYNN P. COSTANTINI, DIRECTOR—INFORMATION TECHNOLOGY, NORTH AMERICAN ELECTRIC RELIABILITY COUNCIL

CRITICAL INFRASTRUCTURE PROTECTION: THE NEED FOR PUBLIC-PRIVATE PARTNERSHIP

My name is Lynn Costantini, and I am the Director of Information Technology for the North American Electric Reliability Council. NERC is a not-for-profit organization formed after the Northeast Blackout in 1965 to promote the reliability of the bulk electric systems that serve North America. NERC comprises ten Regional Reliability Councils that account for virtually all of the electricity supplied in the United States, Canada, and a portion of Baja California Norte, Mexico.

In addition to its job of "keeping the lights on," NERC serves as the electric industry's contact and coordinator in the United States and Canada for bulk electric system security matters and operates the Electricity Sector's Information Sharing and Analysis Center (ES-ISAC).

As the Director of Information Technology, it is my responsibility to ensure NERC's information assets and the environment in which they operate are secure. I serve on NERC's Critical Infrastructure Protection Advisory Group and I am a member of the ES-ISAC team.

Generally NERC supports the Administration's proposed Department of Homeland Security. NERC appreciates the recognition in this proposal of the role of the private sector in protecting critical infrastructures. Furthermore, NERC believes it is imperative to national security to refine and strengthen the public-private partnership. Organizing the authority and responsibilities for critical infrastructure protection under the Department of Homeland Security supports that goal.

In this testimony, I will discuss the need to keep information flowing between the public and private sectors, the barriers to information sharing, what can be done to overcome those barriers, and, finally, the electricity sector's experience in these areas.

Background

The information age dawned with little thought to security. We were in awe of the power at our fingertips (information!) and we rushed to find new ways to gather and use more and more information through an increasing array of new techniques. A computer on every desktop, complete with tools to improve efficiency and productivity, networked together so we could share precious resources. How could something so positive, so beneficial, be used against us? Never!

Today we know better. The silver cloud had a black lining. First "script kiddies" exploited vulnerabilities in our computing armor for fun. Then committed hackers exploited us for profit. Now we are faced with the prospect of nation-states exploiting us to rain terror. The need for security was never clearer or more urgent.

We now also understand that security is multi-faceted. It is guards, gates, and guns. It is firewalls and intrusion detection systems. It is policy statements and disaster planning. It is also about understanding the spectrum of threats we face so we can accurately assess risk in the context of our industries, our operating environments. Ultimately, security is about awareness, preparedness, and action.

The Need for Partnership

Security, then, demands cooperation and coordination between the public and private sectors. In fact, the public-private relationship is vital. It is true that more than 80% of assets that drive our economy are privately held. However, without the assistance of the United States government to help the owners of these assets understand their threat environment and warn them when they are called out as targets, these assets may be vulnerable.

Moreover, the public-private partnership is crucial to helping us understand such complicated potential vulnerabilities as the interdependencies between and among different infrastructures, such as telecommunications, electricity, transportation, and natural gas.

Barriers to Public-Private Partnership

Although the idea of information sharing seems so simple, it raises serious concerns. Except in special circumstances, information provided to the government is subject to disclosure to the citizenry and others via the Freedom of Information Act (FOIA). Furthermore, information sharing among members of private industry is subject to anti-trust regulations. Trust is as much an issue as anti-trust.

Freedom of Information Act

Participants in critical infrastructure industries repeatedly cite the inability of the federal government to assure them that any sensitive information they supply will not fall into inappropriate hands as a significant barrier to information flow between the public and private sectors. The effect of these private-sector concerns is that some valuable information necessary to fully analyze vulnerabilities and risks to critical national interests is not being reported. This will likely remain the case until the government can offer such assurances of protection from disclosure.

Of course, legitimate market participants, regulators, and others need to obtain information in a timely manner, but truly sensitive information must be protected.

The existing FOIA disclosure exemptions do not provide the necessary levels of assurance.

Exemption 4 asserts that information voluntarily given to the government will be protected if the provider customarily treats such information as confidential. This language leaves the door open to legal challenges and thus, to the possibility of disclosure of sensitive information. Rather than risk disclosure, the private sector may decide not to release information to the government.

Exemption 1 protects sensitive information from disclosure by classifying it in the interest of national defense or foreign policy. This is strong, assuring language; however, only a small percentage of the personnel working in critical infrastructure industries have security clearances. The flow of information from the public sector back to the private sector would be jeopardized if sensitive information were classified.

FOIA disclosure concerns are not simply theoretical. The United States Department of Energy, working with the Office of Homeland Security, has asked the electric utility industry to provide the government with a list of nationally critical electric facilities. We understand how this information would be useful. Indeed, NERC has maintained a critical equipment database since the mid-1980s, to which strict access controls are applied. NERC and its members are unwilling to hand over even a small part of any such database without adequate assurance that such information will receive appropriate protection.

Anti-trust Regulations

Anti-trust regulation is another serious private-sector concern and goes beyond the potential problems caused by merely sharing information about threats. Entire industries must decide whether and how to share spare parts or other finite resources to repair major, widespread damage and prevent worse calamities due to cascading failures. The issue of sharing also involves potential allocations of scarce commodities—both supplies for repair and products for customers. Further, entire industries may determine security-related requirements to ask of their suppliers and business partners. At the least, entire industries may discuss the security-related shortcomings of existing products, suppliers and partners. Each of these actions is ripe for anti-trust allegation. The risk of allegation seriously dampens the willingness to share information, which, in turn, jeopardizes the ability to adequately analyze cross-sector dependencies and develop effective protection strategies.

Trust

As noted by the General Accounting Office last October, one issue critical "to establishing, developing, and maintaining effective information-sharing relationships [to] benefit critical infrastructure protections efforts, [is to] foster . . . trust and respect . . ."[1] Without a trust relationship between government and private industry, information sharing stands little chance of success.

A report by the President's Commission on Critical Infrastructure Protection (PCCIP) in October 1997 specifically commended NERC as a model for information sharing, cooperation, and coordination between the private sector and government. Clearly, the successful relationship between NERC and its government partners (the FBI and its National Infrastructure Protection Center, the Department of Energy, and others (has been a benefit to the electricity sector.

OVERCOMING THE BARRIERS TO PUBLIC-PRIVATE PARTNERSHIP

Clarify the Freedom of Information Act disclosure exemptions.

FOIA disclosure exemptions do not provide the necessary levels of assurance to the private sector that its sensitive information will be protected. Congress should clarify the exemptions to create indisputable, consistent rules for the non-disclosure of sensitive critical infrastructure protection information. Alternatively, create new statutes stipulating non-disclosure of specific, sensitive data voluntarily provided to the United States government for the purposes of critical infrastructure protection.

Because of the FOIA concerns, participants in the electricity sector are asking federal regulators, agencies, and states to reconsider what information they request of utilities, especially market information that identifies system constraints and the availability of critical facilities. Our industry has especially asked that they reconsider how they share that information once they obtain it. In fact, the Federal Energy Regulatory Commission (FERC) is beginning to address those issues. FERC recently asked for advice and suggestions on how to prevent sensitive information from being disclosed despite the requirements of FOIA. However, there is no clear process or timeline for any final decision by FERC. Congress is in the best position to mitigate the security risks inherent in information-sharing activities.

Grant security clearances for personnel in critical infrastructure industries so that the flow of information between the public and private sectors remains intact and secure.

The owners of critical infrastructure assets need access to more specific threat information and analysis from the public sector in order to develop adequate protection strategies. This may require either more security clearances or treatment of some intelligence and threat information and analysis as sensitive business information, rather than as classified information.

Provide limited anti-trust exemptions.

The possibility of anti-trust allegations inhibits cross-sector information sharing. The private sector wants clarity as to what information it can share and the extent to which information can be exchanged without risking anti-trust allegations. A legislative action similar to the 1998 Y2K Information and Readiness Disclosure Act would provide the necessary level of clarity.

Build Trust

Infrastructure security requires a healthy, trusting public-private relationship. Overlapping and inconsistent roles and authorities may have hindered development of productive working relationships. Clarification of roles and responsibilities both within the government and the private sector is an important factor in building a trust model. Centralizing leadership, authority, and responsibility under the Department of Homeland Security is a step forward in building trust. Recognizing a voluntary system of information sharing between the public and private sector as an effective means of promoting critical infrastructure assurance is another. Helping the private sector overcome barriers to effective participation by clarifying FOIA and

[1] *Information Sharing—Practices That Can Benefit Critical Infrastructure Protection,* GAO Report to Senator R. F. Bennett, Joint Economic Committee (October 2001)

providing anti-trust protection will allow the trust relationship to grow and be fruitful.

THE ELECTRICITY SECTOR EXPERIENCE

NERC has a long history of coordination with the federal government on grid security. It began in the early 1980s when NERC became involved with the electromagnetic pulse phenomenon. Since then, NERC has worked with the federal government to address the vulnerability of electric systems to state-sponsored, multi-site sabotage and terrorism, Year 2000 rollover impacts, and most recently the threat of physical and cyber terrorism. At the heart of NERC's efforts has been a commitment to work with various federal agencies including the National Security Council (NSC), the Department of Energy (DOE), the Nuclear Regulatory Commission (NRC), and the Federal Bureau of Investigations (FBI) to reduce the vulnerability of interconnected electric systems to such threats.

NERC maintains a close working relationship with the FBI's National Infrastructure Protection Center (NIPC) and the Department of Energy's Emergency Operations Center (DOE EOC), and participates in and hosts several related critical infrastructure protection programs, the Indications, Analysis, and Warnings Program (IAWP); the Electricity Sector Information Sharing and Analysis Center (ES-ISAC); and the Partnership for Critical Infrastructure Security (PCIS).

On at least two occasions, Congress has asked the General Accounting Office (GAO) to study the practices of organizations that successfully share sensitive information. GAO report B-247385, April 1992, "Electricity Supply, Efforts Under Way to Improve Federal Electrical Disruption Preparedness," and GAO report GAO-02-24, October 15, 2001, "Information Sharing: Practices That Can Benefit Critical Infrastructure Protection," outline and report on many of the ways in which NERC coordinates industry response activities.

Information Sharing and Analysis Center for the Electricity Sector (ES-ISAC)

Presidential Decision Directive (PDD-63), issued in May 1998, called for government agencies to become involved in the process of developing a National Plan for Information Systems Protection, and to seek voluntary participation of private industry to meet common goals for protecting the country's critical systems through public-private partnerships. In September 1998, then Secretary of Energy Richardson sought NERC's assistance in developing a program for protecting the nation's critical electricity sector infrastructure and NERC agreed to participate as the electricity sector coordinator.

In its role as the ES-ISAC, NERC performs the following functions:

- Receives incident data from electricity sector entities
- Assists the National Infrastructure Protection Center to analyze electricity sector events
- Disseminates threat and vulnerability assessments
- Liaisons with other ISACs
- Analyzes sector interdependencies
- Participates in infrastructure exercises

Critical Infrastructure Protection Advisory Group

NERC created its Critical Infrastructure Advisory Group (CIPAG) to evaluate sharing cyber and physical incident data affecting the bulk electric systems in North America. The CIPAG, which reports to NERC's Board of Trustees, has Regional Reliability Council and industry sector and associations representation as well as participation by the Critical Infrastructure Assurance Office in the Department of Commerce (CIAO), DOD, DOE, NIPC, and FERC.

Participation in CIPAG represents all electricity sector segments, which is an essential ingredient to its success. The participants include the dedicated experts in the Electricity Sector who represent physical, cyber, and operations security. NERC is recognized as the most representative organization of the Electricity Sector for this coordination function, as demonstrated by NERC's performance as project coordinator for the Electricity Sector for the Y2K transition. The security committees and communities associated with industry organizations (American Public Power Association, Canadian Electricity Association, Edison Electric Institute, and National Rural Electric Cooperative Association) provide the expertise for security in the electricity sector to compliment NERC's existing operational and cyber security expertise. The CIPAG relies on small self-directed working teams, a proven and effective method for developing detailed processes and practices by subject matter experts, concluding with peer review in the forum environment, and approval by NERC's Board of Trustees.

CIPAG activities are targeted to reducing the vulnerability of the North American bulk electric system to the effects of physical and cyber terrorism. The CIPAG's activities include developing recommendations and practices related to monitoring, detection, protection, restoration, training, and exercises.

CONCLUSIONS

NERC believes it is imperative to national security to refine and strengthen the public-private partnership. Building a strong trust relationship is essential to the success of this partnership. Overcoming the hurdles to effective communications and information sharing as described in this testimony will enable cooperation for the ultimate goal of protecting our nation's critical infrastructures, its economy, and the well-being of all its citizens. Thank you.

Mr. WHITFIELD. Mr. Sullivan, you're recognized for 5 minutes.

TESTIMONY OF JOHN P. SULLIVAN, JR.

Mr. SULLIVAN. Thank you, Mr. Chairman and members of the subcommittee.

My name is John Sullivan. I'm the Chief Engineer of the Boston Water and Sewer Commission and the President of the Association of Metropolitan Water Agencies on whose behalf I'll testifying today.

AMWA is an organization of the Nation's largest publicly owned drinking water agencies. In 1998, AMWA was designated the Water Sector's liaison to the Federal Government on critical infrastructure protection. In this role, AMWA has served as a community coordinator of security activities.

Governor Ridge said 2 weeks ago that the DHS would focus the resources of the Federal Government on critical infrastructure protection. Giving the Cabinet-level agency the authority to coordinate and consolidate the Federal Government's vast resources will better protect consumers from bioterrorism and life-threatening disruption from water sources.

We recognize the importance of engaging in a new and unique partnership with the Federal Government. We have been working with the National Infrastructure Protection Center, EPA and the Critical Infrastructure Assurance Office. We have also been working with the Office of Homeland Security to develop a national physical infrastructure protection plan, and we will be working with that office to develop a report on cyber security leads. We have also engaged the Departments of Defense, Energy and Interior and the national laboratories in discussions relating to security.

AMWA serves as the first-ever Water Information Sharing and Analysis Center which became operational last September. The ISAC issues advisories and warnings and EPA security-related notices directly to approximately 1,000 major drinking water or wastewater systems, State drinking water administrators and several associations. Through the water ISAC, systems can also submit incident reports to be analyzed by NIPC. In the future, a more sophisticated ISAC that will be operational by the end of this year is being developed with seed money from the EPA grant.

Relating to the ISAC is the issue of information sharing by the Federal Government. Title II of the President's proposal directly relates to the water sector's need for credible and timely intelligence. It is imperative that the information gathered by law enforcement and the intelligence agencies be shared with the water sector by way of the water ISAC. This data is necessary not only to prevent or reduce damages from a potential attack but also to better understand the type of disruptions that could occur, to analyze trends and to build protections into the design of our systems.

Protecting security risk and other information is another top priority of the water sector. As part of the partnership between the government and water sector, AMWA is hopeful that the highest possible protection of information will be assured.

As always, water utilities stand ready to share key information with Federal law enforcement and intelligence agencies as we would with the new DHS help them conduct their mission for protection of Americans in critical infrastructure. We look forward to engaging in a dialog on this important topic because it is essential that we avoid educating the terrorists. It is why the system vulnerability assessment program works.

In 2000, more than a year before the September 11 attacks, the water sector began development of the vulnerability methodology the Boston water and sewer systems have since used. Now thousands of water systems are engaged in this process.

Within the mission of the proposed department is the comprehensive assessment of the vulnerabilities of America's critical infrastructures, including water systems. Two weeks ago, Governor Ridge asked Congress to give the new Department the responsibility for the water system vulnerability assessment program. We strongly support this idea. If DHS is going to be a primary assessor of critical infrastructure responsibilities, then to separate water structures from the other sectors would undermine the ability to consolidate, coordinate and streamline homeland security. This is important given the interdependencies among the various sectors such as the reliance on electricity supplies to treat and distribute water and the need for reliable water supply by hospitals and industry.

Research is another priority for the water sector. Under title III of the President's proposal, DHS would help fill in gaps in research with a national scientific research and development program. We believe that DHS should specifically be authorized to conduct research in methodologies and technologies to detect, prevent and respond to acts of terrorism, including acts of cyber terrorism against drinking water systems. The need for new, sophisticated science in water technology is vital.

Thank you for holding this important hearing and for inviting us today. We anticipate a close and mutually beneficial relationship with the Department of Homeland Security and look forward to further discussions with Congress.

[The prepared statement of John P. Sullivan, Jr. follows:]

Prepared Statement of John P. Sullivan Jr., President, Association of Metropolitan Water Agencies

INTRODUCTION

Chairman Greenwood, Ranking Member Deutsch and members of the subcommittee, thank you for inviting me to testify. My name is John Sullivan. I am the Chief Engineer of the Boston Water and Sewer Commission. I am also the President of the Association of Metropolitan Water Agencies, or AMWA, on whose behalf I am testifying today.

AMWA is an organization of the nation's largest publicly owned drinking water agencies, collectively serving more than 110 million people across the country. In 1998, AMWA was designated the Water Sector's liaison to the Federal government on critical infrastructure protection. In this role, AMWA has served as a coordinator of security activities across the Water Sector, which includes both drinking water and wastewater systems, the vast majority of which are publicly owned. We provide a single point of contact for the government to both gather important information about the Water Sector and communicate data from the government back to water systems across the United States.

Water utilities are especially sensitive to maintaining the public's health, as well as its trust and confidence in a safe and reliable supply of water. We operate both in small towns and in the nation's largest cities and have a significant responsibility to the communities we serve. We are on the front line for defending critical water facilities here in the homeland, and we are acutely aware of this responsibility.

Given these leadership responsibilities, we recognize the importance of engaging in a new and somewhat unique partnership with the Federal government. We are in the midst of a War on Terrorism and must view this partnership in new and creative ways to adapt to the evolving risk environment.

THE DEPARTMENT OF HOMELAND SECURITY

The proposed Department of Homeland Security must provide a vital link between the Federal government and the Water Sector. Like other critical infrastructures, the Water Sector is dependent on the continuous supply of timely information on threats, warnings and other security risks to fulfill our responsibilities to the nation.

There are a number of key areas within the enabling legislation that should be strengthened to ensure that the new department relates directly to the Water Sector. Four key provisions include:

- Critical infrastructure protection.
- Intelligence and information sharing.
- Vulnerability assessments.
- Science and technology development.

CRITICAL INFRASTRUCTURE PROTECTION

Governor Ridge said here two weeks ago that DHS would focus the resources of the Federal government on critical infrastructure protection. He also recommended that Congress provide the new department with the responsibility for the Water Sector's vulnerability assessment program—a proposal that we support.

AMWA, in its security role, has been working with a number of Federal entities, such as the National Infrastructure Protection Center (NIPC), the Environmental Protection Agency (EPA), and the Department of Commerce's Critical Infrastructure Assurance Office (CIAO). We have also been working with the Office of Homeland Security to develop of a national physical infrastructure protection plan and, we will be working with that office to develop a report on cyber security needs. We have also engaged the Department of Interior, the Department of Energy, the National Laboratories and the Department of Defense in discussions related to security. Having a Cabinet-level agency with the authority to coordinate and consolidate the Federal government's vast resources will better protect consumers from bioterrorism and life-threatening disruption of water services.

INTELLIGENCE AND INFORMATION SHARING

AMWA has undertaken a leadership role in organizing and coordinating the flow of information and cooperation across the Water Sector and with the government. AMWA is developing the first-ever Water Information Sharing and Analysis Center, or Water ISAC, which will provide water systems with alerts of potential terrorism and other security-related services and information. The Water ISAC is being developed to incorporate multiple pathways for communicating. It is essential that these pathways run both ways—local to Federal and Federal to local.

Title II (Information Analysis and Infrastructure Protection) of the President's proposal directly relates to the Water Sector's need for credible and timely intelligence, and it is particularly relevant to the security of water systems and the effectiveness of the Water ISAC. The ISAC, which AMWA is developing in close cooperation with NIPC and EPA, will provide the nation's drinking water and wastewater utilities with a secure forum for gathering, analyzing and sharing security-related information. In addition, the Water ISAC will:

- Serve as a single point-of-contact for the Water Sector;

- Feed incident and trend information to the Federal government;
- Facilitate the assessments of water systems' vulnerabilities (required under the bioterrorism bill);
- Analyze threats and risks unique to the Water Sector; and
- Serve as a delivery vehicle for water security research, as authorized under the bioterrorism bill.

Although the ISAC is not yet functional, the Water Sector has developed an informal process for distributing threat information to utilities and, in collaboration with NIPC, an interim mechanism to collect utility security incident information in order to analyze trends and imminent or ongoing threats.

Regardless of which Federal agencies oversee critical infrastructure protection, it is imperative that information gathered by law enforcement and intelligence agencies be shared with the Water Sector, via the Water ISAC. This data is necessary not only to prevent or reduce damages from a potential attack, but also to better understand the types of disruptions that could occur, to analyze trends and to build protections into the design of our systems.

Furthermore, as part of the partnership between the government and the Water Sector, AMWA is hopeful that the highest possible protection for security, risk and other information will be assured. AMWA is taking on responsibility for complex critical infrastructure responsibilities. We are focused on nothing less than promotion of the public's trust and confidence in the communities where we operate. Sensitive information that is either voluntarily shared by utilities, required by the government or is produced by the government must not fall into the hands of those who wish to harm the nation. Likewise, sensitive information developed by the government to assist water systems in deterring threats and protecting their systems must also be protected. Non-disclosure requirements and an exemption to the Freedom of Information Act are solutions, but there are others. We look forward to engaging in a dialogue on this important topic, because it is essential that we avoid educating the enemy.

VULNERABILITY ASSESSMENTS

Within the mission of the proposed department is the comprehensive assessment of the vulnerabilities of America's critical infrastructures, including water systems. Two weeks ago, Governor Ridge asked Congress to give the new department the responsibility for the water system assessments program—a proposal that we strongly support. If DHS is going to be the primary assessor of critical infrastructure vulnerabilities, then to separate water systems from the other sectors would undermine DHS's goal to coordinate, consolidate and streamline homeland security. This is particularly relevant given the interdependencies among the various sectors, such as the reliance on electricity supplies to treat and distribute water and the need for a reliable water supply by hospitals and industry.

In the context of DHS legislation, we also urge the subcommittee to revisit other provisions in the bioterrorism statute relating to the assessments. Assessing vulnerabilities is the first step in securing a water system, and many water utilities have already completed their assessments. The drinking water community does not object to being required to conduct vulnerability assessments. In fact, in mid-2000—more than a year before the September 11 attacks—the Water Sector began development of the vulnerability assessment methodology that Boston Water and Sewer and other large systems have since used. But under the bioterrorism law, EPA is required to collect hardcopies of these vulnerability assessments—more than 8,000 of them. In spite of non-disclosure provisions, the Water Sector is concerned that these extremely sensitive documents could wind up, intentionally or inadvertently, in the hands of malicious people. To avoid this, we recommend that the government not be required to collect the assessments. Instead, utilities could be subject to audits to ensure compliance.

SCIENCE AND TECHNOLOGY DEVELOPMENT

Under Title III (Chemical, Biological, Radiological, and Nuclear Countermeasures), DHS would help fill in the gaps in research with a national scientific research and development program. We believe that DHS should be specifically authorized to conduct research into methodologies and technologies to detect, prevent and respond to acts of terrorism against drinking water systems. The need for new, sophisticated science and technologies in water security is inarguable. Congress and the President recognized this need in the recently enacted bioterrorism law, which not only directed EPA to initiate a research program, but also authorized EPA to disseminate research results via the Water ISAC.

We also encourage the inclusion of cyber terrorism prevention and response in DHS's research program. Water utilities increasingly rely on information systems to control many aspects of water treatment and distribution. It is essential that resources be invested now to design information systems with fewer vulnerabilities, rather than spend limited resources patching up those systems after installation.

This research must be funded, and the Water Sector has requested the $15 million that Congress has authorized in the bioterrorism bill, to initiate this all-important research program.

CONCLUSION

Thank you for holding this important hearing and for inviting us to testify. We would be happy to work with you on changes to the DHS legislation that would further focus efforts to protect the nation's water supply from terrorist attack—whether domestic or international. We anticipate a close, mutually beneficial relationship with the Department of Homeland Security, and we look forward to further discussions with Congress.

Mr. WHITFIELD. Thank you, Mr. Sullivan.

Mr. Watson, you're recognized for 5 minutes.

TESTIMONY OF KENNETH C. WATSON

Mr. WATSON. Thank you, Mr. Chairman and distinguished committee members. I'm honored to testify before you today for PCIS in support of the President's proposal for a Homeland Security Department. A single Department with a clear line of authority would not only consolidate efforts currently spread over a hundred organizations but also provide needed national emphasis to improve our preparedness.

Because networks are now integral to core business and government practices, security has become the top or next-to-top requirement for CEOs and corporate boards. Both the cyber and physical aspects of security must be integrated into core networking practices and environments, especially now that we read in the Washington Post that al Qaeda is exploring the Internet as a means for attack, mapping our vulnerabilities in cyberspace and had detailed information on digital control systems on a laptop recovered in Afghanistan.

Four years prior to the attacks of 9/11, the President's Commission on Critical Information Protection identified eight infrastructure sections critical to national and economic security and the health and safety of American citizens. Because there are no boundaries in cyberspace and because the vast majority of the Nation's critical infrastructures are privately owned and operated, the Commission recommended an unprecedented partnership between private industry and government. The PCIS was launched in December, 1999, in the World Trade Center to fulfill that need. The private sector portion of the PCIS was incorporated as a 501(c)6 nonprofit organization in January, 2001.

We have eight member companies, representing all the critical infrastructure sectors. In the cyber dimension, private sector infrastructure companies represent the front lines of defense against attacks that take an average of 1½ minutes to traverse multiple jurisdictions and countries at the speed of light and cost the anonymous attacker no more than a personal computer and downloaded free software.

The mission of PCIS is to coordinate cross-sector initiatives and complement public-private efforts to promote and assure reliable provision of critical infrastructure services in the face of emerging risks to economic and national security. This involves more than either physical or cyber security alone, and it spans actions from prevention, planning and preparation to business continuity recovery and reconstitution.

Our top six initiatives this year are to coordinate the private input at the National Strategy for Critical Infrastructure Assurance; to serve as a clearinghouse for security efforts to the public; to publish an Effective Practices compendium in collaboration with CIAO; to provide critical infrastructure awareness materials and references on our website; to develop a risk assessment guidebook for use by any region or sector; and facilitate cross-sector information exchange.

As a public service to promote awareness of the need to secure home and small business computers, another public-private partnership was incorporated as a 501(c)3 within PCIS earlier this year. The website www.staysafeonline.info, has experienced over 5 million page views since February, and we believe this campaign is helping to lower the risk that America's growing broadband user base could be used to stage attacks against our infrastructures.

I'd like to concentrate the remainder of my remarks on two key areas we believe still need work: First, additional emphasis on critical infrastructure assurance activities and, second, the removal of barriers to help with private information sharing.

Critical infrastructure services are interlinked and 85 percent of them are owned and operated by the private sector. The line between physical and cyber assets is becoming even more blurred by the widespread use of digital control systems; as Sam Barco said, electronically controlled devices that report on kilowatt hours transmitted, gallons per hour, cubic feet of natural gas, traffic on smart roadways and can actually control physical assets like floodgates, oil, gas and water valves and flood controllers, ATM machines and the list keeps growing.

After over 20 years as a marine officer, it is second nature for me to relate everything I do to mission. Title II of the Homeland Security Act establishes an Under Secretary for Information Analysis and Infrastructure Protection. We believe that these are two all-encompassing functional areas. The information and analysis and warning function alone will be a full-time job. The job of critical infrastructure assurance is too vital to American commerce to be subsumed by the intelligence gathering reporting mission.

However, similar to a corporate chief executive officer, the Secretary should have the latitude to organize the department to meet both the information analysis and warning requirements and those needed to protect America's critical infrastructures.

Information sharing is the key to solving problems together. Both the private sector and the government agree that the exchange of timely and e-cyber vulnerability and countermeasure information will greatly benefit the cause of protecting our critical infrastructures, and the private sector wants to share this kind of information with the government.

Most critical infrastructures have established information-sharing analysis centers to share information on cyber threats, vulnerabilities, countermeasures, best practices and other solutions. Some of these are strictly in the private sector, while others include public and private participation. Some have been sharing critical information for a number of years and ISAC-type information to other normal reporting information or exchange vulnerabilities established. As ISACs mature, their effectiveness in sharing countermeasures within their industries dramatically improve in both quality and timeliness.

However, even with all of the efforts toward public-private information exchange, only rarely is the private sector sharing most sensitive cyber vulnerability information with the government. The main reason for this is that companies do not believe Federal agencies can protect the information from Freedom of Information Act requests.

Critical infrastructure threat and vulnerability information voluntarily shared with the government should be given similar protection as government classified information. The PCIS supports a narrowly written exemption for infrastructure threat and vulnerability information shared with the government.

The other side of——

Mr. WHITFIELD. Mr. Watson, if you'll excuse me, you're over about a minute. So if you could move ahead and summarize, we'd appreciate it.

Mr. WATSON. There's still much opportunity to work together to remove redundancy and improve communication and clarify roles. On behalf of the PCIS and our 80 member companies, I would like to thank you for your time today. I'll be glad to answer any questions you may have.

[The prepared statement of Kenneth C. Watson follows:]

PREPARED STATEMENT OF KENNETH C. WATSON, PRESIDENT, PARTNERSHIP FOR CRITICAL INFRASTRUCTURE SECURITY

INTRODUCTION

Chairman Greenwood and distinguished Committee Members, I am honored to testify before you today in support of the President's proposal for a Homeland Security Department. A single Department with a clear line of authority would not only consolidate efforts currently spread across over 100 Federal organizations, but also would provide needed national emphasis to improve our preparedness.

Internet-based technologies are driving unprecedented productivity increases and dependencies. As you know, the US government reported that productivity in this country rose 8.4 percent in the first quarter this year, even with the sluggish market.[1] This is unprecedented. In the past, productivity has been in the 1.5- to 2-percent range during down market conditions. Emerging high-growth "tornado" markets such as IP telephony, storage networking, wireless, optical, virtual private networking, and cable integration of voice, video, and data are sweeping business sectors worldwide, bringing about both evolutionary and revolutionary changes in the way businesses and governments do business. These changes—increasing bandwidth, exploding connectedness, integration of all types of applications into multipurpose devices, distribution of both processes and storage, and erosion of physical boundaries—bring old and new vulnerabilities with them. Because networks are now integral to core business and government practices, security has become the top or next-to-top requirement of CEOs and Boards. Both the cyber and physical aspects of security must be integrated into core networking practices and environments, especially now that we read in the Washington Post that al-Qaeda is exploring the Internet as a means for attack, mapping our vulnerabilities in cyberspace, and had detailed information on digital control systems on a laptop recovered in Afghanistan.[2]

[1] US Bureau of Labor Statistics, "Productivity and Costs, First Quarter 2002, Revised," USDL 02-318, May 31, 2002.

[2] Barton Gellman, "Cyber-Attacks by Al Qaeda Feared: Terrorists at Threshold of Using Internet as Tool of Bloodshed, Experts Say," Washington Post, Thursday, June 27, 2002; Page A01

Four years prior to the attacks of 9-11, the President's Commission on Critical Infrastructure Protection (PCCIP) identified eight infrastructure sectors as critical to national and economic security and the health and safety of American citizens. Securing the nation's critical infrastructures goes well beyond the government's traditional role of physical protection through defense of national airspace and national borders. Because there are no boundaries in cyberspace, and because the vast majority of the nation's critical infrastructures are privately owned and operated, the commission recommended an unprecedented partnership between private industry and government. The Partnership for Critical Infrastructure Security (PCIS) was launched in December 1999 in the World Trade Center to fill this need. The private-sector portion of the PCIS was incorporated as a 501(c)6 non-profit organization in January 2001, and I was elected its first President and Chairman of the Board in March of that year.

The PCIS Board and I fully support the President's plan and look forward to working with the Administration and the Congress to further cement the public-private relationships we have forged to assure the delivery of critical services to our citizens and customers. In the cyber dimension, private-sector infrastructure companies represent the front lines of defense against attacks that take an average of one and one-half minutes, traverse multiple jurisdictions and countries at the speed of light, and cost the anonymous attacker no more than a personal computer and downloaded free software.

PARTNERSHIP FOR CRITICAL INFRASTRUCTURE SECURITY

The mission of the PCIS is to coordinate cross-sector initiatives and complement public-private efforts to promote and assure reliable provision of critical infrastructure services in the face of emerging risks to economic and national security. This involves more than either physical or cyber security alone, and it spans actions from prevention, planning, and preparation to business continuity, recovery, and reconstitution.

Presidential Decision Directive 63 followed the PCCIP recommendations by establishing Sector Liaison officials in the pertinent Federal Lead Agencies involved in critical infrastructure assurance, to work with Sector Coordinators who were industry leaders in the private sector in each of the critical sectors. We structured the PCIS Board so that those Sector Coordinators always represent a majority of Directors to ensure that the PCIS continues to meet the needs of all the infrastructure sectors. The PCIS currently has over 80 corporate members from all the critical infrastructure sectors, plus ad hoc representation from all pertinent Federal lead agencies and the National Association of State Chief Information Officers.

To illustrate the level of support in industry for the PCIS, the Board members are either presidents or chief operations or information security officer equivalents in their organizations: **Presidents:** Airports Council International—North America; Association of American Railroads; Association of Metropolitan Water Agencies; Information Technology Association of America; North American Electric Reliability Council; and The Institute of Internal Auditors. **COO/CISO or Equivalent:** Bank of America; BellSouth; Cellular Telecommunications & Internet Association; Conoco; Consolidated Edison of New York; Microsoft; Morgan Stanley; Union Pacific Corporation; US Telecommunications Association; and Telecommunications Industry Association.

Lead agencies, coordinated by the Critical Infrastructure Assurance Office (CIAO) of the Department of Commerce, fully participate in PCIS working groups and its public-private coordinating committee. Our current "top six" initiatives are:

- Coordinate private-sector input to the National Strategy for Critical Infrastructure Assurance, especially those areas of cross-sector interest and dependency;
- Serve as a clearinghouse for digital control systems security efforts, including research and development, exercises and tests, and awareness;
- Publish an "Effective Practices" compendium, in collaboration with the CIAO, starting with lessons learned during the recovery from the 9-11 attacks;
- Provide critical infrastructure assurance awareness materials and references for all PCIS members and the public;
- Develop a risk assessment guidebook for use by any region or sector, concentrating on cross-sector dependencies; and
- Facilitate cross-sector information exchange, augmenting efforts by the industry Information Sharing and Analysis Centers (ISACs) and government cyber warning and information organizations.

As a public service to promote awareness of the need to take steps to secure home and small business computers, another public-private partnership, the National Cyber Security Alliance, was incorporated as a 501(c)3 educational foundation within the PCIS earlier this year. The web site, www.staysafeonline.info, has experienced over 5 million page views since February, and we believe this campaign is helping to lower the risk that America's growing broadband user base could be used to stage denial of service attacks against our infrastructures.

THE PRESIDENT'S PROPOSAL

After reviewing the President's proposal, we believe it provides a clearer and more efficient organizational structure to accomplish homeland security missions than currently exists in the Federal government. Consolidating information analysis and warning; chemical, biological, nuclear, and radiological countermeasures; emergency preparedness and response; border and transportation security; and critical infrastructure assurance is a much-needed, logical response to the continuing threats of terror against the United States.

Additionally, Section 732 shows foresight in taking advantage of current business practices such as "other transactions" for research and development and prototyping, creation of employer-employee relationships for contracting, authorization to invoke 40 U.S.C. 474, and flexible acquisition and disposition of property. These practices should encourage innovation, rapid procurement, advanced research, and beneficial contracting relationships with industry, but will require discipline and oversight.

I'd like to concentrate the remainder of my remarks on two key areas we believe still need work: first, additional emphasis on critical infrastructure assurance activities; and second, the removal of barriers to public-private information sharing.

After over 20 years as a Marine officer, it is second nature for me to relate everything I do to mission. In business as well as in government, those organizations that structure themselves and order their actions around their missions are the most successful. The mission of critical infrastructure assurance is imbedded within the overall mission of Homeland Security, but needs additional organizational emphasis.

As critical infrastructure assurance has matured over the last five years, those of us intimately involved recognize its strong suits: public-private partnership, interdependency, and the recognition that physical business operations of our critical infrastructures depend on information systems and networks, far more so than in any other country in the world.

The PCIS defined critical infrastructure assurance two years ago as: "efforts to promote and assure reliable provision of critical infrastructure services in the face of emerging risks to economic and national security."

Economic and national security are important to assuring our critical infrastructures, but the essence of the mission is assuring the delivery of services over the infrastructures. Those services are what our citizens and customers expect and need, especially in time of crisis, and they include accurate and uninterrupted financial transactions, on-time and safe transportation, reliable electric power, available and dependable information and communications, safe and clean drinking water, safe and available oil and natural gas, and timely emergency services. All these services are interlinked in the Internet Economy; they depend more and more on networks to carry out basic business; and 85 percent of them are owned and operated by the private sector. The line between physical and cyber assets is becoming even more blurred by the widespread use of digital control systems—electronically controlled devices that report on kilowatt hours transmitted, gallons per hour of oil and water, cubic feet of natural gas, traffic on "smart roadways," and can actually control physical assets like flood gates; oil, gas, and water valves and flow controllers; ATM machines; and the list keeps growing.

Industry defines critical infrastructure assurance to include both physical and cyber assets, but by "physical" we mean those assets essential to the delivery of each infrastructure's critical services. Cyber security also includes physical threats to critical infrastructures such as intentional or unintentional interruptions of the high-technology support to the infrastructures, like a backhoe cutting a key fiber-optic line.

AN EFFECTIVE CRITICAL INFRASTRUCTURE ASSURANCE ORGANIZATION

Title II of the Homeland Security Act establishes an Under Secretary for Information Analysis and Infrastructure Protection. We believe these are two all-encompassing functional areas. The information analysis and warning function alone will be a full-time job, especially considering the monumental task of merging the 100-plus intelligence and law enforcement databases in order to effectively administer national threat correlation and support the Homeland Security Advisory System. The job of critical infrastructure assurance is too vital to American commerce to be subsumed by the intelligence gathering and reporting mission. Similar to a corporate Chief Executive Officer, the Secretary should have the flexibility to organize the Department to meet the requirements needed to protect America's critical infrastructures.

The mission of Critical Infrastructure Assurance includes:

- Coordinating vulnerability assessments of key resources and critical infrastructures;
- Development and maintenance of the National Strategy for Critical Infrastructure Assurance;
- Facilitating true partnerships with private industry and state and local government to address critical infrastructure issues;
- Taking or influencing measures necessary for securing key resources and critical infrastructures;
- Facilitating and defining requirements for cutting-edge research and development to enhance long-term critical infrastructure assurance;
- Facilitating cross-sector and public-private sharing of critical infrastructure threat, vulnerability, and countermeasure information;
- Promoting awareness and education at all levels of critical infrastructure assurance issues, including public and private roles and responsibilities; and
- Coordinating with other executive agencies, state and local governments, and the private sector regarding critical infrastructure assurance.

COORDINATION WITH NON-FEDERAL ORGANIZATIONS

Section 701 of the proposal requires the Secretary of Homeland Security to coordinate with state and local officials and the private sector in carrying out the mission of the Department of Homeland Security. Since most of the critical infrastructures are owned and operated by the private sector, coordination with the private sector has become an established norm, led by the efforts of the Critical Infrastructure Assurance Office (CIAO). The CIAO has developed working, productive relationships with the infrastructure leaders, the audit and other risk management industries, and now the National Governors' Association and the National Association of State CIOs. It also has facilitated the development of the PCIS and the various industry Information Sharing and Analysis Centers (ISACs). The various Under Secretaries should be given responsibility for coordinating with state and local governments and the private sector in their respective areas of responsibility, although it is understood and useful for the office of the Secretary of Homeland Security to coordinate activities across the entire Department.

REMOVING INFORMATION SHARING BARRIERS

Information sharing is key to solving problems together. The best leaders know that the more their people know about the problems they're trying to solve, the better they will be able to use their intellect, creativity, and drive to solve them most effectively. Most critical infrastructure sectors have established Information Sharing and Analysis Centers (ISACs) to share information on cyber threats, vulnerabilities, countermeasures, best practices, and other solutions. Some of these are strictly in the private sector, while others include public and private participation. Some have been sharing critical information for a number of years, and some organizations added ISAC-type information to other normal reporting or information exchange responsibilities previously established. As ISACs mature, their effectiveness in sharing both warnings and countermeasures within their industries is dramatically improving, in both quality and timeliness. They are developing a depth of knowledge that enables analysis and trending, beneficial to their industries and member companies. To date, these include: Financial Services ISAC, Telecom ISAC, Information Technology ISAC, Energy ISAC (oil and gas), Electric Power ISAC, Emergency Law Enforcement Services, and Surface Transportation ISAC.

The water, food safety, chemical and manufacturing, aviation, and firefighting sectors are in the process of establishing ISACs.

Several government organizations have cyber information sharing missions: FedCIRC (GSA), DoDCERT (DoD), NSIRC (IC), and NIPC (FBI).

The ISACs are developing an Inter-ISAC Information Exchange Memorandum of Understanding, and some ISACs have signed MOUs with the NIPC. PCIS is facilitating cross-sector information exchange by developing a common taxonomy and co-hosting multi-ISAC and public-private action meetings in conjunction with the President's Office of Cybersecurity. Both the private sector and the government agree that the exchange of timely cyber vulnerability and countermeasure information would greatly benefit the cause of protecting our critical infrastructures, and the private sector wants to share this kind of information with the government.

However, even with all the efforts toward public-private information exchange, only rarely is the private sector sharing its most sensitive cyber vulnerability information with the government. The main reason for this is that companies do not believe Federal agencies can protect the information from Freedom of Information Act (FOIA) requests. Under the current law, companies have no assurance that information they share with a government agency will be treated confidentially, and agencies are not required to commit to confidentiality at the time of disclosure. Agencies are not even required to initiate the FOIA exemption process until a FOIA request is received. When it is received, the agency is asked to defend the information's confidentiality, and is not required to inform the originator if it believes it has enough information to proceed.

Critical infrastructure threat and vulnerability information voluntarily shared with the government should be given the same protections as government classified information. The PCIS supports very narrowly written exemptions for infrastructure threat and vulnerability information shared with the government.

Detractors claim that these new exemptions would provide walls behind which companies could hide environmental accidents and hazards, or that companies would use them to violate citizens' or employee privacy. Neither claim is true. Industry wants the exemption language written narrowly so as to cover only infrastructure threat and vulnerability information, and welcomes specific exclusions covering spills or other environmental accidents. Industry wants to share critical information with the government in a trusted working environment. Let's remove the exemption ambiguity in the current law and start sharing information with each other so that we can deter a digital 9-11 before it happens.

The other side of the information-sharing coin is information from the government to the private sector. This process also needs work. Industry is generally dissatisfied with the quality and timeliness of cyber security information flowing from the government. One example will serve to illustrate the problem. The Klez.H worm began proliferating on April 17 this year. The IT-ISAC issued an advisory on that day, and the Computer Emergency Response Center Coordination Center at Carnegie Mellon University posted its alert on April 19. The NIPC advisory was not issued until April 29, 12 days later, and there was no new information in that alert. This does not mean that the NIPC isn't doing everything it can to release information. On the contrary, they participate in daily conference calls with at least two ISACs, and strive to overcome their intelligence classification and law enforcement sensitivity problems that are not present in the private sector. Delays in NIPC reporting may be due to protecting intelligence sources and methods, or because they decide not to repeat information already disclosed by the private sector or CERT/CC. Removing the FOIA barrier to information exchange will open up the private sector as an unclassified source of valuable information for NIPC and others working hard to protect the country.

Regarding intelligence and law enforcement agencies, the proposal does not clarify jurisdiction issues between CIA, FBI, Secret Service, and other organizations that could be involved in cyber investigations. Private industry appreciates choice in its service suppliers. However, many companies do not know under what circumstances nor whom to call when they suspect cybercrime in their networks. Industry needs clear information about the various agencies regarding their programs, jurisdictions, competencies, and points of contact.

CONCLUSION

The PCIS and I think the proposed Homeland Security Department is vital to providing needed focus to the area of Critical Infrastructure Assurance for America. There is still much opportunity, as we move forward together, to remove redundancy, improve communication, and clarify roles—organizing to support commerce is vital to our economic and national security. It is vitally important to make progress in developing processes and providing legislative support to facilitate sharing of security information and alerts between government and the private sector. It is also important to improve information sharing from the government to industry, and to clarify jurisdiction among the myriad intelligence and law enforcement agencies involved in cyber security and cyber investigations. Finally, I encourage you to leverage existing expertise in the National Security Telecommunications Advisory Committee, the ISACs, and the PCIS as you shape this new, much-needed Department. However the government organizes itself, we in the private sector stand ready to assist any way we can.

On behalf of the PCIS and our 80 member companies, I would like to thank you for your time today. I'll be glad to answer any questions you may have.

Mr. WHITFIELD. Mr. Watson, thank you very much, and Mr. Baumann, you're recognized for 5 minutes.

TESTIMONY OF JEREMIAH D. BAUMANN

Mr. BAUMANN. Thank you, Mr. Chairman, and members of the committee, for the opportunity to testify before you today on the important issue of the proposed Department of Homeland Security. The U.S. Public Interest Research Group is the Federal advocacy office for the State Public Interest Research Groups, or State PIRGs, a network of advocacy organizations with a 30-year record of working to protect public health and safety and work with good government reforms.

My testimony will focus primarily on the issue of chemical security, what needs to be done about this critical gap in security improvements to date, and ways the Department of Homeland Security as currently proposed could become an obstacle rather than an asset in addressing this important issue, specifically through its unclear designation of authority addressing safety of particular industry sectors, its lack of focus on protecting safety by reducing hazards and vulnerabilities, and through its preference for secrecy rather than safety.

While my testimony focuses on chemical plants, I think many of these themes apply across industry sectors that face significant vulnerabilities. Let me first talk about the need for a Federal chemical security program.

Across America thousands of industrial facilities are using and storing hazardous chemicals in quantities that put large numbers of Americans at risk. The best summary of this risk is that at almost 125 facilities, each of them put at least a million people at risk in the event of a chemical release.

Unfortunately, security at these facilities ranges from poor to nonexistent. A series in a Pittsburg Tribune Review 6 months after September 11 reported that an intruder could freely enter and walk through more than 60 chemical facilities not only in Pennsylvania but in Houston, Chicago and Baltimore as well.

The Army Surgeon General has identified this threat as second only to that of bioterrorism. However, unlike bioterrorism, virtually nothing has been done to address this issue since September 11.

A Federal chemical security program requires three basic components, a vulnerability assessment, a hazard reduction plan and increases in site security where significant threats remain. At chemical plants the need for a focus beyond mere assessment and even beyond traditional security is necessary, because fortunately there are well-established measures for reducing hazards at facilities using safer materials and processes that could eliminate these terrorist targets in communities.

The bill as currently proposed not only does not establish any chemical security program, but could in some ways confuse or delay progress on the issue of chemical security. First, the bill does not clearly define what the new department's authorities regarding critical infrastructure should be generally, and chemical plant security specifically is left completely unclear.

The committee should clarify two critical points regarding the new department's authority. First, the role for the Department of

Homeland Security is one of coordinating security programs and advising agencies whose functions are not being transferred to the new department, not that their authority is either being reestablished in a new department or being transferred to the new department. This is particularly important for an agency like EPA where chemical plants have a much broader risk than just a security risk.

There's also a significant chemical accident risk. Accident prevention as well as security could be undermined by removing the authority from this long history of expertise and experience in addressing this threat.

The committee should also clarify that the creation of the new department does not delay, hinder or otherwise affect the ability of other regulatory agencies to exercise their authority, particularly regarding security and safety threats. EPA has already been pressured not to move forward with any chemical security program until the creation of a new department has been addressed. Such a delay would be irresponsible and potentially dangerous.

The critical infrastructure and research and development sections of the bill as proposed have another potential problem, which is they focus almost entirely on securing infrastructure, with little attention to making infrastructure safer in order to protect public health and safety.

Congress should ensure that the new department prioritizes reducing hazards and reducing vulnerabilities, not simply assessing them and not relying only on the traditional security strategies and perimeter security access control, surveillance and related measures.

As discussed above, public health and safety can best be secured against a deliberate chemical release by reducing the hazard itself and eliminating the chance that any chemical release could harm the surrounding community.

Congress should direct the new department to establish public health and safety as a priority and reducing hazards and vulnerabilities as a priority strategy and working with existing agencies to make sure that happens.

Finally, I'll address what is perhaps the proposed bill's most threatening measure when it comes to protecting public health and safety, which is the surprisingly broad loophole proposed in the Freedom of Information Act. The public's right to know about public health and safety and the ensuing public accountability are safety tools that have a long record of protecting public safety, and the new Department of Homeland Security should treat information as such rather than undermining current protections.

Restricting the public's right to know about hazards in communities and industry or government actions to remedy those hazards could hurt safety rather than help it. This information in a lot of cases has been shown to help enable the public communities, local emergency responders and other important constituencies to understand, prepare for and respond to not only accidents but potential terrorist attacks. It's also one of the most effective incentives for public safety improvements. Public disclosure has a long record of reducing risk.

I'll wrap up briefly by just pointing out that the proposed bill goes against the tradition of the Freedom of Information Act. FOIA typically requires a concrete reason in the public interest to withhold specific documents and a specific definition of what documents need to be withheld in order to protect public safety. The proposed bill doesn't even define what documents could be exempt, explain why they wouldn't be covered by current FOIA exemptions, much less explain why they need to be exempt. The requirements are so vague that in theory some currently mandated public information could be removed from public view, because there are no definitions of voluntary information or exactly what critical infrastructure vulnerabilities could be included.

In concluding, I would recommend that Congress only create FOIA exemptions for specific information and types of information being required in the private sector by the government, and since this bill does not do that, I would recommend that the FOIA exemption be removed from this bill and considered separate.

[The prepared statement of Jeremiah D. Baumann follows:]

PREPARED STATEMENT OF JEREMIAH D. BAUMANN, ENVIRONMENTAL HEALTH ADVOCATE

Thank you, Mr. Chairman and members of the Committee on Energy and Commerce, for the opportunity to testify before you today on the proposed Department of Homeland Security. My name is Jeremiah Baumann and I am the Environmental Health Advocate for the U.S. Public Interest Research Group (PIRG). U.S. PIRG is the federal advocacy office of the state PIRGs, a network of state-based public interest advocacy organizations with a 30-year history of advocacy for environmental and public health protection, consumer protection, good-government reforms, and other public interest issues.

My testimony will focus on the issue of chemical security, what needs to be done about this critical gap in security improvements to date, and ways that the Department of Homeland Security—as proposed—could become an obstacle rather than an asset in addressing this issue. In advance, however, I would like to make a few general observations that I think pervade the proposed bill creating a Department of Homeland Security beyond the realm of chemical security:

- The proposed bill, and particularly the sections addressing critical infrastructure, and chemical, biological, radiological, and nuclear countermeasures, lacks a focus on protecting public health and safety. Instead, the focus is on securing infrastructure and protecting assets. While these are often closely related to public health and safety, they need to be put in this context, and the new Department of Homeland Security needs a mandate from Congress to make public health and safety its priority.
- The proposed bill tends to focus on securing existing infrastructure when the first priority should be making infrastructure safer. Some attributes of critical infrastructures are inherently hazardous, but could be made inherently safer. Making our infrastructure safer will require changes to the infrastructure and investment in the near term. However, making infrastructure safer will be less expensive in the long term because the up-front investment will reduce or eliminate the significant costs of making inherently dangerous facilities and operations more secure, and of preparing for or responding to attacks on infrastructure.
- The proposed bill indicates a dangerous preference for secrecy. This could undermine basic mechanisms of public accountability, the public's right to know about threats to health and safety, and is likely to hinder, rather than help, safety.

Examining the threats posed by the use and storage of highly hazardous chemicals in facilities through out nation's industrial infrastructure demonstrates why these three concepts are important. Protecting against terrorist attacks on a chemical-using industrial site requires a focus on protecting public health and safety using the most effective strategies, not just securing industrial facilities and protecting their assets. Furthermore, simply securing facilities as they are, without making them inherently safer, will not protect public health and safety from terrorist-related chemical incidents. Finally, new secrecy measures will be an obstacle to protecting public health and safety from chemical incidents, a category of hazard where a long record of public safety improvements has demonstrated the value of openness and of the public's right to know.

THE NEED FOR AN AGGRESSIVE FEDERAL CHEMICAL SECURITY PROGRAM

The Threat of Chemical Terrorism

Across America, thousands of industrial facilities use and store hazardous chemicals in quantities that put large numbers of Americans at risk of serious injury or death in the event of a chemical release. One hundred twenty-five facilities each put at least 1 million people at risk; 700 facilities each put at least 100,000 people at risk; and 3,000 facilities each put at least 10,000 people at risk.[1] According to a 1998 report by U.S. PIRG, 1 in 6 Americans lives within a vulnerable zone—the area in which there could be serious injury or death in the event of a chemical accident—created by a nearby industrial facility.[2]

The threat of terrorism has brought new scrutiny to the potential for terrorists to deliberately trigger accidents that until recently the chemical industry characterized as unlikely worst-case scenarios. Such an act could have even more severe consequences than the tens of thousands of chemical accidents that kill 150 Americans and injure 5,000 every year.[3]

Frederick L. Webber, president of the American Chemistry Council, has said "No one needed to convince us that we could be—and indeed would be—a target at some future date . . . If they're looking for the big bang, obviously you don't have to go far in your imagination to think about what the possibilities are."[4] The Agency for Toxic Substances and Disease Registry said in 1999 that chemicals at industrial sites provide terrorists with ". . . effective and readily accessible materials to develop improvised explosives, incendiaries and poisons."[5]

Unfortunately, that report found security at these facilities ranging from poor to nonexistent. More recent investigations since September 11th tell the same story. Just months ago, a series in the Pittsburgh *Tribune-Review* reported that an intruder could freely enter and walk through more than 60 chemical facilities in Pennsylvania, Houston, Chicago, and Baltimore—completely unchallenged.[6] A recent report by the Department of Justice, made secret for unexplained reasons, apparently confirms these findings.

An Opportunity to make Communities Safe

Fortunately, there are well-established measures for reducing hazards at facilities—and making communities safer. Reducing chemical hazards at industrial facilities means making process changes that reduce or eliminate the possibility of a chemical release by reducing chemical use or switching to safer chemicals and processes. For many chemicals and processes, there are readily available and safer alternatives. A few examples demonstrate this simple concept:

- In New Jersey, 553 water treatment facilities have stopped using chlorine gas because of its notorious potential for disastrous chemical releases.[7]
- Here in Washington, DC, the city's Blue Plains Sewage Treatment Plant has long recognized that a release of chlorine gas or sulfur dioxide could blanket the downtown area, as well as Anacostia, Reagan National Airport, and Alexandria.[8] Over the course of eight weeks after September 11th, authorities quietly removed up to 900 tons of liquid chlorine and sulfur dioxide, moving tanker cars at night under guard. The city switched to a hypochlorite process that dramatically reduces the safety risk, virtually eliminating the chance of any off-site impact.[9]

---

[1] James Belke, U.S. Environmental Protection Agency. "Chemical accident risks in U.S. industry—A preliminary analysis of accident risk data from U.S. hazardous facilities," September 25, 2000.

[2] U.S. Public Interest Research Group and National Environmental Law Center. *Too Close to Home*. July 1998.

[3] Mannan, Gentile, and O'Connor. "Chemical Incident Data Mining and Application to Chemical Safety Trend Analysis," Mary Kay O'Connor Process Safety Center, Texas A&M University, 2001.

[4] Eric Pianin. "Toxic Chemicals' Security Worries Officials," *Washington Post*, November 12, 2001.

[5] Pianin 2001 *Ibid.*

[6] Carl Prine. "Lax Security Exposes Lethal Chemical Supplies," Pittsburgh *Tribune-Review*, April 7, 2002; and "Chemicals Pose Risks Nationwide," Pittsburgh *Tribune-Review*, May 5, 2002.

[7] Information provided by R. Baldini, Bureau of Release Prevention, New Jersey Department of Environmental Protection, September 2001.

[8] Radian Corporation. "Air Dispersion Model Assessment of Impacts From a Chlorine Spill at the Blue Plains Wastewater Treatment Plant," 1982; See also Chlorine Institute, Pamphlet 74, April 1998.

[9] Carol D. Leonnig and Spencer S. Hsu. "Fearing Attack, Blue Plains Ceases Toxic Chemical Use," *Washington Post*, November 10, 2001.

- In response to the Pittsburgh *Tribune-Review* series on the danger of chemical plants' lax security, Bethlehem Steel in Pennsylvania is switching from hazardous sulfur dioxide to safer materials and processes.[10]

The threat of terrorism requires looking for ways to make industrial facilities inherently safer when it comes to chemical use. If terrorists continue to use airplanes or truck bombs, add-on security measures such as safety guards and physical barriers cannot prevent a chemical release. Similarly, secondary prevention or mitigation measures, such as safety valves, would be decidedly inadequate in the event of an attack like those seen on September 11th.

Inherent safety is an opportunity for policymakers to *remove* a terrorist threat in many cases. This is an option that is not available for all terrorist risks. Airline passengers have to rely on increased security to make flying safer. For American industry, however, many chemicals have readily available safer alternatives and many facilities could re-design processes to be inherently safer.

Inaction on Chemical Security

Since September 11th, the Senate has introduced, held hearings, and scheduled mark-up on a bill. But at this late date, little else has occurred to address chemical security. The administration developed a proposal on chemical security, but appears to have backed away from it. An EPA presentation in May outlined an aggressive legislative proposal, but later reports indicated that the proposal had been scaled back in scope and potentially reduced to agency guidance with little enforceability. News reports indicate that progress on the proposal slowed in response to resistance from the industry and from within the administration.

The Department of Justice has released its "Sandia methodology," guidance on assessing site security at chemical facilities. Unfortunately, this guidance has been issued with no indication that facilities will be required to implement it. Also, the guidance relies primarily on site security with only minimal mention of making facilities inherently safer. Additionally, the guidance is quite complicated and relies on sophisticated judgments on the relative risk of different security threats; it is unlikely that the average plant manager would have the expertise to implement this plan without assistance from security experts.

The American Chemistry Council touts a voluntary program being developed to increase site security at chemical plants. While the American Chemistry Council is doing the right thing by beginning to address the security risks at their facilities, their program is not and cannot be sufficient, for three reasons:

1. *The program is voluntary.* In the wake of September 11th, airline security, water supply security, and nuclear security have not been allowed to happen on a voluntary basis. It makes no sense to allow thousands of facilities with hazardous chemical stockpiles to increase security on a voluntary basis. Furthermore, other voluntary programs, particularly the industry's "Responsible Care" program, to which the new security code is closely linked, have too often been heavy on public relations and promotional campaigns and light on substantive safety improvements. A 1998 survey of American Chemistry Council members showed that, despite their on-paper commitment to the right-to-know principles of the "Responsible Care" program, citizens could not get basic information about toxic chemical use and accidents at 75% of the facilities.[11]
2. *The program applies only to American Chemistry Council members,* which comprise 11% of the 15,000 industrial facilities that store and use high enough quantities of hazardous materials to be subject to EPA's chemical accident prevention program. With nearly 125 facilities in the country each putting 1 million Americans at risk, increasing security at 10% of them is not enough.
3. *The program focuses primarily on increasing site security and only peripherally mentions reducing hazards.* Reducing the hazards themselves—potentially eliminating terrorist targets—must be at the core of any program to make communities safer from a terrorist attack on a chemical plant.

A Federal Chemical Security Program

The threat of chemical use and storage at thousands of industrial facilities deserves the same attention to security as water treatment facilities and nuclear plants. Three basic components are required: a vulnerability assessment, a hazard reduction plan, and increases in site security where significant threats of off-site consequences remain.

---

[10] Carl Prine. "Companies Respond to Infiltration of Facilities." The Pittsburgh *Tribune-Review,* May 5, 2002.
[11] U.S. PIRG Education Fund. *Trust Us, Don't Track Us.* January 1998.

The vulnerability assessment can follow methodologies laid out for other industry sectors and by various federal agencies and industry experts on a voluntary basis to date, with one critical difference: accountability. EPA's proposals have not included a requirement that vulnerability assessments be submitted to the federal government. This basic accountability is critical to government's ability to increase safety and protect against terrorist risks. Without the basic requirement that facilities submit their vulnerability assessments (and plans for reducing hazards and increasing site security) to the government, a federal program would be hardly an improvement over a voluntary program.

Requiring facilities to submit hazard reduction plans must be the heart of a federal chemical security program. Reducing hazards means reducing or eliminating terrorist targets in communities nationwide—the most effective protection possible. A program can take two approaches:

1. *Mandate specific process changes to reduce the inherent dangers at industrial plants.* A federal security program could identify technologies or materials that are highly hazardous and have available alternatives and require that any facility using those technologies or materials adopt the alternative. Examples include chlorine used at wastewater treatment facilities and hydrogen fluoride used at many oil refineries.
2. *Require facilities to look for inherently safer technologies and implement available alternatives.* This approach allows more flexibility to accommodate the significant differences between plants. For this planning-based model to work, facilities must be required to report to the government specifically what safer alternatives were identified, which alternatives they plan to implement and on what timeline, and the reasons for rejecting any safer alternatives that were identified. The reasons permitted should be strictly limited.

A federal chemical security program should be led by EPA. The agency has the expertise and history with chemical plant safety, as well as, appropriately, the regulatory authority. The new Department of Homeland Security should play an advisory or coordinating role, particularly on the site security components. Additionally, research-and-development funding could be directed toward identifying and promoting inherently safer technologies. It is critical that the new Department help improve chemical safety and security, but it is equally critical that the development of the new Department not stand in the way of swiftly establishing a federal chemical security program.

As noted above, EPA's attempts to establish a chemical security program have met obstacles in recent weeks. Congress should mandate a chemical security program to ensure that the program moves forward without delay. The Chemical Security Act, S. 1602, introduced in the Senate, provides a good model. That legislation should be passed by Congress, either as an amendment to the Homeland Security bill, or separately on a similar or shorter timeline.

THE PROPOSED DEPARTMENT OF HOMELAND SECURITY

The bill, as currently proposed, does not establish any chemical security program and moreover could confuse or delay progress on chemical security. It could do so because of its lack of clarity on the Department's role in chemical plant security and because of its lack of clear vision for how to address chemical security. Additionally, the proposed bill could undermine existing chemical safety programs by creating a sweeping exemption from the Freedom of Information Act that could reduce government and industry accountability and limit public access to information that could prove critical to protecting communities.

Ambiguous Authority and Responsibility

The bill does not clearly define what the new Department's authorities regarding critical infrastructure generally, and chemical plant security specifically, would be. Section 201 provides the Under Secretary for Information Analysis and Infrastructure Protection with "primary responsibilities" including:

- "comprehensively assessing the vulnerabilities" (paragraph (2)) and "developing a comprehensive national plan for securing" (paragraph (4)) "the key resources and critical infrastructure" (paragraphs (2) and (4));
- "integrating relevant information—to identify protective priorities and support protective measures by the Department, by other executive agencies—and by other entities" (paragraph (3));
- "taking or seeking to effect necessary measures to protect the key resources and critical infrastructures . . . in coordination with other executive agencies and . . . other entities" (paragraph (5)).

Section 301 provides the Under Secretary for Chemical, Biological, Radiological, and Nuclear Countermeasures with "primary responsibilities" including:Q02

- "securing the people, infrastructures, property, resources, and systems" from acts of terrorism involving "chemical, biological, radiological, or nuclear weapons or other emerging threats" (paragraph (1));
- "conducting a national scientific research and development program" including efforts to "identify, devise, and implement scientific, technological, and other countermeasures" to the same threats (paragraph (2)); and
- "establishing priorities for, directing, funding, and conducting national research, development, and procurement of technology and systems . . . for detecting, preventing, protecting against, and responding to terrorist attacks that involve [chemical, biological, radiological, nuclear, and related] weapons and material" (paragraph (3) and (3)(B)).

These sections, examined together, create confusion and contradictions about where various authorities and responsibilities lie:

1. There are internal contradictions and confusion. What is the difference (or relationship) between the responsibility of the Under Secretary for Information Analysis and Infrastructure Protection for "taking or seeking to effect measures necessary to protect" critical infrastructure and the responsibility of the Under Secretary for Chemical, Biological, Radiological, and Nuclear Countermeasures for "securing" the people and infrastructures? Similarly, what is the difference (or relationship) between the former and latter Under Secretaries' responsibilities for identifying and establishing priorities?

2. It is unclear how these new Under Secretaries' "primary responsibilities" relate to those of other agencies whose functions are not transferred to the new Department. In some cases the new Department's responsibility seems to include "securing" people and infrastructure, but in other cases "taking or seeking to effect" measures "in coordination" with other executive agencies.

Clarifying Authority, Assuring Effective Security

Congress should clarify that the role for the Department of Homeland Security is one of coordinating security programs and advising agencies whose functions are not transferred to the new Department, but that new authority in these cases is not being transferred or otherwise given to the new Department. EPA has the expertise and experience to address chemical safety and security. Moreover, EPA has the authority to address chemical safety and security, granted by the 1990 Clean Air Act Amendments.[12]

The agency which has the substantive expertise on safety protections for the affected industry should retain the authority. This is particularly true for the chemical industry, because deliberate or criminal efforts to trigger chemical releases are only one of many reasons a chemical release could threaten health and safety in a community (as noted above, there are thousands of accidental and non-terrorist related spills and releases every year), and because safety improvements through hazard reduction must be the primary strategy for securing public health and safety from chemical releases related to terrorism.

Making Public Health and Safety a Priority

The bill, as proposed, focuses almost entirely on securing infrastructure and resources with little attention to protecting public health and safety. In fact, one of the few mentions of the public int the bill is to the Department having primary responsibility for "securing the people" (Sec. 301, paragraph (3)). "Securing" people hardly implies sound protection for public health and safety. While protecting public health and safety are presumably an end for which protecting critical infrastructure is a means, it is important that the new Department's mandate reside explicitly in this context. Without a clear mandate to protect public health and safety, the new Department could expend time and resources on measures that are in the short-term interest of protecting infrastructure and property but not in the long-term interest of protecting public safety, or could expend time and resources on security programs without a clear public benefit.

Without a clear definition of "critical infrastructures" or "key resources," there are few limits on or clear characteristics of what types of industrial facilities or other private properties represent resources whose protection is sufficiently in the public interest to justify expending considerable public funds. Protecting undefined assets could result in programs to protect private property and resources without any requirement that such protections merit the use of public resources. To help clarify what types of facilities or properties may merit protection, Congress should make

[12] Clean Air Act Section 112(r)'s general duty clause, definitions, and particularly 112(r)(7)(a).

protecting public health and safety a clear priority of the Department, its Secretary, and each of the relevant Under Secretaries.

Beyond Assessment: Reducing Hazards and Vulnerabilities

Congress should ensure that the new Department prioritizes reducing hazards and reducing vulnerabilities, not simply assessing them and not relying only on traditional security strategies of perimeter security, access control, and surveillance. As discussed above, public health and safety can best be "secured" against a deliberate chemical release from an industrial facility by reducing the hazard such that off-site impacts of a release are reduced or eliminated.

Site security—perimeter security, access control, and surveillance—are band-aid fixes that should only be relied on where there is no way to reduce the inherent danger. The first question that the new Department (and other agencies with whom it coordinates) should ask is: Can the infrastructure be made safer? Reducing or eliminating the possibility of a chemical release is the most effective and long-term protection for public health and safety and also reduced (or eliminates) the need for security measures, reducing costs to the government and the affected industry.

Inherent safety can be applied across industry sectors. For example, transporting nuclear waste throughout our country to move it to the Yucca Mountain site will dramatically increase the inherent dangers in our infrastructure. Since nuclear facilities will continue to generate highly hazardous nuclear waste on site, regularly moving waste across our highways and rails will expand, not reduce, the amount of highly hazardous "infrastructure" in our country. This increased hazard will require more costs for security than would leaving the waste on site, where a fixed facility would be easier to secure than a moving vehicle.

Fossil fuel energy offers another example. Securing the length of a vulnerable pipeline would likely be extraordinarily expensive and questionably effective. Removing pipelines from densely populated areas would not only be less costly, but also dramatically (and inherently) safer. Moving toward renewable energy, such as solar and wind, and particularly distributed generation, such as on-site solar or wind generators, would be inherently safer than expanding production from fossil-fuel based energy sources that rely on highly vulnerable systems for transporting energy.

Congress should make reducing hazards and vulnerabilities the national policy of the United States, as the most effective threat reduction strategy, and should direct the new Department to work toward this end with the agencies that have current authority, rather than providing new or redundant authority.

THE PUBLIC'S RIGHT TO KNOW AND PUBLIC ACCOUNTABILITY AS SAFETY TOOLS

The proposed bill shows a troubling request for secrecy by proposing a sweeping and unprecedented exemption from the Freedom of Information Act. Restricting the public's right to know about hazards in communities and industry or government actions to remedy them could hurt safety rather than help it. By restricting our right to know, even through a well-intentioned effort to protect safety, government is abandoning its duty to warn the public if a community is at risk. It is limiting the ability of the public and communities to understand, prepare for, and respond to threats to safety. And it is also removing one of the most effective—and in a democratic society, substantively important—incentives for public safety improvements: public information.

There are three primary ways that restricting public access to information can decrease public safety. First, secrecy without safety provisions—which is the strategy proposed by the bill—does nothing to address the threats except make them secret. Since September 11th, the administration has regularly employed public warnings. Allowing new secrecy could undermine efforts to provide due warning. Restricting public access also makes the community less safe because the ability of individuals and communities to participate in safety decisions ranging from chemical management and hazard reduction to site security and emergency response planning, is reduced and potentially eliminated. It is for exactly this reason that the Congress has for several decades used public disclosure and right-to-know laws—not secrecy provisions—to protect public safety, particularly from chemical hazards.

Public disclosure is a strategy with a long record of reducing risk. Public information empowers individuals and communities to work for measures that will reduce risk by working directly with a company locally or by advocating for policy changes to require risk reductions. As importantly, right-to-know programs provide a public incentive for relevant parties to be accountable to public values. The Toxics Release Inventory, established under the 1986 Emergency Planning and Community Right-to-Know Act, has been credited with contributing to a nearly-50% reduction in toxic chemical releases. More robust right-to-know programs have seen proportionally greater impacts. In Massachusetts, where companies report not just chemical releases but also chemical use, in products or in the workplace, chemical use is down approximately 40% and chemical releases are down nearly 90%. Restricting public access to information restricts opportunities for these kinds of protections of public safety and health and removes accountability for government and corporate actors.

Section 204 of the proposed bill would contradict these lessons by creating an unprecedented and unwarranted loophole in the Freedom of Information Act. This section runs counter to the fundamental principle of FOIA: a presumption that the people of the United States have wide-ranging access to their government and that a government of, by, and for the people requires an open government. In the rare cases where a compelling public interest requires secrecy, FOIA allows carefully limited exceptions for specific documents.

The proposed bill runs almost exactly counter to this approach. It does not even define what documents would be exempt from FOIA that could not be covered by current FOIA exemptions (which already exist for national security, trade secrets, and certain voluntarily provided information), much less explain what compelling public interest necessitates this exemption. The requirements for what information could be made exempt are so vague that virtually any information on American industry, including information required to be public under other laws, could potentially be submitted to the new Department, certified as "relating" to critical infrastructure vulnerabilities, and permanently removed from public access. This would be a colossal step backwards for open government, public accountability, and the public's right to know about safety threats.

When Congress addressed the security of water supplies, it was first determined that for vulnerability assessments being submitted to the government, current FOIA law may require public disclosure and that such disclosure could be a security threat. Congress then exempted only these documents from disclosure under FOIA. This should be the model for considering any exceptions from FOIA.

Because this bill creates no new vulnerability assessments and requires no new information to be submitted to the government, Congress should not consider creating any new FOIA exemptions. Section 204 should be struck from the bill.

ENVIRONMENTAL DEFENSE, GREENPEACE, NATIONAL ENVIRONMENTAL TRUST, NATURAL RESOURCES DEFENSE COUNCIL, OMB WATCH, U.S. PUBLIC INTEREST RESEARCH GROUP

July 8, 2002

DEAR CONGRESSMAN,

While almost ten months have passed since September 11, a significant vulnerability has yet to be addressed. Across the U.S., thousands of industrial facilities use and store hazardous chemicals in quantities that put large numbers of Americans at risk of serious injury or death in the event of a chemical release.

Unfortunately, the administration's Homeland Security Act fails to address these critical safety issues. Moreover, EPA efforts to address the problem have encountered resistance within the administration as well as from some Members of Congress. Under current law, EPA has the expertise and legal authority[1] to address threats posed by major chemical releases at industrial facilities. EPA should act immediately and aggressively to require facilities that store toxic chemicals to assess and reduce their vulnerabilities by eliminating targets (for example, by converting to safer chemicals or processes) and enhancing security. Congress must make it clear that immediate action is expected from EPA to reduce this threat and should amend the Homeland Security Bill to require oversight to ensure that EPA implements a comprehensive hazard assessment and reduction program.

In its current form, the Homeland Security Act not only fails to address chemical safety, but instead proposes to create new, far-reaching secrecy provisions. These restrictions have the potential to keep the American public in the dark about potential risks from chemical facilities and hamper efforts to make communities safer. Congressional precedent has been to establish only very limited exemptions to the Freedom of Information Act (FOIA) for specific documents (for example, the recent exemption in the Bioterrorism Response Act of 2001 for water system vulnerability assessments). Section 204 of the administration's HoomHmeland Security Act contains

[1] Section 112(r) of the Clean Air Act (CAA) authorizes EPA to issue regulations "to prevent accidental releases of regulated substances," defining such a release as "an unanticipated emission of a regulated substance or other extremely hazardous substance into the ambient air from a stationary source." Likewise, the CAA imposes a "general duty" of precaution on sources, directing them "to design and maintain a safe facility taking such steps as are necessary to prevent releases..."

an overly broad exemption from FOIA, not tied to any specific document or mandate. This section should be dropped from the bill.

The lack of any action to address risks at chemical plants in communities around the nation is an irresponsible omission. EPA's proposed actions are long overdue—the agency should use its existing expertise and authority to act immediately. Efforts to further delay EPA action is unacceptable and contradicts the Administration's promise to quickly address priority threats with existing resources.

We urge you to call on EPA to act immediately to require chemical facilities to assess and reduce their vulnerabilities and to eliminate the overly broad secrecy provisions in Section 204 of the Homeland Security Act of 2002.

Sincerely,

CAROL ANDRESS
Environmental Defense
RICK HIND
Greenpeace
ANDY IGREJAS
National Environmental Trust
ALYS CAMPAIGNE
Natural Resources Defense Council
SEAN MOULTON
OMB Watch
JEREMIAH BAUMANN
U.S. Public Interest Research Group

Mr. WHITFIELD. Thank you, Mr. Baumann. Mr. Sobel, you're recognized for 5 minutes.

TESTIMONY OF DAVID L. SOBEL

Mr. SOBEL. Thank you, Mr. Chairman, for providing me with the opportunity to appear before this subcommittee to discuss the administration's proposed legislation to create a new Department of Homeland Security. I will discuss proposals that would ironically limit public access to crucial data in the name of information sharing.

My comments will focus on proposals to create a new Freedom of Information Act exemption for information obtained by the Department of Homeland Security concerning infrastructure protection and counterterrorism efforts, but I would also like to share with the subcommittee some general observations that I have made as the debate over critical infrastructure information has unfolded over the last few years. I believe it is essential to understand the broader context in which the FOIA exemption proposal arises.

First, there appears to be a consensus that the government is not obtaining enough information from the private sector on vulnerabilities that could adversely affect the infrastructure. It is equally clear that citizens, the ones who will suffer the direct consequences of infrastructure failures, are also receiving inadequate information about these vulnerabilities.

Second, there has not yet been a clear vision articulated defining the government's proper role in securing the infrastructure. Despite the emphasis on finding ways to facilitate the government's receipt of information, it remains unclear just what the government will do with the information it receives. The administration's homeland security proposal does not clearly define the new department's role in protecting the infrastructure.

Third, rather than seeking ways to hide information, Congress should consider approaches that would make as much information as possible available to the public, consistent with the legitimate interests of the private sector. This is particularly critical in the context of the new department, which will assume an unprecedented range of responsibilities involving public safety.

A broad coalition of organizations has serious concerns about various proposals, such as section 204 of the administration's bill to create a broad new FOIA exemption for information relating to security flaws and other vulnerabilities in the infrastructure.

Section 204 would cast a shroud of secrecy over one of the new department's critical functions, removing any semblance of meaningful public accountability. If section 204 or a similar secrecy provision such as Representative Davis' bill is enacted, the public will be unable to hold the department accountable should it fail to make effective use of the information it obtains. What did DHS know and when did it know it is a question that will go unanswered.

While section 204 is, in my view, exceedingly broad, I would urge the subcommittee to approach more circumspect exemption proposals with skepticism as well. Any new exemption, unless extremely limited, is likely to remove important information from public view and restrict public oversight of critical government operations. Perhaps most importantly, any new exemption designed to protect the voluntarily submitted private sector information is simply not needed. Established case law makes it clear that existing exemptions contained in the FOIA provide adequate protection against harmful disclosures of the type of information we are discussing.

Exemption 4, which covers confidential private sector information, provides extensive protection. As my written statement explains in detail, Exemption 4 extends to virtually all of the infrastructure material that properly could be withheld from disclosure.

In light of the substantial protections provided by FOIA Exemption 4 and the case law interpreting it, I believe that any claimed private sector reluctance to share important data with the government grows out of at best a misperception of current law. The existing protections for confidential private sector information have been repeatedly—have been cited repeatedly over the past 2 years by those of us who believe that a new exemption is unwarranted.

Exemption proponents respond that the FOIA creates a perceived barrier to information sharing. They have not cited a single instance in which a Federal agency has disclosed voluntarily submitted data against the express wishes of an industry submittal.

It should be noted that we are discussing the desire of private companies to keep secret potentially embarrassing information at a time when the disclosure practices of many in the business world are being scrutinized. If a company is willing to fudge its financial numbers to maintain its stock price, it would be similarly inclined to hide behind a critical infrastructure FOIA exemption in order to conceal gross negligence in its maintenance and operation of a chemical plant or a transportation system.

In summary, overly broad new exemptions could adversely impact the public's right to oversee important and far-reaching governmental functions and remove incentives for remedial private sector action.

I urge the Congress to preserve the public's fundamental right to know as it considers the establishment of a Department of Homeland Security, and I thank the subcommittee for considering my views.

[The prepared statement of David L. Sobel follows:]

PREPARED STATEMENT OF DAVID L. SOBEL, GENERAL COUNSEL, ELECTRONIC PRIVACY INFORMATION CENTER

Mr. Chairman and Members of the Subcommittee: Thank you for providing me with the opportunity to appear before the Subcommittee to discuss the Administration's far-reaching proposed legislation to create a new Department of Homeland Security. I will discuss the role that the exchange of information plays in protecting our nation's infrastructure and preventing terrorism, and focus on proposals that would, ironically, limit public access to crucial data in the name of "information sharing." The Electronic Privacy Information Center (EPIC) has a longstanding interest in computer and network security policy and its potential impact on civil liberties, emphasizing full and informed public debate on matters that we all recognize are of critical importance in today's inter-connected world.

My comments will focus primarily on proposals to create a new Freedom of Information Act (FOIA) exemption for information obtained by the Department of Homeland Security concerning infrastructure protection and counter-terrorism efforts. But I would also like to share with the Subcommittee some general observations that I have made as the debate over "critical infrastructure information" has unfolded over the past few years. I believe it is essential to understand the broader context in which the FOIA exemption proposal arises.

• There appears to be a consensus that the government is not obtaining enough information from the private sector on security risks and vulnerabilities that could adversely affect the critical infrastructure. I hasten to add that citizens—the ones who will suffer the direct consequences of infrastructure failures—are also receiving inadequate information about these vulnerabilities.

• There has not yet been a clear vision articulated defining the government's proper role in securing the infrastructure. While there has been a great deal of emphasis on finding ways to facilitate the government's receipt of information, it remains unclear just what the government will do with the information it receives. In fact, many in the private sector advocate an approach that would render the government virtually powerless to correct even the most egregious security flaws. Despite its ambitious reach, the Administration's homeland security proposal does not clearly define the new Department's role in protecting the infrastructure.

• The private sector's lack of progress on security issues appears to be due to a lack of effective incentives to correct existing problems. Congress should consider appropriate incentives to spur action, but secrecy and immunity, which form the basis for many of the proposals put forward to date, remove two of the most powerful incentives—openness and liability. Indeed, many security experts believe that disclosure and potential liability are essential components of any effort to encourage remedial action. [1]

• Rather than seeking ways to hide information, Congress should consider approaches that would make as much information as possible available to the public, consistent with the legitimate interests of the private sector. This is particularly critical in the context of the new Department, which will assume an unprecedented range of responsibilities involving public safety.

As indicated, I would like to focus my comments on proposals to limit public access to information concerning critical infrastructure protection. EPIC is a strong advocate of open government, and has made frequent use of the FOIA to obtain information from the government about a wide range of policy issues, including (in addition to computer security) consumer privacy, electronic surveillance, encryption controls and Internet content regulation. We firmly believe that public disclosure of this information improves government oversight and accountability. It also helps ensure that the public is fully informed about the activities of government.

I have personally been involved with FOIA issues for more than twenty years and have handled information requests on behalf of a wide range of requesters. In 1982,

[1] *See, e.g.*, "Counterpane CTO Says Insurance, Liability to Drive Security," InfoWorld (February 20, 2002), <http://www.inforld.com/articles/hn/xml/02/02/20/020220hncounterpane.xml> (According to security expert Bruce Schneier, "[t]he challenges and problems of computer and network security won't be adequately addressed until companies can be held liable for their software and the use of their computer systems").

I assisted in the preparation of a publication titled *Former Secrets,* which documented 500 instances in which information released under the FOIA served the public interest. I am convinced that an updated version of that publication would today yield thousands of examples of the benefits we all derive from the public access law that has served as a model for other nations around the world.

EPIC and other members of the FOIA requester community have, for the past several years, voiced concerns about various proposals to create a broad new FOIA exemption, such as those contained in the Cyber Security Information Act (H.R. 2435) and the Critical Infrastructure Information Security Act (S. 1456), for information relating to security flaws and other vulnerabilities in our critical infrastructures. Section 204 of the Administration's proposed legislation, as I will discuss in more detail, contains an exemption provision that appears to be even more far-reaching than those previously proposed. We collectively believe these exemption proposals are fundamentally inconsistent with the basic premise of the FOIA, which, as the Supreme Court has recognized, is "to ensure an informed citizenry, vital to the functioning of a democratic society, needed to check against corruption and to hold the governors accountable to the governed."[2] To accomplish that end, "[d]isclosure, not secrecy, is the dominant objective of the Act."[3]

It is clear that, as we simultaneously move further into the electronic age and confront the risks of terrorism, the federal government increasingly will focus on the protection of critical infrastructures. It is equally apparent that government policy in this emerging field will become a matter of increased public interest and debate. The proposal to create a vast Department of Homeland Security raises that debate to a new level of urgency. While reasonable observers can disagree over the merits of specific initiatives, I believe we all agree that infrastructure protection and counter-terrorism activities raise significant public policy issues that deserve full and informed public discussion.

The issue is perhaps best illustrated by examining the latest iteration of the "critical infrastructure information" exemption approach—Section 204 of the Administration's proposed Homeland Security Act. In what is surely among the most far-reaching one-sentence statutory provisions ever drafted, Section 204 provides:

> Information provided voluntarily by non-Federal entities or individuals that relates to infrastructure vulnerabilities or other vulnerabilities to terrorism and is or has been in the possession of the Department [of Homeland Security] shall not be subject to [the FOIA].

It should be noted that this provision would conceal from public scrutiny a major component of the Department's statutory mission—the information analysis and infrastructure protection functions set forth in Title II of the Administration's proposed legislation. Indeed, "information analysis and infrastructure protection" is the first of the Department's "primary responsibilities" enumerated in Section 101(b)(2).

Section 204 would cast a shroud of secrecy over one of the Department's critical functions, removing any semblance of meaningful public accountability. The tragic events of September 11th illustrate the importance of such accountability mechanisms; the Congress, the media and the public are currently engaged in an examination of possible failures of intelligence or analysis that may have contributed to the tragedy. Indeed, the legislation we are discussing today is a direct outgrowth of that review process and public debate. If Section 204, or a similar secrecy provision, is enacted, the news media and the public will be unable to hold the new Department accountable should it fail to make effective use of information it obtains. "What did DHS know and when did it know it?" is a question that will go unanswered. Such insulation from accountability is clearly the wrong way to go as we seek to create an effective new entity.

While Section 204 is, in my view, exceedingly broad, I would urge the Subcommittee to approach more circumscribed exemption proposals with skepticism as well. Any new exemption, unless extremely limited, is likely to remove important information from public view and restrict public oversight of critical government operations. And, perhaps most importantly, any new exemption designed to protect voluntarily-submitted private sector information is simply not needed.

It is clear that government activities to protect the infrastructure will be conducted in cooperation with the private sector and, accordingly, will involve extensive sharing of information between the private sector and government. To facilitate the exchange of information, some have advocated enactment of an automatic, wholesale exemption from the FOIA for any information concerning potential vulnerabilities to the infrastructure that may be provided by a private party to a federal agency. Given the breadth of the proposed definitions of the categories of information to be

---

[2] *NLRB v. Robbins Tire & Rubber Co.,* 437 U.S. 214, 242 (1978).

[3] *Department of the Air Force v. Rose,* 425 U.S. 352 (1976).

exempted, I believe such an exemption would likely hide from the public essential information about critically important—and potentially controversial—government activities undertaken in partnership with the private sector. It could also adversely impact the public's right to know about unsafe practices engaged in by the private operators of nuclear power plants, water systems, chemical plants, oil refineries, and other facilities that can pose risks to public health and safety. In short, critical infrastructure protection is an issue of concern not just for the government and industry, but also for the public—particularly the local communities in which these facilities are located.

If the history of the FOIA is any guide, a new exemption would likely result in years of litigation as the courts are called upon to interpret its scope. The potential for protracted litigation brings me to what I believe is the most critical point for the Subcommittee to consider, which is the need for a new "critical infrastructure" FOIA exemption. FOIA caselaw developed over the past quarter-century makes it clear that existing exemptions contained in the Act provide adequate protection against harmful disclosures of the type of information we are discussing. For example, information concerning the software vulnerabilities of classified computer systems used by the government and by defense contractors is already exempt under FOIA Exemption 1. A broad range of information collected for law enforcement purposes may be (and routinely is) withheld under Exemption 7. Most significantly, Exemption 4, which protects against disclosures of trade secrets and confidential information, also provides extensive protection from harmful disclosures. Because I believe that Exemption 4 extends to virtually all of the "critical infrastructure" material that properly could be withheld from disclosure, I would like to discuss briefly the caselaw that has developed in that area.

For information to come within the scope of Exemption 4, it must be shown that the information is (A) a trade secret, or (B) information which is (1) commercial or financial, (2) obtained from a person, and (3) privileged or confidential.[4] The latter category of information (commercial information that is privileged or confidential) is directly relevant to the issue before the Subcommittee. Commercial or financial information is deemed to be confidential "if disclosure of the information is likely to have either of the following effects: (1) to impair the government's ability to obtain the necessary information in the future; or (2) to cause substantial harm to the competitive position of the person from whom the information was obtained."[5] The new FOIA exemption that has been proposed seeks to ensure that the government is able to obtain critical infrastructure information from the private sector on a voluntary basis, a concern which comes within the purview of Exemption 4's "impairment" prong. The courts have liberally construed "impairment," finding that where information is voluntarily submitted to a government agency, it is exempt from disclosure if the submitter can show that it does not customarily release the information to the public.[6] In essence, the courts defer to the wishes of the private sector submitter and protect the confidentiality of information that the submitter does not itself make public.

In addition to the protections for private sector submitters contained in FOIA Exemption 4 and the relevant caselaw, agency regulations seek to ensure that protected data is not improperly disclosed. Under the provisions of Executive Order 12600 (*Predisclosure Notification Procedures for Confidential Commercial Information*) issued by President Reagan in 1987, each federal agency is required to establish procedures to notify submitters of records "that arguably contain material exempt from release under Exemption 4" when the material is requested under the FOIA and the agency determines that disclosure might be required. The submitter is then provided an opportunity to submit objections to the proposed release. The protections available to private sector submitters do not end there; if the agency determines to release data over the objections of the submitter, the courts will entertain a "reverse FOIA" suit to consider the confidentiality rights of the submitter.[7]

In light of the substantial protections against harmful disclosure provided by FOIA Exemption 4 and the caselaw interpreting it, I believe that any claimed private sector reticence to share important data with the government grows out of, at best, a misperception of current law. The existing protections for confidential private sector information have been cited repeatedly over the past two years by those of us who believe that a new FOIA exemption is unwarranted. In response, exemption proponents have not come forward with any response other than the claim that the

---

[4] *Getman v. NLRB,* 450 F.2d 670, 673 (D.C. Cir. 1971), stay denied, 404 U.S. 1204 (1971).

[5] *National Parks and Conservation Association v. Morton,* 498 F.2d 765, 770 (D.C. Cir. 1974).

[6] *Critical Mass Energy Project v. Nuclear Regulatory Commission,* 975 F.2d 871 (D.C. Cir. 1992) (en banc), cert. denied, 113 S.Ct. 1579 (1993).

[7] *See GTE Sylvania, Inc. v. Consumers Union,* 445 U.S. 375 (1980).

FOIA creates a "perceived" barrier to information sharing.[8] They have not cited a single instance in which a federal agency has disclosed voluntarily submitted data against the express wishes of an industry submitter. Nor have they provided a single hypothetical example of voluntarily submitted "critical infrastructure" information that would not fall within the broad protection of Exemption 4.

Frankly, many in the FOIA requester community believe that Exemption 4, as judicially construed, shields far too much important data from public disclosure. As such, it is troubling to hear some in the Administration and the private sector argue for an even greater degree of secrecy for information concerning vulnerabilities in the critical infrastructure. As I have noted, shrouding this information in absolute secrecy will remove a powerful incentive for remedial action and might actually exacerbate security problems. A blanket exemption for information revealing the existence of potentially dangerous vulnerabilities will protect the negligent as well as the diligent. It is difficult to see how such an approach advances our common goal of ensuring a robust and secure infrastructure.

It should not go unnoticed that we are discussing the desire of private companies to keep secret potentially embarrassing information at a time when the disclosure practices of many in the business world are being scrutinized. If a company is willing to fudge its financial numbers to maintain its stock price, what assurance would we have that it was not hiding behind a "critical infrastructure" FOIA exemption in order to conceal gross negligence in its maintenance and operation of a chemical plant or a transportation system?

In summary, the Freedom of Information Act has worked extremely well over the last 36 years, ensuring public access to important information while protecting against specific harms that could result from certain disclosures. After monitoring the development of critical infrastructure protection policy for the last several years, I have heard no scenario put forth that would result in the detrimental disclosure of information under the current provisions of the FOIA. Overly broad new exemptions could, however, adversely impact the public's right to oversee important and far-reaching governmental functions and remove incentives for remedial private sector action. I urge the Subcommittee and the Congress to preserve the public's fundamental right to know as it considers the establishment of a Department of Homeland Security.

Mr. WHITFIELD. Thank you, Mr. Sobel. The Chair at this time will recognize the gentleman from North Carolina for 5 minutes.

Mr. BURR. I thank the Chair, and I will be briefer than 5 minutes. Let me ask you, Mr. Sobel, since you just finished, I think we've heard testimony today calling for FOIA exemptions for sensitive information and others have indicated the concerns that currently there's too much information in the public domain, and that that would be useful from the standpoint of targeting and prioritizing—targeting the manufacturing facilities for terrorist attacks. Is there any information that is currently in the public domain via FOIA disclosures or otherwise about private sector critical infrastructure assets that you believe should not be in the public domain because it provides too much information that could be used by terrorists? And if so, what would it be?

Mr. SOBEL. Well, Congressman, the answer is very simple. I do not believe that as a result of the FOIA any material that could create potential harms or problems has been released, and as I indicated, the proponents of the FOIA exemption have not in the last 2 years that this issue has been debated come forward with a single example of such a disclosure. Their entire case is hypothetical and, as I said, as far as I'm concerned based on a misunderstanding and a misperception——

Mr. BURR. So there's nothing that you would remove today?

[8] *See, e.g.*, Letter from Daniel P. Burnham, Chair, National Security Telecommunications Advisory Committee to the President, June 28, 2001 ("*Real or perceived*, barriers to [information] sharing must be removed. Among those barriers are the Freedom of Information Act and potential legal liabilities") (emphasis added).

Mr. SOBEL. That is correct, not information released as a result of the FOIA.

Mr. BURR. Mr. Smith, what steps can we take to improve information sharing between the Federal Government and the private sector?

Mr. SMITH. Before I answer that question, let me just respond to a different point of view than an earlier answer. One of the reasons why that material has not been presented is because we haven't—we've refused to provide it. We have been asked, for example, to provide a list of the 100 most critical buildings in our network. Now, that would be tantamount to providing a road map to terrorists to say if you really want to hurt us, if you really want to take the telecommunications infrastructure down, here are a hundred buildings to target. So we refuse to do that.

Now, the problem is, that limits our ability to work with agencies that we might cooperate with in developing preplanned response in the event of that kind of instance such as what we saw in Manhattan, but that is the reason why you don't see that information released. We have chosen not to do that.

But to respond to your particular question, we think that with respect to H.R. 4598, recently passed, the information from the government to the private sector must really be actionable, that it supports the sharing of information, but sharing classified information or very limited sharing that we can act on is only part of the way. So we think there is work to do to improve that. And then certainly, as we've just said, on the other direction we think that FOIA protections really are critical for us to be able to share more information with government agencies in order to respond to those kind of threats.

Mr. BURR. Well, it clearly is an issue that I don't think we will come to consensus based upon those who have sat and testified today. But this body, along with this administration, will be asked to move some very significant legislation that in the end won't be perfect, but hopefully will give us a greater degree of comfort in knowing that tools are in place to allow whoever the Secretary is of Homeland security the ability to carry out the job, to make assurances, but more importantly, to make sure that the security of this country is in fact intact in more ways than one.

I'd like to thank all of you for your willingness to come, and I'd yield back.

Mr. WHITFIELD. The gentleman from Florida is recognized for 5 minutes.

Mr. DEUTSCH. Thank you, Mr. Chairman. You know, Mr. Smith, I guess your comments were directly on point, and I guess this is really the differences on this panel in terms of the vulnerability assessments.

Mr. Sullivan, in terms of the area of water in the West 1 month ago the President signed the bioterrorism bill which passed the House 425 to 1 and the Senate 98 to 0, and it contained specifically this whole context of water issues in terms of safety and security and the vulnerability assessments. You testified that you believe the vulnerability assessments provided to EPA, you're concerned that these documents could wind up inadvertently in the hands of malicious people. The bioterrorism law establishes criminal penalties for releasing these documents. Do you have a basis for your concern or just more information out there, not really understanding the controls that they would have on that information?

Mr. SULLIVAN. Well, first of all, the water industry has no objections at all to preparing vulnerability assessments. You've got to understand that these are absolute blueprints of how to take down a water system. It shows you the way. Now, these vulnerability assessments currently are going to be supplied, over 8,000 of them, in hard copy, but they're going to be smaller books and smaller systems in volumes of pages. And first, we are concerned that EPA, which is a regulatory agency, where are they going to secure these? It's not truly the ambition to be doing security of critical infrastructure. So as we all learned, and Congress has done it many times, as new information comes across, new opportunities come across, there may be a better opportunity here to have DHS provide this repository, if needed.

Our preference, of course, is that these vulnerability assessments be kept in each utility and that access to them through an audit process or DHS audit, local FBI, would be dealt with in each utility itself so that we wouldn't be sending hard copies to some spot that we don't know where they go.

Mr. DEUTSCH. If I can just follow up on this, I understand what you're saying, but my understanding of the law as passed is you have to do it, and it's sort of our burden to try to keep it secure. But it's your determination that you don't like the way EPA is securing these vulnerability assessments?

Mr. SULLIVAN. We don't know how they're securing them yet. What we're suggesting is that there's another location—there's an improvement available to Congress now to house them under DHS. It's a suggestion by the water industry.

Mr. DEUTSCH. There's nothing specific about EPA, but just the way—your experience with EPA, I mean, you just don't feel they have the ability to secure these documents?

Mr. SULLIVAN. They're a regulatory agency. We're looking at DHS as the new secure critical infrastructure, that they can house it. So we think it is a better opportunity.

Mr. DEUTSCH. Mr. Sobel, Mr. Watson seems to be raising a salient point about sharing highly sensitive information regarding cyber security owned and operated by the private sector. The overwhelming bulk of critical cyber systems in the United States today are usually not government owned. He says the private companies are not sharing this information with the public because of concerns that it's not protected from public disclosure under FOIA.

Is he correct that information would not be protected under current law?

Mr. SOBEL. Congressman, I think clearly in my view that is not a correct assessment of current law. Basically what the courts have said is that if a disclosure of voluntarily submitted information against the wishes of the submitter would impede the government's ability in the future to get that kind of cooperation, it's not going to be disclosed, and that is precisely what we're talking about here.

If there would be any disincentive that resulted from disclosure of voluntarily provided information, it is not disclosable. So virtually by definition, information is not disclosed against the wishes of the submitter of the voluntary information. It's really as simple as that, and as I've pointed out repeatedly, the proponents of this exemption have not pointed to one disclosure that has been against the wishes of an industry submitter.

Mr. DEUTSCH. Thank you.

Mr. WHITFIELD. Mr. Copeland, in your testimony, you call for a greater emphasis on cyber security in the legislation, and you even urge the creation of a Bureau of Cyber Security. I was wondering if you would just elaborate on why you believe that's important and why do you believe it should be separated from the physical security issues and what the primary mission of the new bureau should be.

Mr. COPELAND. Thank you, Mr. Chairman. I would like to clarify to the—I wasn't suggesting that it be separated from the physical security issues and in fact pointed out that it should be attending to physical attacks on cyber facilities. So, for example, physical attacks on network nodes that might cause a telecommunications outage are a serious concern. In fact, today by far the most common form of outage that networks experience, the back operators do more damage to our network than terrorists do today, and that is a common physical experience and fires and hurricanes and earthquakes caused the kind of physical problems they are accustomed to dealing with.

One of the observations that our association has made and many of its members have made is that since September 11 there's been an extremely enhanced level of awareness of physical security issues and attention to them, and unfortunately, we have been building a significantly high level of interest and awareness in some of the developing cyber security issues, and there was an attendant fall-off after September 11, and the focus shifted to physical security.

Our concern is that the cyber security issues frequently are more esoteric, more complex, more difficult to understand because of the complexity, and if they tend to get mixed in with other activities that are focusing on the physical aspects, they tend to get lost in the shuffle.

We think it's important that they get the degree of visibilities that are needed, that the resources necessary are applied and that a single executive be accountable for those particular aspects of the security and have to report to the Congress with that accountability.

One needs only look at some of the experiences in the Federal Government where they should be leading by example to see that cyber security unfortunately has not been getting traditionally the attention it should in the departments and agencies. The score that has been given to information security by Congressman Horn in his oversight is one example of that. The reports that were submitted to the Congress under GISRA, the Government Information Security Requirements Act, indicate that most of the departments are at best achieving only barely passing grades in that area. So anything that can help to focus attention on that area is extremely important.

Mr. WHITFIELD. Thank you, Mr. Copeland.

Mr. Watson, some have argued that this public-private partnership created by Presidential Directive 63 to build a strong business model for ensuring the security and reliability of our Nation's critical infrastructures is not an effective model, and primarily because it does not include additional regulatory directives to compel the private sector owners to take additional steps. Do you agree with that argument, or do you feel like the directive has been successful?

Mr. WATSON. Mr. Chairman, I don't agree with the argument, because the PCIS is represented by presidents and chief operating officers, chief information security officers from the companies and trade associations that represent the critical infrastructures. There's enthusiastic participation. It's more enthusiastic because the government has approached industry eschewing the new regulation. So it creates an atmosphere of trust.

One of the initiatives we took on last year was to look at research and development requirements with the idea of defining what the market could provide, identifying the gap between market security provision and national security and then going back to the government and saying we can come up to 80, 90 percent, whatever the market might produce, and then let the government provide incentives for direct funding for research to fill the remaining gap. We think that kind of participation and partnership is new, and so we have a lot of work to do, but we think the model is very sound and working well.

Mr. WHITFIELD. Thank you, Mr. Watson.

Well, I'm going to thank this panel very much for your patience today and for your testimony. The committee really appreciates it. We appreciate the time and effort you put into it, and at this time I'll dismiss this panel. Thank you very much.

The Chair at this time recognizes himself for a unanimous consent request and to offer a motion. Because of the sensitive nature of this hearing, particularly its implications for national security and after consultation with the minority, I will soon offer a motion that the subcommittee go into executive session. I yield to Mr. Deutsch for any comments on this procedure.

Mr. DEUTSCH. I would agree with the Chair.

Mr. WHITFIELD. Thank you. The Chair moves that pursuant to clause 2(g) of rule XI of the Rules of the House, the remainder of this hearing will be concluded in executive session to protect the information that might endanger national security.

Is there discussion on the motion? If there is no discussion, pursuant to the rule, a recorded vote is ordered. Those opposed say nay. The ayes appear to have it. The ayes have it, and the motion is agreed to. So we'll go into executive session at this time.

[Whereupon, at 2 p.m., the subcommittee proceeded in Executive Session.]

[Additional material submitted for the record follows:]

PREPARED STATEMENT OF JAYSON P. AHERN, ASSISTANT COMMISSIONER, FIELD OPERATIONS, U.S. CUSTOMS SERVICE

Chairman Tauzin, Chairman Greenwood, members of the Subcommittee, thank you for this opportunity to testify.

Mr. Chairman, I know that the Subcommittee has a great deal of interest in the Administration's proposal for a new Department of Homeland Security and the inclusion of the U.S. Customs Service in that Department. I will tell you what Commissioner Bonner has told the employees of the Customs Service: "I fully support the President's proposal and strongly believe that the new Department of Homeland Security will play a key role in safeguarding the American people."

For over 200 years, the U.S. Customs Service has defended our country's borders and facilitated international trade and travel. Since September 11th, at the direction of the President, the top priority of Customs has been responding to the continuing terrorist threat at our land borders, seaports, and airports. I would like to describe for you some of our most significant efforts and initiatives on that front.

Since September 11th, Customs has been at a Level One alert across the country—at all ports of entry. Level 1 requires sustained, intensive anti-terrorist questioning, and includes increased inspections of travelers and goods.

To help ensure that Customs forms a coordinated, integrated counter-terrorism strategy for border security, Customs established a new Office of Anti-Terrorism within the agency to coordinate Customs' role within our national security architecture.

Customs agents are also working diligently under Project Shield America to monitor exports of strategic weapons and materials from the U.S. They are seeking to prevent international terrorist groups from obtaining sensitive U.S. technology, weapons and equipment that could be used in a terrorist attack on our nation.

To help Customs officers in the field, the Commissioner also established the Office of Border Security. The mission of that office is to develop more sophisticated anti-terrorism targeting techniques for passengers and cargo in the seaport, airport, and land border environments.

Customs has also created the Customs-Trade Partnership Against Terrorism,"C-TPAT", which is a partnership with some of the largest U.S. importers to improve security along the entire supply chain, from the loading docks of foreign vendors to our land borders and seaports. We were very pleased to have Governor Ridge, Secretary O'Neill and Commissioner Bonner announce C-TPAT at the Ambassador Bridge in Detroit, Michigan on April 16, 2002. To date, there are over 250 signatories to this initiative.

To complement C-TPAT, Customs developed the Container Security Initiative which places Customs enforcement personnel in major foreign shipping ports. The Customs officers will establish international security criteria for identifying high-risk cargo containers that potentially pose a risk of containing terrorists or terrorist weapons. In addition to having U.S. Customs officers in Halifax, Montreal and Vancouver, Customs has recently signed an agreement that will place our officers in Rotterdam, Antwerp and Le Havre. We anticipate other ports will sign up in the near future.

Customs continues to deploy technology necessary to rapidly and comprehensively inspect arriving and departing people, cargo and in all port environments and across all modes of transportation. To date Customs has deployed 87 large-scale non-intrusive inspection systems along with other technologies that will assist inspectors in conducting high-confidence, non-intrusive inspections quickly and efficiently.

In 1998, Customs began deploying technology to detect radiological sources. Since that time, we have deployed over 4,000 personal radiation detectors and over 200 x-ray van mounted radiation detection units. This year we ordered over 4,000 additional personal radiation detectors and have funding for 172 portal radiation detectors and 128 isotope identifiers for our ports of entry.

Customs is working closely with the Department of Energy to investigate systems and technology to detect radiological and nuclear materials to enhance our detection capabilities. Specifically, we are working with the Pacific Northwest National Laboratory, the Lawrence Livermore National Laboratory and the Special Technology Laboratory. In addition, Customs is engaged with the Department of Transportation in the Container Working Group, with the U.S. Coast Guard for targeting sea containers and with the Federal Aviation Administration for detection technology for cargo and baggage.

We are currently conducting operational field tests of portal radiation detection systems to determine system capabilities and to develop procedures and response protocols. A challenge will be our ability to differentiate between the numerous consumer goods such as cement, porcelain, potash, and bananas that may give off radiation, as well as medical isotopes given to humans for detection and treatment of disease and the attempt to smuggle and/or conceal a second radioactive source.

Concerning other possible weapons of mass destruction, Customs, in partnership with Johns Hopkins University, is working to establish a chemical/biological project to investigate systems and technologies to augment and enhance our existing chemical/biological detection capabilities.

The effective use of technology depends on good targeting, for which we require advance information. The Automated Manifest System, in conjunction with our advanced targeting systems allow Customs to sort through the cargo manifests provided by shippers and carriers, and pick out those that appear unusual, suspect, or high-risk.

Legislation currently under consideration mandates the advance electronic transmission of cargo manifest information. This will significantly increase the amount and timeliness of information input into the Customs database, thus enhancing our ability to identify anomalies. We appreciate the support the House and Senate have shown for making the advance filing of electronic cargo manifest information mandatory.

Thank you again, Mr. Chairman and the members of the Subcommittee, for this opportunity to testify. We look forward to working with your Subcommittee on this important legislation. I would be happy to answer any questions you may have.

PREPARED STATEMENT OF LINTON F. BROOKS NATIONAL NUCLEAR SECURITY ADMINISTRATION, U. S. DEPARTMENT OF ENERGY

INTRODUCTION

Thank you, Mr. Chairman for having me here today. This is an important topic: the establishment of a new Government Agency that will have sweeping responsibilities. The new Department of Homeland Security will enable us to more effectively respond to today's threats, through a streamlined and dynamic institution that will greatly enhance our ability to respond quickly, decisively, and where necessary, before threats against our homeland materialize. We are on the verge of making history. It's critical that we get it right.

The Department of Energy and the National Nuclear Security Administration are fully committed to the homeland security mission, and the successful establishment of the Department of Homeland Security. We recognize that this will require restructuring and relocation of critical assets now under the stewardship of the NNSA. We are prepared to support these shifts in responsibilities, and indeed, to do what is necessary to make any transfer of responsibilities as smooth and painless as possible.

There is an enormous amount of experience and expertise now residing in DOE/NNSA that will be vital to the success of the new Department. Our Technology Research and Engineering assets have been applied to homeland security problems long before last September; since then, such contributions became even more focused and accelerated.

We've conducted the PROTECT subway demonstration, which will help provide chemical protection to the U.S. population. We deployed a prototype biodetection capability at the winter Olympics. We have greatly increased our work with the U.S. Customs and US Coast Guard with radiation and nuclear technology—specific technical support that will directly benefit the new Department. DOE/NNSA is committed to ensuring that its assets can continue to provide enabling science and technology to support homeland security and counter-terrorism mission needs.

There are a number of capabilities currently residing in the Department of Energy that will support or be transferred to the new Department. Today I want to focus on those relevant to Title III of the legislation—those germane to technology research and development in support of the Homeland Security mission.

Before beginning that discussion, let me briefly mention a few things that the Homeland Security Act does *not* do. It will not affect our ability to conduct our principal missions of stockpile stewardship, nuclear nonproliferation, naval nuclear propulsion, and, just coming to NNSA, emergency response. NNSA will retain all of its programs and responsibilities that contribute to our ability to assure the safety, security, and reliability of the nation's nuclear weapons stockpile.

With respect to nuclear nonproliferation, the Administration proposes to transfer the core of our chemical-biological WMD work and certain nuclear programs related to the domestic threat. This is largely self-contained work and primarily supports *domestic* preparedness programs.

NNSA has unique assets and capabilities, developed primarily from our work with nuclear weapons and with nonproliferation, that have been applied to homeland security problems long before last September.

Some of these initiatives have long timelines; Long before 9/11, DOE has led USG efforts to support "first responders" with our chemical, biological, and nuclear research programs. We've worked closely with the FBI and other agencies to ensure that cutting edge detection and identification technologies are available to those that would need them first. And we began this work long before there was a recognized need to do so—we took the initiative because we anticipated the requirement. It is as good an example as any of why long-range research is so critical to the security of this country.

We have aggressively pursued these efforts since last 9/11. But it's time for a more focused organization and we are committed to that change *and* to continuing to provide enabling science and technology in support of homeland security and counterterrorism mission needs.

NONPROLIFERATION AND VERIFICATION RESEARCH AND DEVELOPMENT

The NNSA Nonproliferation and Verification Research and Development Program conducts applied research, development, testing, and evaluation of technologies that lead to prototype demonstrations and resultant detection systems. As such, the program strengthens the U.S. response to current and projected threats to national security worldwide posed by the proliferation of nuclear, chemical, and biological weapons and the diversion of special nuclear material. The R&D program provides operational organizations with innovative systems and technologies to satisfy their nonproliferation and counter-terrorism mission responsibilities. The program's three main elements are:

- Nuclear explosion monitoring, which will remain within the Department of Energy
- Chemical and Biological National Security, which will be transferred in its entirety to the Department of Homeland Security
- Proliferation Detection

Proliferation Detection sponsors a high-risk research on detection technologies that can support both nonproliferation and homeland security. Those elements that can be disaggregated and identified as supporting homeland security will be transferred to the new Department. At a minimum, we will transfer our research and development to counter nuclear smuggling. Where the activity supports both the homeland security and non-proliferation functions, we will examine arrangements as joint programs. The Administration's proposed legislation gives the President the necessary flexibility to provide for joint operation.

Let me describe those functions that will be transferred, after which I will return to the subject of long-term coordination.

Major Activities Identified for Transfer

Within, the Nonproliferation and Verification Research and Development Program, the Chemical and Biological National Security Program and the nuclear smuggling detection activity fall squarely into the Homeland Security mission and thus have been designated for transfer in their entirety.

Chemical and Biological National Security Program

The Chemical and Biological National Security Program works to develop technologies and systems to improve the U.S. capability to prepare for and respond to domestic chemical and biological threats against civilian populations, complementing DOD's focus on the battlefield and military installations. As part of its primary nuclear science and technology mission, NNSA and the National Laboratories have developed extensive capabilities in chemistry, biology, and materials and engineering sciences that form the basis for the NNSA chemical and biological national security program. We have conducted research on the biological foundations necessary to establish signatures of biological threat agents and develop assays certified by the Centers for Disease Control for those agents, which are applied to develop detectors.

NNSA has conducted demonstration projects of prototype detector capabilities in partnership with other agencies to support their operational missions, such as the systems I just mentioned that have been developed and applied for the Olympics and the Washington Metro, to illustrate possible system approaches for population protection. We are now working to expand the number of signatures and assays of biological agents that we can detect with increased sensitivity, and to improve public health response through the CDC. The next generation of bio-detectors will detect a much wider range of agents, which will enable public health agencies to more rapidly treat affected people.

Homeland Security Nuclear Smuggling Activities

The nuclear smuggling component of our proliferation detection program also squarely fits within homeland security and will be transferred. NNSA and the National Laboratories have unique insight into nuclear proliferation activities—the facilities and infrastructure, as well as the observable signatures of nuclear weapon development activity. We also have the capability to develop technical solutions for the U.S. government to detect and characterize such proliferation activities in their early stages. NNSA has worked closely with homeland security agencies, including U.S. Customs, U.S. Coast Guard, and the Departments of Transportation and Justice to apply this technical base to detection of nuclear weapons and materials at U.S. borders. With these agencies, we have previously conducted demonstrations of radiation detection methods at international border crossings, including a port, a rail yard, and airport personnel and baggage handling facilities. With many of these agencies becoming part of the new Department, it is a good fit for the R&D applications to counter nuclear smuggling to be transferred to the Department of Homeland Security.

Nuclear Threat Assessment and Trafficking in Nuclear Materials

In addition to the transfer of research and development, Title III of the proposed legislation provides for the transfer of the Department of Energy's Nuclear Assessment Program to the new Department of Homeland Security. This program provides a national capability to assess accurately and swiftly the credibility of communicated threats of nuclear terrorism. The Lawrence Livermore National Laboratory (LLNL) leads this unique effort. Since September 1978, the Nuclear Assessment Program has been used to assess the credibility of over 60 nuclear extortion threats, 25 nuclear reactor threats, 20 nonnuclear extortion threats and approximately 650 cases involving the reported or attempted illicit sale of nuclear materials.

When activated, DOE-based threat credibility assessment teams perform comprehensive technical, operational and behavioral assessments of communicated nuclear threats at the start of an actual or perceived emergency. Since communicated nuclear threats are a serious violation of federal law, the FBI is the lead federal agency. Since the Program's inception in 1977, the Nuclear Assessment Program has developed close and working relationships with its counter-terrorism counterparts in Customs, State, FBI, DIA, CIA, and others in the nonproliferation community. The Program also provides expert technical support to law enforcement and others for Special Event Preparedness, on-scene technical support, and national and international training.

Since 9/11 the Nuclear Assessment Program has performed approximately 70 assessments involving communicated nuclear threats, reports of illicit trafficking of nuclear materials, and special analysis reports for law enforcement and intelligence components. This national asset provided immeasurable support to all government agencies tasked with separating critical from non-critical information in the aftermath of 9/11.

Observations

With the transfer of these programmatic responsibilities to the Department of Homeland Security, it will be critically important that the new Department assume the leadership to maintain the technical base at the National Laboratories. Upon this foundation is built our future technical capability. The multidisciplinary scientific environment of a national laboratory is ideally suited to pursue high risk, long-term research, in spite of the need to focus on short-term requirements for homeland security. It is the ability to pursue such research that makes our national laboratories a national treasure—and a unique asset with unmatched capabilities. Only through such investment will the scientific and technical capability exist to meet the needs for innovative solutions to future homeland security problems.

With respect to the remainder of the proliferation detection program, no matter how the responsibilities are finally apportioned, the research will be of value to both departments. For that reason, it is critical that we work together closely. By so doing, our nonproliferation and homeland security efforts will continue to benefit from the unparalleled capabilities of the National Laboratories.

I support fully the concept of locating the new Department's main research facility at Lawrence Livermore, with satellite centers of excellence located at other national laboratories. It will create a campus-like environment where scientists will be dedicated, full-time, to thinking about homeland security, and it will allow for direct interaction with the expertise that resides at the other DOE labs as well as other labs throughout the federal government. It's good for DOE and it's good for the Department of Homeland Security.

CONCLUSION

I want to reiterate in no uncertain terms: The National Nuclear Security Administration supports fully the transfer of the programs noted in Section 302(2) of the bill under discussion. The details of what would be included in the legislative package were worked out directly with my office. These programs are a natural fit for the Department of Homeland Security, whose primary mission is the critical task of protecting the United States from catastrophic terrorism. DOE/NNSA will also work to ensure that its assets can continue to contribute enabling science and technology in support of DHS mission needs.

Obviously, that is a goal that I am pleased to support wholeheartedly. I believe that the Administration's proposed legislation represents a major step toward its realization.

Thank you, and I look forward to any questions you may have.

PREPARED STATEMENT OF ROBERT A. BRYDEN, STAFF VICE PRESIDENT OF SECURITY, FEDEX CORPORATION

Mr. Chairman and Members of the Committee,

My name is Robert A. Bryden, Staff Vice President of Security for FedEx Corporation, the parent corporation of FedEx Express. It is an honor for me to address this committee and speak about the very important topic of cargo security, particularly the prevention of the importation of unauthorized radiological materials into the United States. As you know, for at least six months FedEx has been in close contact with members of the staff of this committee on this subject. We have had several meetings with Ray Shepard and Chris Nauer, and have allowed them to tour our facilities at Charles DeGaulle Airport in Paris, and our National Hub at Indianapolis. We have fully cooperated with staff, and trust that they have provided you with the information you need and desire.

I would like to discuss some of the measures that FedEx utilizes in the detection and handling of radioactive materials. FedEx has a multi-layered security program in place to detect the presence of unapproved dangerous goods, including radioactive materials, and prevent their movement in the FedEx worldwide system. These include, but are not limited to, employee training and awareness programs, physical screening of packages originating at certain security-sensitive areas of the world, and radioactive monitoring devices aboard FedEx aircraft and on FedEx employees. In addition, our ability to track and trace shipments back to the place of tender constitutes a significant deterrent to utilizing the FedEx system for the shipment of illegal materials. While we are confident that these measures exceed those of any other transportation company in the world, the events of 9/11 have shown us that the state-of-the-art must be advanced. To that end, we have deployed, on a trial basis, advanced radioactive monitoring sensors at our Indianapolis Hub. This is a much more difficult endeavor than one might initially think. Because of the normal volume of radioactive shipments that FedEx routinely transports, such as pharmaceuticals, it is difficult to calibrate the sensors to detect undeclared shipments while at the same time not creating "false alarms" for legal shipments. The Indianapolis sensing equipment is very sophisticated, and can be finely tuned. In addition, it measures different types of radiation, and appears to be up to the task of adding a significant, additional layer of security to the FedEx system. We would be happy to update you periodically as we gather more data. If the trials continue to be successful, we intend to deploy the equipment at significant FedEx facilities throughout the world.

Again, I would like to thank you for the opportunity to address the committee. Security is a shared responsibility, and I want you to know that FedEx has committed its time, expertise, and money to ensure that its worldwide express delivery system is not employed as a tool of wrongdoers. I would be happy to answer any questions that you may have at this time.

PREPARED STATEMENT OF FRANK PANICO, MANAGER, INTERNATIONAL NETWORKS AND TRANSPORTATION, UNITED STATES POSTAL SERVICE

Good morning, Mr. Chairman and members of the Subcommittee. Thank you for this opportunity to speak with you today about in-bound international mail.

We share your concerns about the possible use of the mail to ship radioactive materials and other potentially dangerous substances. As events of the past year have demonstrated, the United States Government must be more vigilant than ever in regard to both international and domestic terrorism.

Needless to say, compromising the U.S. Mail system has the potential to adversely impact the entire nation through a single terrorist act, so we take threats to this system very seriously.

Last year, Congress provided the Postal Service with $500 million in the Fiscal Year 2002 Department of Defense Appropriations bill. The conference report required that the Postal Service prepare a comprehensive Emergency Preparedness

Plan. The Emergency Preparedness Plan was submitted to Congress on March 6, 2002.

The Postal Inspection Service provided the Postal Service with a threat assessment, which serves as a basis for our Emergency Preparedness Plan. The Postal Inspection Service maintains a continuous liaison with all appropriate federal law enforcement agencies and monitors threats to the nation and its mail.

The threat assessment concluded, "Accordingly, the Postal Service believes, and is acting on the assumption that the threat for the inappropriate use of the mails continues." The threat assessment also notes, "The greatest opportunities to limit the damage of covert NBC [nuclear, biological, or chemical] attacks, or prevent them entirely, exist during the first phases of the incident."

Therefore, our Emergency Preparedness Plan places a premium on threat identification, combined with protection to both employees and customers of the Postal Service at the earliest feasible point in our distribution system.

Unfortunately, the Postal Service has had to deal with the issue of bombs in the mail for a number of years. The most widely reported case was the Unabomber, but there have been other incidents over the years. As a result, we continue to educate our employees about identifying suspicious packages, particularly package bombs. Beyond education, we make responsible changes to our processes as necessary to meet new threats.

For example, in 1996, the Postal Service revised our procedures for accepting domestic and international parcels for mailing. Since that time, all domestic *stamped* parcels and *all* international and military mail weighing 16 ounces or more must be presented in person to a postal retail clerk or letter carrier. As a result, the number of package bombs in the mail system dropped from 18-20 per year to an average of about 3-5 per year.

The Postal Service is looking at a variety of process changes and technology initiatives that could be applied to the threat of chemical, biological and radiological hazards in the mail.

As described by the plan, the Postal Service is currently testing bio-detection technology on the automated processing equipment at one of our mail processing plants. This equipment has already passed tests at Edgewood Arsenal. In addition, we are testing filtration devices to improve our employee safety and to minimize cross-contamination of the mail. We anticipate a contract award by the end of September.

Careful review and consideration is being given to all currently available processes and technologies. The paramount conclusion is that no single solution exists to solve the complex problem of using the mail as a tool of terrorism. Further, no solution or even series of solutions can totally eliminate the threat.

To assist us in this review, we have contracted with Mitretek Systems to perform a comprehensive threat analysis. Mitretek is a well-respected, nonprofit systems-engineering company that provides programmatic and technical support on chemical, biological, radiological, and conventional weapons threats to the U.S. defense and intelligence communities.

In fact, Mitretek's President and CEO, Dr. Lydia W. Thomas, has been appointed to serve on the recently established Homeland Security Advisory Council by President Bush.

This assessment will review threats that may impinge on the mail, including the full spectrum of biological, chemical, explosive, and radiological threats. The assessment considers threats that may be directed at the Postal Service or may use the Postal Service as a vehicle.

As a result of this assessment, we will propose steps that may be taken to counter the threats and develop an overall risk/cost/benefit analysis, including an estimate of system effectiveness for protecting employees and customers, and for ensuring the continuity of postal operations in the event of a terror attack.

The viability of the Postal Service, and its value to the American people, is dependent upon an open and accessible system. Extreme procedural changes could reduce threats, but would significantly damage the usefulness of the mail to the American people—and the American economy.

Since the anthrax attacks, the Postal Service has worked closely with both the Office of Homeland Security and the President's Office of Science and Technology Policy. We provided both of these organizations with copies of our Emergency Preparedness Plan and followed up with briefings to their staffs.

Building upon our Emergency Preparedness Plan, we worked with Homeland Security in the development of a national Critical Infrastructure Plan. The Office of Science Technology and Policy has established the Inter-Agency Working Group for the protection of vulnerable systems, a group on which our Vice President of Engineering sits. He chairs the Mail and Package Working Group. This group is evaluating existing technology, as well as providing guidance as to where research and development efforts should be best directed.

We also continue to coordinate with all appropriate agencies about mail security, including the US Customs Service. The USPS and the Inspection Service have met with Customs and discussed the potential for radiological, chemical and biological hazards in foreign-originating mail.

Most of our previous discussions have centered on the examination of outbound mail, as the Customs Service has the authority and responsibility for the examination of all in-bound mail and cargo.

In addition, the Dangerous Goods Subgroup of the Universal Postal Union's Postal Security Action Group (PSAG) is working closely with the International Atomic Energy Agency (IAEA) to address the issue of radioactive materials in the mail. PSAG has worked with some of the international posts that have experimented with screening mail.

While these experiments have identified low-level radioactive materials, such as smoke detectors and medical equipment, there have been no instances of suspicious radioactive mailings.

As for international mail, we believe that the Customs Service is the agency with the responsibility and authority to detect radiological material imported into the country. And Customs has assured us that it is working on the issue. We believe Customs would be the most effective and efficient agency to perform this duty and we will continue to work with them, and all appropriate agencies at home, and abroad, to assure the safety of America's mail system.

To that end, we would be pleased to work with this Committee in any way possible to preserve the security and the usefulness of the United States mail.

Thank you, Mr. Chairman. I would be happy to respond to any questions or suggestions you might have.

PREPARED STATEMENT OF WAYNE J. SHOTTS, ASSOCIATE DIRECTOR FOR NONPROLIFERATION, ARMS CONTROL, AND INTERNATIONAL SECURITY, LAWRENCE LIVERMORE NATIONAL LABORATORY

Mr. Chairman and members of the committee, thank you for the opportunity to appear before you today. I am Wayne Shotts, the Associate Director for Nonproliferation, Arms Control, and International Security at the Lawrence Livermore National Laboratory (LLNL). I am responsible for managing the work being conducted at the Laboratory that pertains to homeland security. The urgency of our efforts has increased dramatically in the wake of September 11. The events of that day tragically make clear that the United States is not immune to the scourge of terrorism, and they call for the nation's leaders and technical community to take dramatic steps to improve homeland security.

Enactment of legislation to form a Department of Homeland Security—an idea supported by the President and the Congress—will fundamentally change for the better the nation's approach to preventing terrorist attacks on the United States, reducing the nation's vulnerability to terrorism, and managing the aftermath of any attack. The mission is complex and daunting in scope. One major challenge for the new department will be effective integration of relevant activities, which are currently dispersed among many government organizations. Another challenge will be focusing the unsurpassed scientific and technical talent of this nation to improve capabilities to deal effectively with threats, those most critical today and as well as those emerging in the future.

I support formation of a Department of Homeland Security and I am here to comment from a technical perspective on both the needs of the new department to pursue a sustained research, development, testing, and evaluation (RDT&E) program and the capabilities available to it to do so. Currently, RDT&E capabilities are dispersed, but there is an important concentration of them—particularly related to chemical, biological, radiological and nuclear threats—in the Department of Energy's National Nuclear Security Administration (NNSA) and its laboratories and other sites. I will discuss relevant capabilities at LLNL and some of the important programs and partnerships we have in place. They illustrate LLNL's approach to developing and deploying technologies and systems to strengthen homeland security and the success we are having in placing the right tools in the hands of the right people.

Effective partnerships among the various sources of expertise and with the users of new capabilities are required to make necessary improvements in homeland defense to cope with today's dangers and prepare for the threats of tomorrow. Focus on the most effective approaches to the highest priority issues is also required. At

LLNL, we are anxious to contribute to homeland security to the best of our abilities and confident that we can help make the Department of Homeland Security a success.

LLNL'S CONTRIBUTIONS TO HOMELAND SECURITY

Lawrence Livermore National Laboratory was established 50 years ago to pursue innovative solutions to the nation's pressing needs to advance nuclear weapons science and technology. Since then, the Laboratory has continually adapted to address the evolving challenges of the day and anticipate future needs, keeping a central focus on national security. As one of NNSA's three national laboratories, LLNL is a principal participant in the Stockpile Stewardship Program to maintain and enhance the safety, security, and reliability of the nation's nuclear weapons stockpile. The Laboratory is also engaged in vital national programs to reduce the threat posed by the proliferation of weapons of mass destruction (WMD) and to provide for homeland security. These complementary missions—stockpile stewardship and countering WMD threats—are integrally connected in terms of their overarching goal of enhancing security, and the research activities largely draw on the same base of scientific and technical capabilities and expertise.

Because Livermore and our sister NNSA laboratories (Los Alamos and Sandia) have long been working to develop technical capabilities to detect, counter, and mitigate WMD proliferation and terrorism, we were able to respond rapidly and effectively to the events of September 11 and its aftermath. Although those investments are paying great dividends in the newly declared war on terrorism, substantial sustained investment is needed to develop vastly improved warning and response capabilities to protect the U.S. against these threats, now and in the future. We are fully committed to this long-term national security endeavor and are well positioned to provide RDT&E support to the Department of Homeland Security.

Lawrence Livermore is contributing widely and effectively to the war against terrorism with capabilities and partnerships and through RDT&E programs directly relevant to the Department of Homeland Security's mission. The provided examples illustrate three major points about the Laboratory:

- LLNL has demonstrated the capability to work problems from end-to-end—starting with an understanding of the threat and the users' needs, devising a systems solution, developing the enabling technology advances, testing both the component technologies and systems solution in cooperation with users, moving the new technologies to U.S. industry, and working with the user community to ensure effective deployment and training.
- LLNL has strong capabilities and active programs in each of the WMD areas—chemical, biological, radiological, and nuclear. In addition, the Laboratory has major programmatic activities in threat assessment and intelligence support as well superb supercomputing capabilities. Accordingly, we have a "critical mass' of programs and capabilities that provides the Laboratory an excellent overall perspective of threats, technical opportunities, and user needs.
- LLNL has many strong ties to research partners and the user community—including sister laboratories, the Nevada Test Site for remote testing, a wide range of universities, and many ties at the local- and state-government level.

THE CAPABILITY TO WORK PROBLEMS FROM END-TO-END—BASIS AS AN EXAMPLE

A research and development program particularly focused on the area of WMD terrorist threats is an integral part the legislative proposal for a Department of Homeland Security for good reason—the nation faces a dire immediate threat that unquestionably will grow more sophisticated over time. The nation's vulnerabilities vary widely in their significance and their potential for being ameliorated by new capabilities and/or changes in operations. What is needed is a comprehensive perspective of the issues, a vision where one wants to go, and a pragmatic approach to problem solving to put products in the field expeditiously.

At LLNL, we take a systems approach to the overall problem and determine what priority items can be dealt with expeditiously with existing equipment or modest improvements in technology and where investments in longer-term research and development will be necessary. In those areas where a new system based on existing or emerging technologies can make a substantial difference, it is important to work the problem comprehensively with the end user in mind.

The development of the Biological Aerosol Sentry and Information System (BASIS) by Livermore and Los Alamos exemplifies this approach and serves as model of how the Department of Homeland Security could most rapidly and effectively take technology from the conceptual stage through to actual deployment. The process is more than R&D, it is RDT&E—research, development, testing, and evaluation.

In late 1999 we were challenged by the Secretary of Energy to develop and field a biological detection system in time for the 2002 Salt Lake City Olympics. At the time, there was no system suitable for civilian use for broad-scale biological environmental detection and monitoring. Early detection and rapid response are the keys to reducing the human health consequences of a biological agent attack. Over the next three years, we and our colleagues at Los Alamos developed and demonstrated a successful system to meet this challenge. BASIS was fielded at Salt Lake City in February 2002 as part of the overall security strategy for the Olympic Games where it performed exactly as designed. The goal-oriented approach used in this program greatly contributed to its outstanding achievement. In particular, BASIS benefited from:

- *A Clear Objective at the Outset.* For BASIS, clear, top-level objective was established at the beginning of the project with respect to the desired cost and performance attributes of the system. The objective was based on an understanding of the threat, technical possibilities, and user needs. After this, the management of the program and the technical details were left to the technical team.
- *Close Interactions between Users and Technology Developers.* There were extensive direct interactions with the Salt Lake Olympic Committee, local, state, and federal response agencies, the public health system, and the technology developers from conception through implementation and operation.
- *Problem-Solving Systems Approach.* The sponsors, users, and technologists recognized the need for a system-level solution, not a single technological widget, and for the system to work in conjunction with other equipment (e.g., medical surveillance systems). LLNL and LANL brought together a team of engineers, biologists, computer scientists, and operations specialists to execute the program.
- *Advanced Technology Developed by Labs, Transferred to and then Procured from Industry.* The system used the most advanced biological detection technologies available (i.e., PCR). The best biological detection instrument for this application was from a commercial entity (Cepheid) that had earlier licensed the technology from LLNL.
- *Testing and Evaluation against Standards by Recognized Authority.* The biological assays were co-developed by LLNL and the Center for Disease Control's (CDC) Bioterrorism Laboratory. The testing regimen was established with law enforcement and public health, assuring a high level of confidence in the system.
- *Transfer of Operations to Contractors.* Local contractors provided the bulk of the staff for all aspects of the system operations at the Olympics. LLNL/LANL staff were used in supervisory roles and for technical support.

STRONG CAPABILITIES AND ACTIVE PROGRAMS—NUCLEAR AND RADIOLOGICAL THREATS

As one of NNSA's three national laboratories, LLNL is fully engaged in the Stockpile Stewardship Program and has a very large science and technology base supportive of work on nuclear weapons, nuclear materials, and nonproliferation that can be leveraged to support homeland security. The Laboratory is home to one of the nation's two research facilities for special nuclear materials. It operates a remote test site and has a close working relationship with the Nevada Test Site where work that requires even greater isolation is carried out. Several activities that contribute to homeland security merit special mention:

Nuclear Threat Assessment Program. The NNSA's Nuclear Assessment Program was established in 1977 to provide a national capability for correctly and expeditiously assessing the credibility of communicated nuclear threats. Shortly after its inception, the Nuclear Assessment Program became the central point of contact and action office within the NNSA for assessing and monitoring illicit nuclear material trafficking incidents worldwide. Selected elements of the program are routinely used to provide NNSA technical support to the law enforcement, diplomatic and intelligence communities. The major support activities include real-time assessments of nuclear threats and black market transactions, participation in FBI designated Special Events, and providing NNSA courses on nuclear crime at various national and international training venues. Since the terrorist attack on September 11, there has been dramatic increase in requests for our services; we have assessed 25 nuclear threats, 90 illicit trafficking cases, and 51 other nuclear related incidents.

The operational capability consists of a small group of professionals who are collectively knowledgeable in nuclear explosives design and fabrication, nuclear reactor operations and safeguards, radioactive materials and hazards, linguistics analysis, behavioral analysis and profiling, as well as terrorist tactics and operations. The assessor teams are organized into specialty teams and operate in secure facilities at the three participating NNSA contractor sites. An Assessment Coordinating Center at LLNL directs credibility assessment operations for the NNSA and provides a single point of contact for federal crisis managers during emergency operations.

Nuclear Incident Response. The Laboratory is a key participant in the national nuclear incident response groups, including the Joint Technical Operations Team (which deals with nuclear terrorism or extortion threats), the Accident Response Group (which responds in the event of an accident involving U.S. nuclear weapons) and the Radiological Assessment Program (which assists state and local agencies). Livermore maintains a deployable response capability, called HOTSPOT, which can be transported to any location by military aircraft to provide local radiological field support.

Specifically, the Radiological Assessment Program (RAP) provides technical and operational expertise to state and local agencies to mitigate the consequences of a radiological incident or emergency. It uses DOE and national laboratory experts with skills in assessing radiological and toxic contamination and the attendant risks to human health. The Livermore RAP teams have primary responsibility for California, Nevada, Hawaii, and the U.S. Pacific Rim territories. They are called upon, on average, three to five times per year. In 2001, they responded to three requests for assistance along with normal exercises and training. Typically, RAP investigates containers suspected of housing radioactive materials, seeks the location of lost industrial or medical radioactive sources, and advises federal, state, and local authorities on the consequences of a radioactive release or personnel contamination. RAP regularly drills with similar teams from other federal agencies, state, local, and tribal governments as well as private companies and organizations.

To deal with the latest emerging threats, LLNL now maintains a home team capability to assist response workers at all levels. The home team is trained to recognize and respond to nuclear terrorism. Included within this umbrella is the ability to supply timely interpretation of signals from field instruments (the so-called "nuclear triage' program being developed at NNSA headquarters).

Search and Inspection Technologies. There is a pressing need for technologies to improve the screening of passengers, baggage, and cargo. Candidate technologies, in various stages of development at Livermore, include computed tomography (CT), x-ray scanning, gamma-ray imaging, neutron interrogation, and ultrasonic and thermal imaging. These efforts build on projects and expertise in the Stockpile Stewardship Program to develop improved sensors for non-destructive evaluation of the condition of weapons and weapon components in the stockpile. NNSA has assigned LLNL the responsibility to establish a national test bed for the inspection of cargo containers (discussed further below).

Two Laboratory-developed search technologies demonstrated their applicability to counterterrorism response when they were deployed to the World Trade Center. The first, a micropower radar, can "see' many feet into concrete rubble and could be a valuable tool for search and rescue operations. The other, a remote monitoring instrument that uses hyperspectral data to detect and identify trace gas emissions, was flown over Ground Zero to characterize hazardous gases emanating from the rubble.

Sensor Networks. Livermore has developed a concept for correlated sensor networks for detecting and tracking ground-delivered nuclear devices or nuclear materials, the Detection and Tracking System (DTS). A novel algorithm integrates data from the various sensors, together with information from other sources (e.g., an intelligent traffic system) to identify sources of concern, track their movement through the road network, and guide responders in intercepting the suspect vehicle. Since September 11, DTS development was accelerated and a prototype system was demonstrated in an urban environment. We are preparing for further, larger scaled demonstrations of this system with added capabilities.

STRONG CAPABILITIES AND ACTIVE PROGRAMS—BIOLOGICAL AND CHEMICAL THREATS

Bioscience research at the Laboratory traces its root to 1963, when a program was established to study how radiation and chemicals interact to produce adverse consequences to humans. Research activities at LLNL and LANL led to a focus on DNA and technology development that led to DOE's decision to launch its Human Genome Initiative in 1987. Both laboratories are part of DOE's Joint Genome Institute, which includes Lawrence Berkeley National Laboratory and is located in nearby Walnut Creek, California, and have contributed to deciphering the human genetic code. We are applying our expertise in genomics to counter the threat of bioterrorism. In addition, in support of Livermore's national security and other programs, the Laboratory also has outstanding capabilities in chemistry and materials science.

Biological Agent Detectors. The biodefense capabilities that have been deployed in the wake of September 11 have, at their core, advances in biological detection instrumentation developed at Livermore. We have made technology breakthroughs in biodetection instrumentation, pioneering the miniaturization and ruggedization of both flow cytometry and DNA identification devices. Our miniature thermal cycler unit makes possible DNA amplification via polymerase chain reaction (PCR) and identification in minutes rather than the hours and days previously required. Livermore's miniaturized PCR technology has been licensed to private industry and forms the basis of today's most advanced commercial biodetection instruments (e.g., Cepheid's Smart Cycler, Environmental Technology Group's handheld biodetector).

Cepheid Smart Cyclers are the heart of the field laboratory of the Biological Aerosol Sentry and Information System (BASIS), developed jointly by Livermore and Los Alamos and previously discussed. In developing BASIS, the two laboratories worked closely with the many law enforcement, emergency response, and public health agencies that would be involved in dealing with a bioterrorism event to develop appropriate sample handling (chain of custody), communications, and response protocols.

DNA Signatures. Biodetectors depend on unique antibodies or DNA sequences to identify and characterize biological pathogens. Livermore is developing gold-standard DNA signatures of top-priority threat pathogens (anthrax, plague, etc.) and are working with the Centers for Disease Control and Prevention (CDC) to validate these signatures and distribute them to public health agencies nationwide. We are also working with the Federal Bureau of Investigation, CDC, Department of Defense, and U.S. intelligence agencies to develop detailed biological "fingerprints' and data to support forensic analysis of any act of biological terrorism.

Chemical Analysis for Forensic Attribution. Timely and complete analysis of suspect chemicals can answer important questions related to nonproliferation, counterterrorism, and law enforcement. Our Forensic Science Center has assembled a unique capability for detecting and characterizing ultratrace levels of virtually any compound in any sample matrix. Expertise and instrumentation are available for complete chemical and isotopic analysis of nuclear materials, inorganic materials, organic materials (e.g., chemical warfare agents, illegal drugs), and biological materials (e.g., toxins, DNA). The Forensic Science Center also develops advanced laboratory and field capabilities for ultratrace analysis, including a portable (55-pound) gas chromatograph/mass spectrometer, field kits for thin-layer chromatography, and novel sample collectors using solid-phase microextraction.

The Forensic Science Center has begun the rigorous testing required to become the second U.S. laboratory certified by the Organization for the Prohibition of Chemical Weapons (OPCW), which is responsible for implementing the Chemical Weapons Convention (CWC). Under the terms of the CWC, all samples collected from inspected facilities must be analyzed at two OPCW-designated laboratories. The U.S. Congress mandates that all U.S. samples be tested in the U.S. Currently, the U.S. has only one designated laboratory, the Edgewood Chemical and Biological Forensic Analytical Center. Livermore will provide the second required facility.

STRONG CAPABILITIES AND ACTIVE PROGRAMS—UNDERPINNING CAPABILITIES AND FACILITIES

Several special capabilities at Livermore merit special mention because they provide broad yet critical support to homeland security: our International Assessments Program, the National Atmospheric Release Advisory Center (NARAC), the Counterproliferation Analysis and Planning System (CAPS), high-performance computations, and the Computer Incident Advisory Capability.

Intelligence Analysis and Threat Assessment. One of the most critical, yet difficult, elements of homeland security and counterterrorism is gaining insight into the capabilities, intentions, and plans of persons, groups, or states hostile to the U.S. Our International Assessments Program (Z Division) is one of the strongest capabilities in the country for analysis and research related to foreign nuclear weapons and other weapons of mass destruction, including early-stage foreign technology development and acquisition, patterns of cooperation, and foreign cyber threats. Such intelligence analyses serve as the foundation for homeland defense against WMD threats. Intelligence provides an essential input to threat analyses that, in turn, provide the basis for defining functional requirements for technical homeland security systems. Furthermore, intelligence can provide "indications and warning' of an imminent attack, thus guiding further deployment of defensive assets. Thus there is a critical need for both long-term, in-depth intelligence analysis and timely, responsive indications and warning.

Z Division regularly provides analysis products to our intelligence, defense and policy-making customers. Our assessments of foreign weapons programs and activities provide important input to policy makers and diplomats as they develop strategies for U.S. responses to events affecting national security. The capabilities in Z Division also support our Nuclear Threat Assessment Program (previously discussed), which analyzes nuclear terrorist threats and smuggling incidents.

In addition to filling a critical niche by providing all-source intelligence analyses of foreign nation-state programs to acquire WMD, we develop data analysis tools and data integration methods to aid intelligence collection and assessment and avoid the pitfalls of information stovepiping. Some of these tools are currently being evaluated by our analysts as well as end-users across the Intelligence Community, while many others are under intense development and will be applied to the counter-terrorism problem. In the aftermath of September 11, we provided intelligence analysts and assessments as well as information-operations tools and expert personnel to the U.S. Intelligence Community.

Atmospheric Modeling for Consequence Management. The National Atmospheric Release Advisory Center (NARAC), located and operated at the Laboratory, is a national emergency response service for real-time assessment of incidents involving nuclear, chemical, biological, or natural hazardous material. NARAC can map the probable atmospheric spread of contamination in time for an emergency manager to decide whether protective actions are necessary. NARAC is on call to respond to real incidents and can also be used to evaluate specific scenarios for emergency response planning, such as optimizing the siting of bioaerosol samplers or determining evacuation routes.

Since it was established in 1979, NARAC has responded to more than 70 alerts, accidents, and disasters and has supported more than 800 exercises. In addition to accidental radiological releases (e.g., Chernobyl, 1986; Three Mile Island, 1979), NARAC has assessed natural and manmade disasters (Mt. Pinatubo volcanic ash cloud, 1991; Kuwaiti oil fires, 1991). NARAC has also provided assessments to state and local responders to toxic chemical accidents (e.g., Richmond sulfuric acid cloud, 1993; Sacramento River Spill, 1991). State and local agencies can request NARAC support for actual releases or planning by contacting DOE's Office of Emergency Response or the NARAC program office at Livermore.

The Counterproliferation Analysis and Planning System (CAPS). Developed continually updated by LLNL, Counterproliferation Analysis and Planning System (CAPS) is a versatile and powerful modeling system for analyzing, end-to-end, a proliferator's WMD production processes and for assessing interdiction options and their corresponding consequences. CAPS is as easy to use as a Web browser, with its powerful and complex science (spectral analysis, toxic release modeling, etc.) invisible to the user. CAPS is widely accepted by the military's mission planners and is the Department of Defense's preferred counterproliferation planning tool.

High-Performance Computing. With supercomputers acquired as part of NNSA's Advanced Simulation and Computing (ASCI) program and additional institutional investments in massively parallel computers, Livermore is an international leader in high-performance computing. Many groundbreaking applications are being developed. An example directly relevant to homeland security is our computational biology work directed at genomics—the development and use of bioinformatics tools and databases.

We have developed computational tools to automatically identify regions of bacterial and viral pathogen genomes that have a high probability of being unique to that genome. We can now process any draft or finished pathogen genome in a few hours and confidently detect all regions that are not "matched' in any other known sequenced genome. This capability has been tested on numerous bacterial and viral pathogens both at LLNL and with collaborators such as the Centers for Disease Control, the U.S. Army Medical Research Institute of Infectious Diseases, and the Department of Agriculture. We are currently using this unique computational capability to satisfy pathogen detection needs of these and other federal and state agencies.

Building on the approach we are taking, we will attempt to tackle more complex problems such as automatically determining all protein signature targets in a genome and determining the "pathomics' of virulence across all pathogens (i.e., the molecular mechanisms of virulence itself). The computational needs to address these problems will require use of cutting-edge supercomputer resources such as those at LLNL.

Computer Incident Response. LLNL is home to DOE's Computer Incident Advisory Capability (CIAC), which was formed in 1989. We assist any DOE facility that experiences a computer security incident with analysis, response, and restoration of operations. CIAC serves as DOE's watch and warning center, notifying the complex of vulnerabilities that are being exploited, specifying countermeasures to apply, and providing a picture of the attack profile. The center also develops science and technology solutions in support of computer network defense and products such as SafePatch, which earned its developers a Government Technology Leadership Award. CIAC's list of clients has grown to encompass other government agencies, and there have been several incidents where the team worked with the Federal Bureau of Investigation.

STRONG TIES TO RESEARCH PARTNERS AND THE USER COMMUNITY

One key attribute of LLNL is the Laboratory's proximity to important assets—potential major partners in RDT&E and commercialization as well as key customers for homeland security. The San Francisco Bay Area is home to three international airports, two seaports, an FBI field office, Customs and INS headquarters, Silicon Valley, area biotechnology firms and health-care providers, mass transit and rail systems, and high-visibility targets (e.g., Golden Gate Bridge). In addition, as part of University of California, LLNL has close ties with the many UC campuses in the area (Berkeley, San Francisco, Davis, and Santa Cruz) as well as Stanford University (and associated medical schools). We are also right next to Sandia-California. Almost every aspect of the homeland security equation is just minutes away from Livermore.

Many of our various research partners are cited throughout my testimony. An often overlooked—yet important—aspect of a successful research and development program is understanding the users' needs. Additional examples of our connections and work with the user community follow.

Expert Personnel Assisting in Homeland Security. Livermore scientists serve on various task forces, committees, and advisory groups dealing with aspects of homeland security and counterterrorism. For example, a Livermore expert on x-ray imaging is a member of the National Academy of Science Committee on Assessment of Technology Deployed to Improve Commercial Aviation Security. Other Laboratory scientists serve as technical advisors to the U.S. Customs Service, the National Guard, and the Los Angeles Emergency Operations Center, and as members or advisors to various Defense Science Board task forces addressing homeland defense. Still others are assisting the California Highway Patrol and the California State Office of Emergency Services (OES) with training related to weapons of mass destruction and serving as members of the California Council on Science and Technology, which is providing technical advice to the OES's State Strategic Committee on Terrorism.

Forensic Science Support to Law Enforcement. Over the years, Livermore's Forensic Science Center (previously discussed) has responded to many requests from law enforcement for assistance in forensic analysis of unique samples. Since September 11 and the subsequent anthrax scare, hundreds of samples of concern have been analyzed for local and federal law enforcement and government officials. Previously, the Center has been brought in to analyze Supernote counterfeit bills, methamphetamine samples, biotoxins, suspect chemical-warfare specimens, and nuclear contraband. It has characterized explosive traces from the 1993 World Trade Center bombing, the Unabomber case, and the Fremont serial bomber; performed forensic sleuthing related to the Riverside "mystery fumes' case; analyzed samples for the Glendale "Angel of Death' case; and analyzed Capitol Hill offices as requested following anthrax decontamination. Locally, the Center assisted Livermore police by rapidly identifying a vapor that sickened response personnel at the scene of a suicide; once the chemical was identified (malathion), law enforcement agencies were able to take appropriate personnel-protection measures and complete their investigation.

LINC for Improved Emergency Preparedness. Through the LINC program (Local Integration of the National Atmospheric Release Advisory Center with Cities), we are currently working with local agencies in the Seattle area. A LINC pilot project is testing and evaluating the effectiveness of an approach to emergency preparedness that offers the potential for dramatic improvements. Sponsored by NNSA's Chemical and Biological National Security Program, LINC integrates capabilities at LLNL's NARAC (previously discussed) with local emergency management and response centers. Ultimately, LINC's goal is to provide continuous operation of an integrated, nationalwide system that aids emergency preparedness and response at all levels of government.

A National Test Bed for Standards, Test, and Evaluation. One key function of the Department of Homeland Security will be the setting of standards for technical homeland security systems. To set such standards will require practical, technical judgment, with consideration of the threats that the technology is intended to address, a concept of operations for its use, and the infrastructure necessary to use it effectively. This process must involve the Intelligence Community, end users in federal, state and local government, and technical experts. Candidate technologies must undergo objective testing and evaluation to determine how well they satisfy the standards, as input to acquisition decisions by those with operational responsibilities.

NNSA has assigned LLNL the responsibility to establish a national test bed for the inspection of cargo containers for chemical, biological, radiological, and nuclear weapons and materials. To meet this responsibility, we have initiated threat analyses to establish the range of threat scenarios that such inspection systems should address. We have also begun a research program, based on calculations and experiments, to characterize the relevant "observables' for successful detection. We have engaged federal, state and local organizations with operational responsibilities in this area to factor in their practical, operational constraints. We have set up a test facility where exemplar containers are loaded with surrogate materials, as well as typical cargo, so that commercial equipment and research prototypes can be tested in meaningful scenarios. We believe that this methodology should be extended to other terrorist scenarios of concern.

Risk and Vulnerability Assessments of Critical Facilities. Through our participation in DOE's Vulnerability and Risk Assessment Program, we have made systematic assessments of the threat environment, cyber architecture, physical and operational security, policies and procedures, interdependencies, impact analysis, risk characterization, and possible mitigation measures for the 2002 Winter Olympic Games in Salt Lake City, eleven electric and gas infrastructures, and several independent service operators (ISOs), including the California ISO during the electrical energy crisis. We have also analyzed the vulnerability of buildings, dams, and other structures to catastrophic damage from earthquakes and explosive events. Projects have included evaluation of the earthquake vulnerability of major bridge structures (including the Golden Gate and San Francisco-Oakland Bay bridges), the structural integrity of nuclear material shipping containers for a variety of impact scenarios, and the likely damage resulting from the explosion of natural gas storage tanks in a suburban environment.

More generally, LLNL has applied risk and decision theoretic methodologies to a wide range of hazardous endeavors, both internal to the Laboratory and for the public sector, and we can be considered a major scientific contributor to the discipline of risk assessment and risk management. We have developed methodologies for and conducted risk assessments of nuclear power generation, nuclear explosive operations, information systems, transportation systems and hazardous material protection (called vulnerability analyses) to identify and enhance safety, safeguards and security. In addition, LLNL has assisted other federal agencies in the application of risk management.

Engineering a Novel Truck-Stopping Device. In October 2001, the Governor of California contacted Livermore requesting assistance to develop a means of stopping tanker trucks, to keep hijacked trucks from becoming motorized missiles. The objective was to make it possible to stop these large trucks using equipment readily available to peace officers, namely their vehicles and their weapons. A retired Livermore engineer and consultant teamed with Laboratory engineers, technicians, and heavy equipment operators to develop a simple mechanical device to accomplish this. It can be readily attached to the back of a tanker truck. When bumped from the rear by the patrol vehicle, the device would cause the trailer braking system to lose air pressure automatically locking the trailer brakes. A prototype was demonstrated in Oakland in late November 2001, and testing at high speeds was conducted at the Nevada Test Site in February and March 2002. We are currently developing a portable remote-controlled system and working with the California Highway Patrol and a major California trucking company on implementing a field trial program.

CLOSING REMARKS

In its efforts to combat terrorism and ensure homeland security, the nation can build on an attribute that has made the United States the world leader that it is—the remarkable capability of the American people to focus extraordinary energy on achieving important objectives in a time of need. Establishing a Department of Homeland Security can fundamentally change for the better the nation's approach to preventing terrorist attacks on the United States, reducing the nation's vulnerability to terrorism, and managing the aftermath of any attack.

As the Administration and many leaders in Congress have already stated, to succeed the new department will need to pursue a sustained RDT&E program—particularly related to chemical, biological, radiological and nuclear threats—that is prioritized to meet prudently established objectives. These threats are significant and will grow more sophisticated over time. At Livermore, we are fully committed to this long-term national security endeavor to improve homeland security and are well positioned to provide effective RDT&E support to the department. LLNL brings to the Department of Homeland Security relevant existing mission responsibilities and programs, experience working with a wide range of research partners and users, and a track record of taking technologies from concept to prototype development and deployment.

PREPARED STATEMENT OF STEVEN W. MARTIN, DIRECTOR, HOMELAND SECURITY PROGRAMS, PACIFIC NORTHWEST NATIONAL LABORATORY

INTRODUCTION

Mr. Chairman, members of the House Energy and Commerce Subcommittee on Oversight and Investigations; my name is Steve Martin, and I am the Director of Homeland Security Programs at the Department of Energy's Pacific Northwest National Laboratory (PNNL). On behalf of the Laboratory Director, Dr. Lura Powell, I am pleased to provide testimony today.

In this statement I begin with a brief overview of Pacific Northwest National Laboratory. This is followed by some comments regarding the nature of our homeland security challenges and some examples of ways in which PNNL is contributing to help meet the needs for securing our homeland. I close with comments on the role of the national laboratories managed by the Office of Science and the National Nuclear Security Administration in the Department of Energy.

PACIFIC NORTHWEST NATIONAL LABORATORY

Pacific Northwest National Laboratory (PNNL) is a Department of Energy (DOE) multi-program laboratory, managed by DOE's Office of Science. Since 1965, the Pacific Northwest Division of Battelle Memorial Institute, a not-for-profit entity based in Ohio, has operated PNNL for the DOE. PNNL employs approximately 3,500 staff and maintains a business volume in excess of $500M annually, $230M of which is related to national security work for a number of government clients in areas such as combating terrorism, homeland security, proliferation detection and monitoring, underground nuclear test detection, nuclear weapon dismantlement, nuclear materials safeguards and security, environmental and waste characterization, and fundamental science.

OUR HOMELAND SECURITY CHALLENGES

Terrorism is not a new phenomenon and for decades PNNL has performed work for government agencies with missions designed to combat terrorism. Recent events serve to remind us of the vulnerabilities to the security of our homeland and it is becoming even more evident that there are terrorist elements with a willingness to deploy weapons of mass destruction against U.S. interests—both abroad and at home.

The threat we face is dynamic and complex. We need to be as flexible and adaptable as are the adversaries who would threaten us. As we organize around the need to manage the risks associated with the threats posed by weapons of mass destruction (WMD), we must do so in a reasonable and systematic manner. The actual financial costs of developing and implementing mitigating strategies and countermeasures are only one consideration of a comprehensive risk management strategy. We must also ensure that the solution is implemented in a manner that considers negative consequences such as reduced operational efficiencies or productivity that currently give U.S. industry and the U.S. economy a competitive advantage.

Finally, it is imperative that organizational and technological standards evolve that ensure solutions can be integrated across the various functions and responsibilities outlined for the new Department of Homeland Security (DHS). Solutions must facilitate integration of operations and functions, information sharing, and interoperability.

PNNL CONTRIBUTIONS TO HOMELAND SECURITY

PNNL participated, along with other DOE and NNSA laboratories, in a demonstration of national laboratory science and technology with potential for application within the Office of Homeland Security. At that demonstration PNNL profiled several of the following technologies. These are but a few examples that demonstrate that capabilities at PNNL span the entire WMD threat spectrum.

• **Millimeter Wave Holographic Imaging System:** This system, developed for the FAA for personal security checkpoint screening, is capable of detecting all threats and contraband.

• **Acoustic Inspection Device:** This handheld system was originally developed by PNNL for inspection of chemical weapon stockpiles in Iraq following the 1991 Gulf War. It can be used by Law Enforcement Officials to *Detect* concealments, hidden compartments or anomalies in liquid-filled containers and solid form commodities; *Sort* material types into groups of like and unlike, and *Identify* liquids and solid materials over a wide range of temperatures. It has recently been commercialized by U.S. Customs as an inspection and screening tool.

• **Biodetection Enabling Analyte Delivery System (BEADS):** It is necessary to process large environmental samples to obtain traces of threat biomaterial and deliver that material in a small volume to a sensor. BEADS enables automated sample preparation for biodetection systems.

• **Plutonium Measurement and Analysis (PUMA):** A radiation monitoring system that uses glass fibers to detect the presence of radionuclides, such as plutonium. This technology offers flexible, lightweight, low-power detection capability.

• **Hazardous Material Chemical Agent Detector (HAZMATCADtm):** This commercially available tool takes advantage of special (sensitive and selective) polymers developed by PNNL and allows faster response times to lower concentrations of hazardous chemicals and agents.

• **WMD Interdiction Training for International and Domestic Border Security Officials:** In 1997, Congress provided for the U.S. training of international border security officers in detecting, identifying, and interdicting the smuggling of WMD materials and items. Since then, Border Officers from 17 nations have been trained as part of the International Border Security Training Program. PNNL is responsible for conducting this highly successful training known as Interdict/RADACAD at the Hazardous Materials Management and Emergency Response (HAMMER) Training Center, a $30M facility located near PNNL at the Hanford Site. The value of this program has been demonstrated by seizures of sensitive materials in Eastern Europe, including nuclear reactor components destined for Iran and a quantity of Uranium-235. The border security officials responsible for both of these seizures attribute their success to the training they received in this program from PNNL at HAMMER.

PNNL initiated training of U.S. Customs Officers this year. Thus far, two 3-day courses in radiation detection and protection and the use of advanced detection equipment have been completed. For the foreseeable future, one U.S. Customs class per month is scheduled.

The practical operational environment of HAMMER is enhanced by props that include a mock border crossing, a Port of Entry building with a loading dock, inspection pit and radiation portal monitor, as well as intermodal shipping containers and transport vehicles with concealment compartments and traps commonly used by smugglers.

• **International Emergency Preparedness for WMD:** PNNL supports a US government-sponsored training program that teaches international first responders how to recognize, respond to and manage an incident involving a WMD. In addition to the operations training at HAMMER, PNNL also supports a course for international mail handlers on Postal Chemical/Biological Incident Management. In the same way the international WMD interdiction training eventually expanded to accommodate U.S. Customs Officers, consideration should be given to leveraging this training capability and facility to accommodate the government's articulated desire to train U.S. first responders to handle WMD incidents.

• **Federal Emergency Management Information System and EMAD VANTAGE:** Decision support and command and control tools have been developed for both emergency managers and emergency responders. These tools provide an automated decision support architecture that applies to situation planning and response capabilities for large multi-user environments.

• **National Counterdrug Center (NCC):** Operational coordination (or interoperability) across multiple agencies, missions, or functions is a known limiting factor impacting interdiction efforts. The NCC is a simulation-based interoperability training system that can improve multi-agency operational planning and execution in a virtual environment. While the current focus is drug interdiction, this national capability can be readily leveraged to accommodate training and planning capability for all-threat interdiction to include weapons of mass destruction. In addition, since the underlying objectives are to support interoperability, it is plausible that the capability and concept of simulation-based interactive environments can support the needs of first responders (police, fire, and emergency medical) as well.

- **Information visualization and knowledge management:** For over a decade PNNL has been conducting research that helps government analysts deal with the overwhelming amount of information they must process. PNNL has developed and successfully deployed tools for exploiting large and diverse sets of information and analysts within a number of government agencies are currently taking advantage of PNNL tools like *SPIRE* and *Starlight* to help them connect the dots.
- **Critical Infrastructure Protection:** PNNL is one of many DOE laboratories tasked to assure the integrity of energy infrastructures by conducting vulnerability assessments and recommending risk-mitigating strategies. The bulk of this work has focused on the electrical power infrastructure, an area wherein PNNL has recognized capability.
- **Radiological Detection Expertise:** Even though PNNL has existed for nearly four decades, there are over 50 years of history related to radiation detection technology development and deployment as a result of the legacy from the Hanford site's involvement in the Manhattan project. Instruments incorporating PNNL radiation detection technologies have been fielded in a number of locations, including: outer space, deep undersea, within the core of both naval and civilian reactors, border crossings, international nuclear test detection networks, high altitude aircraft, nuclear accident sites such as Three Mile Island and Chernobyl, U.S. nuclear complex sites, and deep underground. In addition, PNNL staff participate in a number of U.S. Government or international policy working groups including the Radiation Detection Panel (DOE), the Nuclear Smuggling Working group (IAEA), and the Radiation Instrumentation Steering Committee (IEEE.) PNNL currently holds leadership positions in the International Nuclear Materials Management Association.
- **Radiation Portal Monitoring Support to US Customs:** The U.S. Customs Service, Office of Information and Technology (OIT), Applied Technology Division (ATD), working with the Department of Energy (Pacific Northwest National Laboratory-PNNL), has established a terrorist radiation/nuclear detection project to investigate systems and technologies to augment and enhance their existing radiological detection capabilities. This project addresses the maritime, aviation, land crossing, and rail USCS inspection environments.

THE ROLE OF SCIENCE AND TECHNOLOGY AND OUR NATIONAL LABORATORIES

The science and technology response to our homeland security challenges must draw broadly on the talent and expertise resident in our research universities, our industry, and in all the government laboratories managed by multiple agencies. The national laboratories managed by DOE's Office of Science and the National Nuclear Security Administration will play a very substantial role, particularly on weapons of mass destruction issues. These laboratories have specialized capabilities in several areas of science and technology, such as the control and detection of nuclear materials, and expertise pertinent to radiological, chemical and biological threats. The national laboratories maintain the interdisciplinary approach and scientific and engineering breadth necessary to take a broad systems view of these problems, and have the ability to deliver solutions in a secure environment.

I very much appreciate the opportunity to provide this testimony and will be pleased to answer questions or provide any additional information that would be helpful.

PREPARED STATEMENT OF BARRY S. HOWE, VICE PRESIDENT, THERMO ELECTRON CORPORATION

Thank you Mr. Chairman and members of the subcommittee for this opportunity to submit testimony on behalf of my company, Thermo Electron Corporation (NYSE:TMO), a technology company based in Waltham, Massachusetts. As a senior executive at a major technology firm, my role here today is to offer some ideas on how companies like Thermo Electron can be a partner with government to apply proven technological solutions to the serious homeland security challenges facing our nation—solutions that are already available and in use successfully at border checkpoints and in multiple other applications around the world today.

Improving security at the nation's borders, airports, and seaports has become a top national priority. Recent events have heightened concerns about the potential use of weapons of mass destruction and so-called "dirty" bombs. Given, for example, that the U.S. Customs Agency is currently equipped to screen a small percentage of the large cargo containers that enter the United States, there are clearly gaps in our nation's security system that threaten our country's safety. But current technology is available to help mitigate these risks.

We at Thermo believe that three points are particularly important to this hearing: First, when properly installed and operated, current radiation-detection systems work very well. They can and have successfully thwarted attempts to illegally transport nuclear material—regardless of the mode of transport. Second, the United States should monitor for radiation in non-traditional locations. Third, we should protect our shipping infrastructure against terrorists.

THERMO ELECTRON BACKGROUND AND QUALIFICATIONS

Thermo Electron offers a comprehensive range of security-related instruments—supporting chemical, explosive, radiological, and biological-detection capabilities—to help ensure the safety of public places and people. Many of these products have and will continue to be critical in the detection and prevention of terrorist acts, as well as for the emergency and forensic response to such events.

Our instruments have played an important role in the aftermath of September 11th. Authorities in New York and Washington D.C. deployed a variety of Thermo instruments to understand the nature and extent of the post-attack hazards. For example, Thermo's gas and particulate monitors were deployed at Ground Zero and at the Pentagon to assess levels of asbestos, carbon monoxide, carbon dioxide, formaldehyde, and ammonia to determine whether it was safe for residents and workers near the disaster sites to return to their homes and offices. In addition, our monitors are in continuous use near the Fresh Kills landfill in Staten Island (where Trade Center debris has been transported) to ensure that environmental conditions remain safe.

In the November 2001 anthrax contamination of the Senate Office Building, officials used Thermo's sophisticated sampling equipment to assess the anthrax threat, monitor the cleanup, and evaluate when it was safe to re-occupy the building.

Thermo also produces the EGIS explosive trace systems, which can detect and identify in seconds plastic, commercial, and military explosives, as well as ICAO taggants—chemical markers added to military explosives in the manufacturing process to assist detection. The EGIS system has been approved by the Transportation Security Authority (TSA) and we are in current discussions as to how it will be deployed in the U.S. to support the Congressional mandate to have 100 percent of checked baggage screened for explosives by the end of the year.

The EGIS has become the trace-detection standard throughout Europe for airport screening of bags and electronic items. It has been used to protect embassies in trouble spots worldwide, deployed to screen British Rail freight traveling through the Channel Tunnel, used in Israel to ensure maximum security at border crossings, and installed in mailrooms to screen suspicious packages.

As part of the unprecedented levels of security at the 2002 Olympic Games, the Federal Bureau of Investigation (FBI) used EGIS to search for explosives to ensure the safety of athletes and spectators. In addition, law enforcement agencies, such as the FBI and the Bureau of Alcohol, Tobacco and Firearms (ATF), and forensic laboratories use EGIS routinely on-site and in the lab for post-blast investigation to determine the type and origin of explosives.

However, it is our radiation, nuclear-material detection and radiological protection products that are most relevant to today's hearing. Thermo produces a full range of monitoring systems from hand-held, mobile, and environmental monitors for first responders to complete systems suitable for use in airports and other large public venues, as well as tunnels, border crossings, and other checkpoints.

THERMO ELECTRON RADIATION MEASUREMENT AND PROTECTION GROUP

Thermo's Radiation Measurement and Protection group is a world leader in its field. Current brand names include Bicron, Eberline, ESM, Harshaw, Mini, NE, and NNC. We have been manufacturing and supplying equipment to customers worldwide for more than 50 years. Included in the Radiation Measurement and Protection staff of approximately 350 are mechanical, electrical and software engineers, health physicists, and nuclear power plant operators. Their experience ranges from research environments, such as NASA, to nuclear submarines. This combination of technical prowess and real-world applications expertise is critical to our mission of designing and manufacturing sophisticated and technologically advanced equipment that is foolproof, yet extremely easy to use.

ISO 9001 certified, Thermo's Radiation Measurement and Protection manufacturing facilities have been tested and have been approved by the Czech Metrological Institute (Prague Test), German TUEV, PTB, CE, International Atomic Energy Agency's (IAEA) Illicit Trafficking Radiation Detection Assessment Program (ITRAP

Test), American National Standards Institute (ANSI), and the Health Physics Instrumentation Committee (HPIC), among others.

In 1987, we began development of a large-scale vehicle radiation-monitoring system at the request of the steel industry, following the inadvertent and widely publicized melting of radioactive sources at various facilities. Since then, we have manufactured and installed more than 1,400 of these state-of-the-art systems, which we continue to update and improve.

We also manufacture small-scale and pedestrian monitors for a variety of applications. End users cover the gambit of national laboratories, nuclear power plants, states (both domestic and international), steel mills, foundries, metal-recycling facilities, solid waste facilities/transfer stations, waste-to-energy plants, and most recently, the International Atomic Energy Agency (IAEA) at the Vienna International Centre (VIC) in Austria.

Our systems have been tested again and again the world-over by various entities for compliance against rigorous standards as well as to evaluate their ultimate level of reliable detection. Through the course of this testing, we have consistently met all requirements.

In 1997, we were invited and successfully participated in the Illicit Trafficking Radiation Assessment Program (ITRAP), sponsored by the IAEA, INTERPOL, and the World Customs Organization. Our Automobile and Personnel Monitor (APM), hand-held FH40G, FieldSpec (isotope identifier), and pocket/pager devices (PM1401GN/PM1703GN), products developed for this application, have not only surpassed the ITRAP requirements for overall performance—sensitivity, usability, and reliability—but have proven to be a very cost-effective solution for the interdiction of radioactive materials. The Weapons of Mass Destruction Civilian Support Team (WMDCST) has equipped their 35 teams around the country with our FieldSpec devices.

SECURITY AT INTERNATIONAL BORDERS

Since 1992, the U.S. government has spent $86 million on radiation-detection equipment and personnel training in 30 countries of the former Soviet Union and Central and Eastern Europe as part of the Second Line of Defense, a Department of Energy initiative. In a study released just last month, the General Accounting Office assessed these efforts to stop the smuggling of radioactive materials. The report detailed a number of problems with the deployment overseas of radiation-detection equipment. However, the GAO also recounted noteworthy success stories. Over the past 10 years, according to the GAO, 181 attempts to smuggle nuclear material have been foiled at international borders.

Thermo Electron has placed 80 radiation-monitoring systems at border crossings in 15 countries around the world, including Argentina, Austria, Canada, China, the Czech Republic, Estonia, Finland, Germany, Latvia, Poland, Slovakia, Spain, United Kingdom, and the United States (summarized in Appendix B). They comply with existing international standards, and are proof that current, readily available technology can make a difference.

We believe it continues to be an important priority to stop smuggling at its source. However, as the GAO report demonstrates, the United States government has had little control over other countries' use of the radiation-detection equipment that our taxpayer dollars have funded. Fortunately, we have designed our equipment to work with remote-monitoring capabilities so that any alarm can notify supervisory personnel as appropriate. The equipment can also be monitored remotely to verify status, review history, and prepare reports detailing any alarms, equipment failures, and downtime—which can be a vital part of any follow-up program to help ensure the Second Line of Defense program increases its effectiveness.

Ultimately, it is clear from the GAO findings that we cannot rely on other nations to prevent the illicit export of radioactive material to our shores. We should control our own destiny by having a comprehensive radiation-detection program in place at our own borders, airports, and seaports.

MONITORING FOR RADIATION IN NON-TRADITIONAL LOCATIONS

Protecting military installations, government buildings, and other government facilities remains a critical radiation-detection priority. However, the obvious aim of these terrorists is to target our symbols and our citizens. Some of the clearest threats today involve large gatherings of everyday people at national holiday celebrations, parades, protests, and sporting events.

We believe the federal government can provide protection at these venues that is flexible, effective, and a prudent use of public funding using mobile systems, including handheld devices, described in Appendix A, the product section of this document.

PROTECTING THE NATION'S SHIPPING INFRASTRUCTURE AGAINST TERRORISTS

Courier services and the postal system are obvious areas of vulnerability. Recent anthrax incidents have taught us that terrorists can send their packages of destruction using our own infrastructure.

Thus, the same detection systems that the United States should put in place at our borders, seaports, and airports should also be installed at courier and shipping locations around the world.

A leading worldwide courier service has been testing a radiation-monitoring program with Thermo Electron equipment at a U.S. facility. We are also in discussions with other commercial vendors to determine their needs at locations around the world. We believe a comprehensive system of radiation-monitors at courier sites would go a long way in defending this nation and others against terrorism.

STANDARDS

One topic that may be discussed today is whether the U.S. should delay deployment of radiation-detection systems to develop new, U.S. standards for these devices. Thermo has participated in the development of existing standards and would be willing to offer expertise again in the ongoing development of any new standards.

Thermo's devices already comply with the existing international standards that were developed with extensive involvement of American experts. An integral part of the standards requires manufacturers to have all products independently tested. These standards were approved by two highly respected, multi-national organizations—the International Atomic Energy Agency and the International Electrotechnical Commission. They are also the basis for American Society for Testing and Materials standards, now in draft form. We believe the existing international standards, and the ASTM draft rules modeled on them, do constitute a well-considered and effective system of protection.

One critical issue for the committee is how our government can fully leverage the capability of the nation's technology providers and expedite the deployment of proven equipment that can effectively detect radiation today.

September 11th and its aftermath have brought the threat of nuclear and radiological terror to the national consciousness. But we at Thermo have been working successfully to develop solutions for a robust and reliable security system for years—in close partnership with entities like the Department of Energy, the Department of Justice, U.S. Customs, the national laboratories, as well as city and state emergency response teams and first responders.

Of course, terrorists will always try to defeat any security system to accomplish their destructive goals. That is why United States government agencies and the nation's technology companies should continue to work in partnership to develop even better technologies for tomorrow. We at Thermo continue to invest R&D resources to leverage our current technologies as platforms for the next generation of advanced products, including looking at ways to integrate our multiple, proven detection capabilities—radiation, explosives, chemical, and biological—into a comprehensive solution.

But products that are effective and available today can be put in place immediately to ensure the security of our nation, and the safety of our citizens.

Thank you Mr. Chairman for this opportunity to testify on behalf of Thermo Electron Corporation.

PREPARED STATEMENT OF JIM HOLSEN, VICE PRESIDENT OF ENGINEERING, UNITED PARCEL SERVICE, INC.

Good morning Mr. Chairman and members of the Subcommittee. My name is Jim Holsen and I am Vice President of Engineering for United Parcel Service.

I have responsibility to oversee the project that UPS has undertaken to place radiation-detection equipment at key international locations. We made this decision, as we do with all of our security measures, based on an ongoing assessment of the risks. We have had discussions with a number of governmental agencies and with staff of this committee as we have assessed the current risks, and we have determined that deploying the equipment at this time is a prudent decision.

We think our experience provides some useful lessons in how the private sector and governmental agencies could improve their cooperation to enhance the efficacy of the security-related activities of both. The new Department of Homeland Security should foster appropriate relationships between government and the private sector that encourage cooperation.

Because UPS operates all over the world, we have been dealing with international security threats for many years. The multi-faceted security measures we employ are developed based on our continuing risk assessment. Certainly, since September 11, our threat assessments have taken on new dimensions. These decisions have to be made in the context of a business that involves the delivery of millions of packages every day, with time commitments that require an extremely efficient and time-sensitive system, which is crucial to the flow of commerce. Our system includes many inherent security measures.

In developing the radiation-detection equipment deployment strategy, we evaluated our system and the available equipment. We discussed the technology with a number of governmental agencies, vendors and other private sector entities. Our plan is based on the best information available to us at this time. As with all threats, we plan to continue to monitor this risk and may modify our approach if new information indicates that such modification is needed.

I want to emphasize that we have had very cooperative and helpful discussions with the various government agencies we have consulted while developing this strategy. However, issues have arisen that have made the process difficult, inefficient, and perhaps less effective than it could otherwise be. These issues go beyond any one of those agencies, and we believe are as frustrating to the governmental officials we spoke to as they are to us.

The first issue relates to the number of different agencies involved in radiation monitoring. We have consulted with the Customs Service, the Department of Energy, and the Transportation Security Administration. In addition, we have met with the White House Office of Science and Technology Policy and the Office of Homeland Security. We have found that each of these agencies approaches the issues differently. We have also found that these agencies are working with different outside labs and experts. While we have found all of these agencies helpful, it would have been preferable to have one authoritative voice speaking for the federal government. We are put in the position of making decisions about our deployment strategy without consistent, definitive knowledge. To wait until such information may be available leaves the risk unabated.

The second issue of concern is related to the first. We need guidance on the nature of the threats we are trying to address and have been unable to obtain such information from the experts within the agencies with whom we have talked. The government experts have been cooperative and helpful, but their ability to respond to our inquiries has been limited by the restrictions on the information we need. Recognition of the appropriate level of information sharing needed is critical.

While we are willing to work appropriately with governmental agencies to address threats—a role we have played for some time—we cannot do so in the absence of appropriate intelligence information. If the government is unwilling to share information with us, we cannot adequately assist the government in addressing the threats.

In conclusion, we believe that the deployment of this equipment is a prudent step in light of the information available to us at this time. We will continue to evaluate the nature of the threats with the best available information. We will continue our cooperation with the interested governmental agencies and hope that our concerns regarding coordination and intelligence sharing will be addressed. The new Department of Homeland Security should foster appropriate relationships between the government and the private sector.

Again, thank you for the opportunity to meet with you today, and I am prepared to answer any questions that you may have.

Prepared Statement of K. David Nokes, Sandia National Laboratories

Mr. Chairman and distinguished members, it is my pleasure to appear again before this committee. I am David Nokes, Director of the Systems Assessment and Research Center and Coordinator for Sandia National Laboratories' homeland security and combating terrorism activities. My statement is an addendum to the one I provided at your June 25 hearing.

I would like to provide Sandia's views on the role of Science and Technology (S&T) within the new Department of Homeland Security (DHS) and some thoughts on how S&T might be organized.

We believe that a robust and comprehensive S&T portfolio within DHS is absolutely essential if this country is to achieve the breakthrough improvements that it must achieve in homeland security performance. Furthermore, the S&T program must address a range of very different needs. It is important to recognize that the

S&T needs of DHS are a continuum ranging from off-the-shelf items to the fundamental research necessary to solve exceptionally difficult problems.

We must first address the urgent, pressing problems that can at least be partially solved by putting existing, known technology into the hands of the people in the field who have the day-to-day responsibility for homeland security. This task is largely one of quickly establishing performance requirements and then transferring the technology to commercial entities for efficient production.

An example of this class of problem is the detection of clandestine nuclear weapons and Radiological Dispersal Devices (RDDs), so-called "dirty bombs," crossing into the United States at legal points of entry. Sandia has demonstrated equipment that, within this constrained environment, has a very high probability of detecting such devices, even when shielded, and alerting officials in real time. We have demonstrated a very low rate of false and nuisance alarms. I believe that we are well-positioned to move beyond the demonstration stage and implement widespread deployment at ports of entry.

Among the challenges that require substantial additional work are detection systems for chemical and especially biological attacks. Although point sensors for some agents exist and limited demonstrations of area sensors have been performed, much developmental work will be required to broaden the spectrum of agents that are detectable, lower the false alarm rate, and ensure continuous operation. In addition, the command and control architecture to network these sensors into an effective and affordable system that can protect large urban areas has not been designed.

Detecting clandestine nuclear weapons or RDDs in large urban areas (as opposed to ports of entry) is a problem that also needs substantial research. Although, unlike chemical or biological devices, radiological weapons all have a detectable signature prior to use, the limitations of physics prevent individual sensors from affording a large detection range. The problem becomes command and control of networks of sensors and developing a strategy that optimizes performance and cost.

An essential first step for the S&T portfolio at DHS will be developing strategic planning and prioritization of the S&T investments of the Department. This must be driven by threat and vulnerability analyses that identify the areas with greatest need.

The S&T needs of the DHS are exceptionally diverse because of the great variety of the individual elements of its mission. Each Under Secretary of Homeland Security will have unique R&D requirements. Clearly, the Under Secretary for Chemical, Biological, Radiological, and Nuclear Countermeasures will need access to a substantially different set of R&D resources than the Under Secretary for Border and Transportation Security.

We recommend that each Under Secretary create a laboratory network tailored for his or her missions by directly tasking existing institutions that possess the required competencies. We call this entity a "Virtual National Laboratory," and it has already been tried and proven as an effective model for multi-institutional programs involving research and technology development. Virtual national laboratories may be of permanent or limited duration and can be reconfigured as necessary for evolving requirements.

To illustrate, the Under Secretary for Chemical, Biological, Radiologcial, and Nuclear Countermeasures may design one or more matrixed laboratory systems specific to his needs that include representation from the National Institutes of Health, some DOE/NNSA labs, leading research universities, and the pharmaceutical industry. The Under Secretary for Border and Transportation Security may design one or more matrixed laboratory systems specific to her needs that include representation from the Naval Research Laboratory and other DoD labs, DOE/NNSA, industry, and universities.

Each of these "virtual national laboratories" would have a defined organizational structure with a laboratory director and program directors, although it would own no real property. The laboratory director would manage a Laboratory Liaison Council (LLC) with representation from the constituent institutions. The LLC would be the Under Secretary's vehicle for direct access to the national laboratory system. He would not have to go through each institution's sponsoring federal agency in a "work-for-others" procurement process. This structure is illustrated in the diagram attached as supplemental material to my statement.

A significant advantage of this concept is that it encourages competition of the right sort—competition of ideas (not direct competition of labs for money)—and cooperation on results, pulling together the right resources for a particular mission focus. It encourages rapid transition of the fruits of research into application, and helps avoid the "valley of death'" that often prevents promising research from being developed and deployed.

Specific suggestions follow:

- Each Under Secretary should have authority for "conducting a national scientific research and development program to support the missions of the Department" for which he or she is responsible, "...including directing, funding, and conducting research and development relating to the same" (as per Sec. 301 (2) of the President's bill).
- In addition, each Under Secretary should appoint a Director of Research and Development with authority to immediately create networked laboratory systems (virtual national laboratories) through cooperative arrangements with federal, academic, and private research institutions. Appropriate funding will be required.
- Directors of Research and Development will be assisted by Laboratory Liaison Councils with representation from the institutions of the virtual national laboratory.
- Directors of Research and Development should have authority and appropriated funding to originate and award Cooperative Research and Development Agreements (CRADAs) and other technology transfer mechanisms between virtual national laboratories and industry on an expedited basis.
- DHS legislation should authorize all relevant federally funded R&D institutions to accept direct tasking from the DHS and should instruct "landlord" agencies to facilitate DHS taskings of institutions under their sponsorship.
- At least initially, DHS should rely on the established great laboratories of the nation rather than creating new ones for its science and technology (S&T) program. There is insufficient time to establish a "green field" laboratory that can make contributions on the scale required in a timely manner.
- Thought must be given to ensure that S&T activities are agile and not encumbered with bureaucratic processes that stifle the imaginative and innovative work required if we are to be successful. New processes will be necessary in some cases, rather than importing existing ones from organizations brought into the new department.
- As recommended by the National Research Council in their recent report, *Making the Nation Safer: The Role of Science and Technology in Countering Terrorism,* an office of "Under Secretary for Technology" should be created, reporting to the Secretary (p. 12-6). This office will manage a strategic, peer-reviewed research program with universities, national laboratories, and industry. Sustained funding at the mission level will be required.
- Also as recommended by the National Research Council (p. 12-7), a Homeland Security Institute should be established as a Federally Funded Research and Development Center (FFRDC) under the direction of the Under Secretary for Technology. This entity should perform policy and systems analysis, help define standards and metrics, and assist agencies with evaluating technologies for deployment.

The creation of the new DHS will be an enormous undertaking, and we appreciate your hard work helping to achieve an effective structure for securing our homeland. Sandia is committed to contributing to this urgent undertaking.

Thank you, Mr. Chairman. I look forward to your questions.

United States General Accounting Office

GAO

Testimony

Before the Subcommittee on Oversight and Investigations, Committee on Energy and Commerce, House of Representatives

For Release on Delivery
Expected at 9:00 a.m.
Tuesday, July 9, 2002

HOMELAND SECURITY

Title III of the Homeland Security Act of 2002

Statement of (Ms.) Gary L. Jones
Director, Natural Resources and Environment

G A O
Accountability * Integrity * Reliability

GAO-02-927T

Mr. Chairman and Members of the Subcommittee:

I appreciate the opportunity to be here today to discuss several aspects of the Homeland Security Act of 2002. The proposed legislation would bring many federal entities with homeland security responsibilities into a Department of Homeland Security in an effort to mobilize and focus assets and resources. Title III of the proposed legislation would task the new department with developing national policy for and coordinating the federal government's research and development efforts for responding to chemical, biological, radiological, and nuclear threats. It would also transfer to the new department responsibility for certain research and development programs and other activities, including those of the Department of Energy (DOE).[1]

In my testimony today, which focuses on Title III of the proposed legislation, I will address (1) the need for clarification of certain roles and responsibilities of the new department and (2) our observations on transferring certain activities of DOE to the new department. Our testimony is based largely on our previous and ongoing work on national preparedness issues,[2] as well as a review of the proposed legislation.

In concept and if properly implemented, this proposed legislation could lead to a more efficient, effective and coordinated research effort that would provide technology to protect our people, borders, and critical infrastructure. However, the legislation does not address many issues that could impact the Department of Homeland Security's potential effectiveness. For example, while it is tasked with coordinating federal "civilian" research, the new department will also need to coordinate with the Department of Defense and the intelligence agencies that conduct research and development efforts designed to detect and respond to weapons of mass destruction. Further, the proposed legislation does not specify that a critical role of the new department will be to establish collaborative relationships with programs at all levels of government and to develop a strategic plan for research and development to implement the national policy it is charged with developing. In addition, the proposed legislation is not clear on the role of the new department in setting standards for the performance and interoperability of new

---

[1] Sections 301, 302, and 303 of the President's proposed legislation primarily cover these changes.
[2] See "Related GAO Products" at the end of this testimony.

technologies so that users can be confident that the technologies they are purchasing will perform as intended. Lacking this, the Department of Homeland Security may not be able to efficiently and effectively focus the research and development resources of the federal government to address the most important terrorist threats.

Regarding the transfer of certain activities of DOE to the new department, we believe that some of the transfers proposed in the legislation are appropriate, such as DOE's nuclear threat assessment program and the Environmental Measurements Laboratory (EML). However, we are concerned that the transfer of certain DOE research and development activities may complicate research currently being performed to accomplish multiple purposes. For example, some research programs, such as Lawrence Livermore National Laboratory's advanced scientific computing research program, have broad missions such as ensuring the reliability of our nuclear weapons stockpile that are not easily separated into homeland security research and research for other purposes. Furthermore, in some cases, such as the energy security and assurance program activities at DOE, the legislation does not clearly indicate exactly what research would be transferred.

Background

In response to global challenges the government faces in the coming years, the creation of a Department of Homeland Security provides a unique opportunity to create an extremely effective and performance-based organization that can strengthen the nation's ability to protect its borders and citizens against terrorism. There is likely to be considerable benefit over time from restructuring some of the homeland security functions, including reducing risk and improving the economy, efficiency and effectiveness of these consolidated agencies and programs. Realistically, however, in the short term, the magnitude of the challenges that the new department faces will clearly require substantial time and effort, and will take additional resources to make it fully effective.

Recently, we testified that Congress should consider several very specific criteria in its evaluation of whether individual agencies or programs should be included or excluded from the proposed department.[4] Those criteria include the following:

- Mission Relevancy: Is homeland security a major part of the agency or program mission? Is it the primary mission of the agency or program?

- Similar Goals and Objectives: Does the agency or program being considered for the new department share primary goals and objectives with the other agencies or programs being consolidated?

- Leverage Effectiveness: Does the agency or program being considered for the new department create synergy and help to leverage the effectiveness of other agencies and programs or the new department as a whole? In other words, is the whole greater than the sum of the parts?

- Gains Through Consolidation: Does the agency or program being considered for the new department improve the efficiency and effectiveness of homeland security missions through eliminating duplications and overlaps, closing gaps and aligning or merging common roles and responsibilities?

- Integrated Information Sharing/Coordination: Does the agency or program being considered for the new department contribute to or leverage the ability of the new department to enhance the sharing of critical information or otherwise improve the coordination of missions and activities related to homeland security?

- Compatible Cultures: Can the organizational culture of the agency or program being considered for the new department effectively meld with the other entities

---

[4]U.S. General Accounting Office, *Homeland Security: Proposal for Cabinet Agency Has Merit, But Implementation Will be Pivotal to Success*, GAO-02-886T (Washington, D.C.: June 25, 2002).

that will be consolidated? Field structures and approaches to achieving missions vary considerably between agencies.

- Impact on Excluded Agencies: What is the impact on departments losing components to the new department? What is the impact on agencies with homeland security missions left out of the new department?

Federally sponsored research and development efforts, a key focus of the proposed legislation, enhance the government's capability to counter chemical, biological, radiological, and nuclear terrorist threats by providing technologies that meet a range of crisis- and consequence-management needs. Research and development efforts for these technologies, however, can be risky, time consuming, and costly. Such efforts also may need to address requirements not available in off-the-shelf products. These factors limit private and public research and development efforts for these technologies, necessitating federal government involvement and collaboration.

Many federal agencies and interagency working groups have recently deployed or are conducting research on a variety of technologies to combat terrorism. Recently deployed technologies include a prototype biological detection system used at the Salt Lake City Olympics and a prototype chemical detection system currently being used in Washington D.C.'s metro system that was developed by DOE. Technologies under development include new or improved vaccines, antibiotics, and antivirals being developed by the National Institutes of Health. In addition, the Centers for Disease Control and Prevention, in collaboration with other federal agencies, are conducting research on the diagnosis and treatment of smallpox. Moreover, the Food and Drug Administration is investigating a variety of biological agents that could be used as terrorist weapons. Other federal agencies such as the Department of Defense and intelligence community are engaged in similar research and development activities, such as research on technology to protect combatants from chemical and biological agents.

Roles and Responsibilities of the Proposed Department of Homeland Security Need to be Clarified

Certain roles and responsibilities of the Department of Homeland Security in managing research and development need to be clarified. Under the proposed legislation, the Department of Homeland Security would be tasked with developing national policy for and coordinating the federal government's civilian research and development efforts to counter chemical, biological, radiological, and nuclear threats. However, while coordination is important, it will not be enough. Federal agency coordination alone may not address the specific needs of state and local governments, such as those of local police and fire departments that will use this technology. In our view, the proposed legislation should also specify that a role of the new department will be to develop collaborative relationships with programs at all levels of government—federal, state, and local—to ensure that users' needs and research efforts are linked. We also believe the legislation should be clarified to ensure that the new department would be responsible for the development of a single national research and development strategic plan. Such a plan would help to ensure that research gaps are filled, unproductive duplication is minimized, and individual agency plans are consistent with the overall goals. Moreover, the proposed legislation, as written, is unclear about the new department's role in developing standards for the performance and interoperability of new technologies to address terrorist threats. We believe the development of these standards must be a priority of the new department.

Shortfalls in Current Research Coordinating Efforts

The limitations of existing coordination and the critical need for a more collaborative, unified research structure has been amply demonstrated in the recent past. We have previously reported that while agencies attempt to coordinate federal research and development programs in a variety of ways, breakdowns occur, leading to research gaps and duplication of effort.[5] Coordination is limited by compartmentalization of efforts

[5] U.S. General Accounting Office, *Chemical and Biological Defense: Coordination of Nonmedical Chemical and Biological R&D Programs,* GAO/NSIAD-99-160 (Washington, D.C.: August 16, 1999), and U.S. General

because of the sensitivity of the research and development programs, security classification of research, and the absence of a single coordinating entity to ensure against duplication. For example, the Department of Defense's Defense Advanced Research Projects Agency was unaware of U.S. Coast Guard's plans to develop methods to detect biological agents on infected cruise ships and, therefore, was unable to share information on its potentially related research to develop biological detection devices for buildings.

Opportunities to Improve Existing Legislative Proposal

Although the proposed legislation states that the new department will be responsible for developing national policy and coordinating research and development, it has a number of limitations that could weaken its effectiveness. First, the legislation tasks the new department with coordinating the federal government's "civilian efforts" only. We believe the new department will also need to coordinate with the Department of Defense and the intelligence agencies that conduct research and development efforts designed to detect and respond to weapons of mass destruction. The proposed transfer of some DOE research and development efforts to the Department of Homeland Security also does not eliminate potential overlaps, gaps, and opportunities for collaboration. Coordination will still be required within and among the 23 DOE national laboratories. For example, our 2001 report noted that two offices within Sandia National Laboratory concurrently and separately worked on similar thermal imagery projects for two different federal agencies, rather than consolidating the requests and combining resources. In addition, local police and fire departments and state and local governments possess practical knowledge about their technological needs and relevant design limitations that should be taken into account in federal efforts to provide new equipment, such as protective gear and sensor systems. To be most effective, the new department will have to develop collaborative relationships with all these organizations to facilitate technological improvements and encourage cooperative behavior.

Accounting Office, *Combating Terrorism: Selected Challenges and Related Recommendations,* GAO-01-822 (Washington, D.C.: September 20, 2001).

The existing proposal leaves a number of problems unaddressed as well. For example, while the proposed legislation is clear that the position of Undersecretary for Chemical, Biological, Radiological, and Nuclear Countermeasures will be responsible for developing national policy for federal research and development, there is no requirement for a strategic plan for national research and development that could address coordination, reduce potential duplication, and ensure that important issues are addressed. In 2001, we recommended the creation of a unified strategy to reduce duplication and leverage resources, and suggested that the plan be coordinated with federal agencies performing research as well as with state and local authorities.[6] The development of such a plan would help to ensure that research gaps are filled, unproductive duplication is minimized, individual agency plans are consistent with the overall goals, and a basis for assessing the success of the research and development efforts.

Also, while the legislation calls for the establishment of guidelines for state and local governments to implement countermeasures for chemical, biological, radiological, and nuclear terrorism threats, it is not clear to us what these guidelines are to entail. In this regard, we believe it will be important to develop standards for the performance and interoperability of new technologies, something that the legislation does not specifically address. For example, we had discussions with officials from the Utah State Department of Health who prepared for the 2002 Winter Olympic Games. These officials said that local police and fire departments had been approached by numerous vendors offering a variety of chemical and biological detection technology for use during the Olympics. However, these state and local officials were unsure of the best technology to purchase and could find no federal agency that would provide guidance on the technologies. They told us that if the science backing up the technology is poor or the data the technology produces are faulty, the technology can do more harm than good.

Further, the legislation would allow the new department to direct, fund, and conduct research related to chemical, biological, radiological, nuclear, and other emerging threats on its own. This raises the potential for duplication of efforts, lack of efficiency, and an

[6]GAO-01-822

increased need for coordination with other departments that would continue to carry out relevant research. We are concerned that the proposal could result in a duplication of capacity that already exists in the current federal laboratories.

Transferring Certain Activities of DOE to the Department of Homeland Security Raises Concerns

Under Title III of the proposed legislation, a number of DOE programs and activities would be transferred to the new department. Some of these transfers seem appropriate. However, in other cases we are concerned about the transfers because of the potential impact on programs and activities that currently support missions beyond homeland security. Finally, in some cases, transfers proposed by the legislation are not laid out in enough detail to permit an assessment. We discuss each of these groups of transfers below.

Transfer of Certain DOE Activities Seems Appropriate

Title III proposes to transfer to the Department of Homeland Security certain DOE activities that seem appropriate. Specifically, Title III proposes to transfer the nuclear threat assessment program and activities of the assessment, detection, and cooperation program in DOE's international Materials, Protection, and Accountability Program (MPC&A). The threat assessment program and activities, among other things, assesses the credibility of communicated nuclear threats, analyzes reports of illicit nuclear material trafficking, and provides technical support to law enforcement agencies regarding nuclear material/weapons. We would agree with officials of the Office of Nuclear Threat Assessment and Detection who view the potential transfer to the Department of Homeland Security positively. We base our agreement on the fact that, according to officials from DOE, the transfer would not have a negative impact on the rest of the MPC&A program because the functions are separate and distinct. Further, the transfer could tie the office in more closely with the other agencies they work with, such as Customs.

Another program that we believe could be appropriately transferred to the new department is the Environmental Measurements Laboratory (EML), located in New York

City. This government-operated laboratory operates under the Office of Science and Technology in the Office of Environmental Management at DOE. EML provides program management, technical assistance and data quality assurance for measurements of radiation and radioactivity relating to environmental restoration, global nuclear nonproliferation, and other priority issues for DOE, as well as for other government, national and international organizations. According to the laboratory director, the laboratory is completely in favor of the transfer to the proposed Department of Homeland Security and would fit in very well with it. We believe the transfer is appropriate because, unlike some other transfers proposed under Title III, the entire laboratory would be transferred. While it is a multiprogram laboratory serving several elements of DOE as well as other organizations, serving multiple clients could continue under a "work for others" contracting arrangement whether the laboratory was housed within DOE or the Department of Homeland Security.

Some Proposed Transfers Give Reasons For Concern

Title III proposes transferring the parts of DOE's nonproliferation and verification research and development program that conduct research on systems to improve the nation's capability to prepare for and respond to chemical and biological attacks. The legislation also proposes transferring a portion of the program's proliferation detection research. This includes work on developing sensors to help the Coast Guard monitor container shipping at ports of entry. These proposed transfers raise concerns because much of the program's research supports both homeland security and international nonproliferation programs. These programs have broad missions that are not easily separated into homeland security research and research for other purposes and the proposed legislation is not clear how these missions would continue to be accomplished. Furthermore, we are concerned that the legislation does not clearly indicate whether only the programmatic management and funding would move or also the scientists carrying out the research. Moving the scientists may not be prudent. This is because the research is currently conducted by multiprogram laboratories that employ scientists skilled in many disciplines who serve many different missions and whose research benefits from their interactions with colleagues within the laboratory.

In addition, we believe transferring control of some scientists within the DOE national laboratories to the Department of Homeland Security could complicate an already dysfunctional DOE organizational structure by blurring lines of authority and accountability. DOE carries out its diverse missions through a network of multilayered field offices that oversee activities at the national laboratories and other DOE facilities widely dispersed throughout the country. The structure inherited by DOE and the different program cultures and management styles within that structure have confounded DOE's efforts to develop a more effective organization. Transferring control of scientists within the national laboratories could complicate the accomplishment of homeland security missions and DOE's other missions by adding additional lines of authority and accountability between the laboratory scientists, DOE, and the Department of Homeland Security. One alternative would be for the new department to contract with DOE's national laboratories to conduct the research under "work for others" contracts. This would allow for direct contact between the Department of Homeland Security and the laboratories conducting the research without creating a new bureaucracy. Many federal agencies such as the Department of Defense and intelligence agencies currently use this contracting arrangement with the national laboratories.

We have similar concerns about transferring two other activities to the new department:

- The advanced scientific computing research program and activities at Lawrence Livermore National Laboratory are developing supercomputer hardware and software infrastructure aimed at enabling laboratory and university researchers to solve the most challenging scientific problems at a level of accuracy and detail never before achieved. Research conducted under the program include; designing materials atom-by-atom, revealing the functions of proteins, understanding and controlling plasma turbulence, designing new particle accelerators and modeling global climate change. This program is an integral part of DOE's efforts to ensure that the nuclear weapons stockpile is safe and secure. This program may be difficult to separate into homeland security research and research for other purposes.

- The Life Sciences Division within the DOE Office of Science's Biological and Environmental Research Program manages a diverse portfolio of research to develop fundamental biological information and to advance technology in support of DOE's missions in biology, medicine, and the environment. For example, it is determining the whole genome sequences of a variety of infectious bacteria, including anthrax strains—a first step toward developing tests that can be used to rapidly identify their presence in the environment.

In both of these instances, the programs serve multiple missions. These dual-purpose programs have important synergies that we believe should be maintained. We are concerned that transferring control over these programs to the new department has the potential to disrupt some programs that are critical to other DOE missions, such as the reliability of our nuclear weapons. We do not believe that the proposed legislation is sufficiently clear on how both the homeland security and these other missions would be accomplished.

Transfer of Some Activities Is Unclear

The details of two other transfers proposed in the legislation are unclear. First, Title III would transfer the intelligence program activities at Lawrence Livermore National Laboratory. These intelligence activities are related to the overall program carried out by DOE's Office of Intelligence. The Office of Intelligence gathers information related to DOE's missions--energy, nuclear weapons, nuclear proliferation, basic science, radiological research and environmental cleanup. To support this overall intelligence program, Lawrence Livermore National Laboratory, like other weapons laboratories, conducts intelligence activities. At Lawrence Livermore, the "Z" division conducts these activities and has special intelligence expertise related to tracking the nuclear capabilities of countries other than Russia and China. Importantly, the "Z" division receives funding from other DOE programs and/or offices as well as funding from other federal agencies (Department of Defense, Federal Bureau of Investigation, Central Intelligence Agency, etc.). According to officials at DOE Headquarters and Lawrence Livermore National Laboratory, only about $5 million of the division's $30-50 million budget comes from DOE's Office of Intelligence. These officials said the transfer would most likely affect only the $5 million that DOE's Office of Intelligence directly provides to the laboratory, but this is not clear in the proposed legislation. As with other DOE programs discussed in this testimony, the staff that carry out these activities are contractor employees and it is not clear how they would be transferred to the Department of Homeland Security. Moreover, DOE headquarters and other laboratories also have a role in intelligence, and the legislation does not propose to transfer any of their intelligence functions.

Another area of Title III where the details are unclear is the transfer of "energy security and assurance program activities." These activities are carried out by the Office of Energy Assurance, which was created in November 2001 to work with state and local government and industry to strengthen the security of the United States through the application of science and technology to improve the reliability and security of the national energy infrastructure. The national energy infrastructure includes (1) physical and cyber assets of the nation's electric power, oil, and natural gas infrastructures; (2) interdependencies among physical and cyber energy infrastructure assets; (3) national energy infrastructure's interdependencies with all other critical national infrastructures. At the time this testimony was being prepared, DOE and the Office of Homeland Security were trying to define the scope of the proposed transfer.

***

Mr. Chairman, this completes my prepared statement. I would be happy to respond to any questions you or other Members of the Committee may have at this time.

Contact and Acknowledgements

For further information about this testimony, please contact Gary Jones at (202) 512-3841. Gene Aloise, Seto J. Bagdoyen, Ryan T. Coles, Darryl W. Dutton, Kathleen H. Ebert, Laurie E. Ekstrand, Cynthia Norris and Keith Rhodes also made key contributions to this testimony.

Related GAO Products

<u>Homeland Security</u>

Homeland Security: Intergovernmental Coordination and Partnership Will Be Critical to Success. GAO-02-901T. Washington, D.C.: July 3, 2002

Homeland Security: Intergovernmental Coordination and Partnership Will Be Critical to Success. GAO-02-900T. Washington, D.C.: July 2, 2002

Homeland Security: Intergovernmental Coordination and Partnership Will Be Critical to Success. GAO-02-899T. Washington, D.C.: July 1, 2002

Homeland Security: New Department Could Improve Coordination but May Complicate Priority Setting. GAO-02-893T. Washington, D.C.: June 28, 2002.

Homeland Security: Proposal for Cabinet Agency Has Merit, but Implementation Will Be Pivotal to Success. GAO-02-886T. Washington, D.C.: June 25, 2002.

Homeland Security: New Department Could Improve Coordination but May Complicate Public Health Priority Setting. GAO-02-883T. Washington, D.C.: June 25, 2002.

Homeland Security: Key Elements to Unify Efforts Are Underway but Uncertainty Remains. GAO-02-610. Washington, D.C.: June 7, 2002.

Homeland Security: Responsibility and Accountability for Achieving National Goals. GAO-02-627T. Washington, D.C.: April 11, 2002.

Homeland Security: Progress Made; More Direction and Partnership Sought. GAO-02-490T. Washington, D.C.: March 12, 2002.

Homeland Security: Challenges and Strategies in Addressing Short- and Long-Term National Needs. GAO-02-160T. Washington, D.C.: November 7, 2001.

Homeland Security: A Risk Management Approach Can Guide Preparedness Efforts. GAO-02-208T. Washington, D.C.: October 31, 2001.

Homeland Security: Need to Consider VA's Role in Strengthening Federal Preparedness. GAO-02-145T. Washington, D.C.: October 15, 2001.

Homeland Security: Key Elements of a Risk Management Approach. GAO-02-150T. Washington, D.C.: October 12, 2001.

Homeland Security: A Framework for Addressing the Nation's Efforts. GAO-01-1158T. Washington, D.C.: September 21, 2001.

Public Health

Bioterrorism: The Centers for Disease Control and Prevention's Role in Public Health Protection. GAO-02-235T. Washington, D.C.: November 15, 2001.

Bioterrorism: Review of Public Health Preparedness Programs. GAO-02-149T. Washington, D.C.: October 10, 2001.

Bioterrorism: Public Health and Medical Preparedness. GAO-02-141T. Washington, D.C.: October 9, 2001.

Bioterrorism: Coordination and Preparedness. GAO-02-129T. Washington, D.C.: October 5, 2001.

Bioterrorism: Federal Research and Preparedness Activities. GAO-01-915. Washington, D.C.: September 28, 2001.

Chemical and Biological Defense: Improved Risk Assessment and Inventory Management Are Needed. GAO-01-667. Washington, D.C.: September 28, 2001.

West Nile Virus Outbreak: Lessons for Public Health Preparedness. GAO/HEHS-00-180. Washington, D.C.: September 11, 2000.

Chemical and Biological Defense: Program Planning and Evaluation Should Follow Results Act Framework. GAO/NSIAD-99-159. Washington, D.C.: August 16, 1999.

Combating Terrorism: Observations on Biological Terrorism and Public Health Initiatives. GAO/T-NSIAD-99-112. Washington, D.C.: March 16, 1999.

Combating Terrorism

National Preparedness: Technologies to Secure Federal Buildings. GAO-02-687T. Washington, D.C.: April 25, 2002.

National Preparedness: Integration of Federal, State, Local, and Private Sector Efforts Is Critical to an Effective National Strategy for Homeland Security. GAO-02-621T. Washington, D.C.: April 11, 2002.

Combating Terrorism: Intergovernmental Cooperation in the Development of a National Strategy to Enhance State and Local Preparedness. GAO-02-550T. Washington, D.C.: April 2, 2002.

Combating Terrorism: Enhancing Partnerships Through a National Preparedness Strategy. GAO-02-549T. Washington, D.C.: March 28, 2002.

Combating Terrorism: Critical Components of a National Strategy to Enhance State and Local Preparedness. GAO-02-548T. Washington, D.C.: March 25, 2002.

Combating Terrorism: Intergovernmental Partnership in a National Strategy to Enhance State and Local Preparedness. GAO-02-547T. Washington, D.C.: March 22, 2002.

Combating Terrorism: Key Aspects of a National Strategy to Enhance State and Local Preparedness. GAO-02-473T. Washington, D.C.: March 1, 2002.

Chemical and Biological Defense: DOD Should Clarify Expectations for Medical Readiness. GAO-02-219T. Washington, D.C.: November 7, 2001.

Anthrax Vaccine: Changes to the Manufacturing Process. GAO-02-181T. Washington, D.C.: October 23, 2001.

Chemical and Biological Defense: DOD Needs to Clarify Expectations for Medical Readiness. GAO-02-38. Washington, D.C.: October 19, 2001.

Combating Terrorism: Considerations for Investing Resources in Chemical and Biological Preparedness. GAO-02-162T. Washington, D.C.: October 17, 2001.

Combating Terrorism: Selected Challenges and Related Recommendations. GAO-01-822. Washington, D.C.: September 20, 2001.

Combating Terrorism: Actions Needed to Improve DOD Antiterrorism Program Implementation and Management. GAO-01-909. Washington, D.C.: September 19, 2001.

Combating Terrorism: Comments on H.R. 525 to Create a President's Council on Domestic Terrorism Preparedness. GAO-01-555T. Washington, D.C.: May 9, 2001.

Combating Terrorism: Accountability Over Medical Supplies Needs Further Improvement. GAO-01-666T. Washington, D.C.: May 1, 2001.

Combating Terrorism: Observations on Options to Improve the Federal Response. GAO-01-660T. Washington, DC: April 24, 2001.

Combating Terrorism: Accountability Over Medical Supplies Needs Further Improvement. GAO-01-463. Washington, D.C.: March 30, 2001.

Combating Terrorism: Comments on Counterterrorism Leadership and National Strategy. GAO-01-556T. Washington, D.C.: March 27, 2001.

Combating Terrorism: FEMA Continues to Make Progress in Coordinating Preparedness and Response. GAO-01-15. Washington, D.C.: March 20, 2001.

Combating Terrorism: Federal Response Teams Provide Varied Capabilities; Opportunities Remain to Improve Coordination. GAO-01-14. Washington, D.C.: November 30, 2000.

Combating Terrorism: Need to Eliminate Duplicate Federal Weapons of Mass Destruction Training. GAO/NSIAD-00-64. Washington, D.C.: March 21, 2000.

Combating Terrorism: Chemical and Biological Medical Supplies Are Poorly Managed. GAO/T-HEHS/AIMD-00-59. Washington, D.C.: March 8, 2000.

Combating Terrorism: Chemical and Biological Medical Supplies Are Poorly Managed. GAO/HEHS/AIMD-00-36. Washington, D.C.: October 29, 1999.

Combating Terrorism: Observations on the Threat of Chemical and Biological Terrorism. GAO/T-NSIAD-00-50. Washington, D.C.: October 20, 1999.

Combating Terrorism: Need for Comprehensive Threat and Risk Assessments of Chemical and Biological Attacks. GAO/NSIAD-99-163. Washington, D.C.: September 14, 1999.

Chemical and Biological Defense: Coordination of Nonmedical Chemical and Biological R&D Programs. GAO/NSIAD-99-160. Washington, D.C.: August 16, 1999.

Combating Terrorism: Use of National Guard Response Teams Is Unclear. GAO/T-NSIAD-99-184. Washington, D.C.: June 23, 1999.

Combating Terrorism: Observations on Growth in Federal Programs. GAO/T-NSIAD-99-181. Washington, D.C.: June 9, 1999.

Combating Terrorism: Analysis of Potential Emergency Response Equipment and Sustainment Costs. GAO/NSIAD-99-151. Washington, D.C.: June 9, 1999.

Combating Terrorism: Use of National Guard Response Teams Is Unclear. GAO/NSIAD-99-110. Washington, D.C.: May 21, 1999.

Combating Terrorism: Observations on Federal Spending to Combat Terrorism. GAO/T-NSIAD/GGD-99-107. Washington, D.C.: March 11, 1999.

Combating Terrorism: Opportunities to Improve Domestic Preparedness Program Focu and Efficiency. GAO/NSIAD-99-3. Washington, D.C.: November 12, 1998.

Combating Terrorism: Observations on the Nunn-Lugar-Domenici Domestic Preparedness Program. GAO/T-NSIAD-99-16. Washington, D.C.: October 2, 1998.

Combating Terrorism: Observations on Crosscutting Issues. GAO/T-NSIAD-98-164. Washington, D.C.: April 23, 1998.

Combating Terrorism: Threat and Risk Assessments Can Help Prioritize and Target Program Investments. GAO/NSIAD-98-74. Washington, D.C.: April 9, 1998.

Combating Terrorism: Spending on Governmentwide Programs Requires Better Management and Coordination. GAO/NSIAD-98-39. Washington, D.C.: December 1, 199'

Disaster Assistance

Disaster Assistance: Improvement Needed in Disaster Declaration Criteria and Eligibility Assurance Procedures. GAO-01-837. Washington, D.C.: August 31, 2001.

Chemical Weapons: FEMA and Army Must Be Proactive in Preparing States for Emergencies. GAO-01-850. Washington, D.C.: August 13, 2001.

Federal Emergency Management Agency: Status of Achieving Key Outcomes and Addressing Major Management Challenges. GAO-01-832. Washington, D.C.: July 9, 2001.

Budget and Management

Budget Issues: Long-Term Fiscal Challenges. GAO-02-467T. Washington, D.C.: February 27, 2002.

Results-Oriented Budget Practices in Federal Agencies. GAO-01-1084SP. Washington, D.C.: August 2001.

Managing for Results: Federal Managers' Views on Key Management Issues Vary Widely Across Agencies. GAO-01-592. Washington, D.C.: May 25, 2001.

Determining Performance and Accountability Challenges and High Risks. GAO-01-159SP. Washington, D.C.: November 2000.

Managing for Results: Using the Results Act to Address Mission Fragmentation and Program Overlap. GAO-AIMD-97-146. Washington, D.C.: August 29, 1997.

Government Restructuring: Identifying Potential Duplication in Federal Missions and Approaches. GAO/T-AIMD-95-161. Washington, D.C.: June 7, 1995.

Government Reorganization: Issues and Principles. GAO/T-GGD/AIMD-95-166. Washington, D.C.: May 17, 1995.

Grant Design

Grant Programs: Design Features Shape Flexibility, Accountability, and Performance Information. GAO/GGD-98-137. Washington, D.C.: June 22, 1998.

Federal Grants: Design Improvements Could Help Federal Resources Go Further. GAO/AIMD-97-7. Washington, D.C.: December 18, 1996.

Block Grants: Issues in Designing Accountability Provisions. GAO/AIMD-95-226. Washington, D.C.: September 1, 1995.

(360244)

United States General Accounting Office

GAO

Testimony

Before the Subcommittee on Oversight and Investigations, Committee on Energy and Commerce, House of Representatives

For Release on Delivery
Expected at 9:00 a.m.
Tuesday, July 9, 2002

HOMELAND SECURITY

New Department Could Improve Biomedical R&D Coordination but May Disrupt Dual-Purpose Efforts

Statement of Janet Heinrich
Director, Health Care—Public Health Issues

GAO
Accountability * Integrity * Reliability

GAO-02-924T

Mr. Chairman and Members of the Subcommittee:

I appreciate the opportunity to be here today to discuss one component—the potential effect on biomedical research—of the proposed creation of the Department of Homeland Security. Since the terrorist attacks of September 11, 2001, and the subsequent anthrax incidents, there has been concern about the ability of the federal government to prepare for and coordinate an effective public health response to such events, given the broad distribution of responsibility for that task at the federal level. Our earlier work found, for example, that more than 20 federal departments and agencies carry some responsibility for bioterrorism research, preparedness, and response and that these efforts are fragmented.[1]

The President's proposed Homeland Security Act of 2002[2] would bring many of the federal entities with homeland security responsibilities, including biomedical research and development, into one department. Title III of the proposed legislation would transfer responsibility for certain chemical, biological, radiological, and nuclear research and development programs and activities to the new department.[3] Much of the research in these areas is sponsored by or conducted at the Department of Health and Human Services' (HHS) National Institutes of Health (NIH). The proposal would also transfer the Laboratory Registration/Select Agent Transfer Program—which controls biological agents with the potential for use in bioterrorism—from HHS's Centers for Disease Control and Prevention (CDC) to the new department.

In order to assist the Subcommittee in its consideration of this extensive reorganization of our government, my remarks will focus on the potential effects of a reorganization on biomedical research under Title III of the President's proposal. My testimony today is based largely on our previous and ongoing work on homeland security,[4] as well as a review of the proposed legislation.

[1] U.S. General Accounting Office, *Bioterrorism: Federal Research and Preparedness Activities*, GAO-01-915 (Washington, D.C.: Sept. 28, 2001).

[2] H.R. 5005, 107th Cong. (2002).

[3] These changes are primarily covered by Sections 301, 302, and 303 of the President's proposed legislation.

[4] See Related GAO Products at the end of this testimony.

In summary, the proposed Department of Homeland Security would be tasked with developing national policy for and coordination of the federal government's civilian research and development efforts to counter chemical, biological, radiological, and nuclear threats. GAO has consistently stated that there is a need for a strategic plan and better coordination of existing research and development programs. The new department could improve coordination of the biomedical research and development efforts. We are concerned, however, that the proposed transfer of control and priority setting for research from the organizations where the research would be conducted could be disruptive to dual-purpose programs,[5] which have important synergies that need to be maintained. Transferring control over these programs, including priority setting, to the new department has the potential to disrupt some programs that are critical to basic public health responsibility. The President's proposal is not sufficiently clear on how both the homeland security and the biomedical research objectives would be accomplished. Because the select agent program's mission fits with homeland security, its transfer to the new department is appropriate.

Background

In response to global challenges the government faces in the coming years, we have a unique opportunity to create an extremely effective and performance-based organization that can strengthen the nation's ability to protect its borders and citizens against terrorism. There is likely to be considerable benefit over time from restructuring some of the homeland security functions, including reducing risk and improving the economy, efficiency, and effectiveness of these consolidated agencies and programs. Realistically, however, in the short term, the magnitude of the challenges that the new department faces will clearly require substantial time and effort, and will take additional resources to make it fully effective.

The Comptroller General has testified that the Congress should consider several very specific criteria in its evaluation of whether individual

[5] In this testimony, dual-purpose programs refer to biomedical research and development programs that are applicable to both bioterrorism and other health research. For example, NIH supports research to expand knowledge of factors that play a decisive role in determining antibiotic resistance, virulence, and invasiveness of pathogens, as well as those events or processes critical to initiating infection or influencing the severity of disease. This knowledge is useful for both intentional and naturally occurring diseases.

agencies or programs should be included or excluded from the proposed department.[6] Those criteria include the following:

- Mission Relevancy: Is homeland security a major part of the agency or program mission? Is it the primary mission of the agency or program?
- Similar Goals and Objectives: Does the agency or program being considered for the new department share primary goals and objectives with the other agencies or programs being consolidated?
- Leverage Effectiveness: Does the agency or program being considered for the new department promote synergy and help to leverage the effectiveness of other agencies and programs or the new department as a whole? In other words, is the whole greater than the sum of the parts?
- Gains Through Consolidation: Does the agency or program being considered for the new department improve the efficiency and effectiveness of homeland security missions through eliminating duplications and overlaps, closing gaps, and aligning or merging common roles and responsibilities?
- Integrated Information Sharing/Coordination: Does the agency or program being considered for the new department contribute to or leverage the ability of the new department to enhance the sharing of critical information or otherwise improve the coordination of missions and activities related to homeland security?
- Compatible Cultures: Can the organizational culture of the agency or program being considered for the new department effectively meld with the other entities that will be consolidated? Field structures and approaches to achieving missions vary considerably between agencies.
- Impact on Excluded Agencies: What is the impact on departments losing components to the new department? What is the impact on agencies with homeland security missions left out of the new department?

In the President's proposal, the new Department of Homeland Security would be responsible for conducting a national scientific research and development program, including developing national policy and coordinating the federal government's civilian efforts to counter chemical, biological, radiological, and nuclear weapons or other emerging terrorist threats. The new department would carry out its civilian health-related biological, biomedical, and infectious disease defense research and development through agreements with HHS, unless otherwise directed by

[6] U.S. General Accounting Office, *Homeland Security: Proposal for Cabinet Agency Has Merit, but Implementation Will Be Pivotal to Success*, GAO-02-886T (Washington, D.C.: June 25, 2002).

the President. As part of this responsibility, the new department would establish priorities and direction for programs of basic and applied research on the detection, treatment, and prevention of infectious diseases such as those programs conducted by NIH.

NIH supports and carries out biomedical research to study, prevent, and treat infectious and immunologic human diseases. Infectious diseases include those caused by new, emerging, and reemerging infectious agents, including those that are intentionally introduced as an act of bioterrorism. The emphasis of antiterrorism research supported by NIH has been in four areas: (1) design and testing of new diagnostic tools; (2) design, development, and clinical evaluation of therapies; (3) design, development, and clinical evaluation of vaccines; and (4) other basic research, including genome sequencing.[7]

The President's proposal also would transfer the select agent program from HHS to the new department. Currently administered by CDC, this program's mission is ensuring the security of those biologic agents that pose a severe threat to public health and safety and could be used by terrorists. The proposal provides for the new department to consult with appropriate agencies, which would include HHS, in maintaining the select agent list and to consult with HHS in carrying out the program.

[7]Genome sequencing reveals the lineup of paired chemical bases that make up a pathogen's DNA, which contains the genetic code and transmits the hereditary pattern. Sequence information can be exploited in many ways, including demarcating genes, locating therapeutic targets, and identifying mutations that contribute to drug resistance.

Proposed Department Could Improve Coordination of Research and Development Programs

The proposed Department of Homeland Security would be tasked with developing national policy for and coordinating the federal government's civilian research and development efforts to counter chemical, biological, radiological, and nuclear threats. The new department also could improve coordination of biomedical research and development efforts. In addition to coordination, the role of the new department would need to include forging collaborative relationships with programs at all levels of government and developing a strategic plan for research and development.

We have previously reported that the limited coordination among federal research and development programs may result in a duplication of efforts.[8] Coordination is hampered by the extent of compartmentalization of efforts because of the sensitivity of the research and development programs, security classification of research, and the absence of a single coordinating entity to help prevent duplication. For example, the Department of Defense's (DOD) Defense Advanced Research Projects Agency was unaware of U.S. Coast Guard plans to develop methods to detect a biological agent on an infected cruise ship and therefore was unable to share information on its research to develop biological detection devices that could have been applicable to buildings infected this way.

The new department would need to develop mechanisms to coordinate and integrate information about ongoing research and development being performed across the government related to chemical, biological, radiological, and nuclear terrorism, as well as harmonize user needs. Although the proposal tasks the new department with coordinating the federal government's "civilian efforts" only, the new department also would need to coordinate with DOD because DOD conducts biomedical research and development efforts designed to detect and respond to weapons of mass destruction. Although DOD's efforts are geared toward protecting armed services members, they may also be applicable to the civilian population. Currently, NIH is working with DOD on biomedical research and development efforts, and it is important for this collaboration to continue. An example of NIH and DOD's efforts is their support of databases to compare the sequences and functions of poxvirus genes. These searchable databases enable researchers to select targets for designing antiviral drugs and vaccines, and serve as repositories for

[8] U.S. General Accounting Office, *Combating Terrorism: Selected Challenges and Related Recommendations,*, GAO-01-822 (Washington, D.C.: Sept. 20, 2001).

information on well documented poxvirus strains to aid in detection and diagnosis.

The President's proposal could help improve coordination of federal research and development by giving one person the responsibility for a single national research and development strategy that could address coordination, reduce potential duplication, and ensure that important issues are addressed. In 2001, we recommended the creation of a unified strategy to reduce duplication and leverage resources, and suggested that the plan be coordinated with federal agencies performing the research as well as with state and local authorities.[9] Such a plan would help to ensure that research gaps are filled, unproductive duplication is minimized, and that individual agency plans are consistent with the overall goals.

Transfer of Control Over Dual-Purpose Research and Development Raises Concern

We are concerned about the implications of the proposed transfer of control and priority setting for dual-purpose research programs. For example, some research programs have broad missions that are not easily separated into homeland security research and research for other purposes. We are concerned that such dual-purpose research activities may lose the synergy arising from their current placement.

The President's proposal would transfer the responsibility for civilian biomedical defense research and development programs to the new department, but the programs would continue to be carried out through HHS. These programs, now primarily sponsored by NIH, include a variety of efforts to understand basic biological mechanisms of infection and to develop and test rapid diagnostic tools, vaccines, and antibacterial and antiviral drugs. These efforts have dual-purpose applicability. The scientific research on biologic agents that could be used by terrorists cannot be readily separated from research on emerging infectious diseases. For example, research being carried out on antiviral drugs in the NIH biodefense research program is expected to be useful in the development of treatments for hepatitis C. NIH biodefense research on enhanced immunologic responses to protect against infection and disease is critical in the development of interventions against both naturally occurring and man-made pathogens.

[9] GAO-01-822.

The proposal to transfer to the new department responsibility for research and development programs that would continue to be carried out by HHS raises many concerns. Although there is a clear need for the new department to have responsibility for setting policy, developing a strategy, providing leadership, and coordinating research and development efforts in these areas, we are concerned that control and priority-setting responsibility will not be vested in those programs best positioned to understand the potential of basic research efforts or the relevance of research being carried out in other, nonbiodefense programs. For example, NIH-funded research on a drug to treat cytomegalovirus complications in patients with HIV is now being investigated as a prototype for developing antiviral drugs against smallpox.

There is the potential that the proposal would allow the new department to direct, fund, and conduct research related to chemical, biological, radiological, nuclear, and other emerging threats on its own. This raises the potential for duplication of effort, lack of efficiency, and an increased need for coordination with other departments that would continue to carry out relevant research. Design and implementation of a research agenda is most efficient at the level of the mission agency where scientific and technical expertise resides. Building and duplicating the existing facilities and expertise in the current federal laboratories needed to conduct this research would be inefficient.

Mission of Select Agent Program Is Aligned with New Department

The proposal would transfer the Laboratory Registration/Select Agent Transfer Program from HHS to the new department. The select agent program, recently revised and expanded by the Public Health Security and Bioterrorism Preparedness and Response Act of 2002,[10] generally requires the registration of persons and laboratory facilities possessing specific biologic agents and toxins—called select agents—that have the potential to pose a serious threat to public health and safety. Select agents include approximately 40 viruses, bacteria, rickettsia, fungi, and toxins. Examples include Ebola, anthrax, botulinum, and ricin. The 2002 act expanded the program's requirements to include facilities that possess the agents as well as the facilities that transfer the agents.

The mission of the select agent program appears to be closely aligned with homeland security. As we stated earlier, one key consideration in

[10]Pub. L. No. 107-188, §§ 201-204, 116 Stat. 594, 637-647 (2002).

evaluating whether individual agencies or programs should be included or excluded from the proposed department is the extent to which homeland security is a major part of the agency or program mission. By these criteria, the transfer of the select agent program would enhance efficiency and accountability.

Concluding Observations

The President's proposal would address some shortcomings noted earlier in this statement. Better coordination could reduce wasteful duplication and increase efficiency. The mission of the select agent program is aligned with the new department and, therefore, the transfer of the program would enhance efficiency and accountability. However, we are concerned about the broad control the proposal grants to the new department for biomedical research and development. Although there is a need to coordinate these activities with the other homeland security preparedness and response programs that would be brought into the new department, there is also a need to maintain the priorities for current dual-purpose biomedical research. The President's proposal does not adequately address how to accomplish both objectives or how to maintain a priority-setting role for those best positioned to understand the relevance of biomedical research. We are also concerned that the proposal has the potential to create an unnecessary duplication of federal research capacity.

Mr. Chairman, this completes my prepared statement. I would be happy to respond to any questions you or other Members of the Subcommittee may have at this time.

Contact and Acknowledgments

For further information about this testimony, please contact me at (202) 512-7118. Robert Copeland, Marcia Crosse, and Deborah Miller also made key contributions to this statement.

80-680 - 02 - 11

Related GAO Products

Homeland Security

Homeland Security: Intergovernmental Coordination and Partnership Will Be Critical to Success. GAO-02-901T. Washington, D.C.: July 3, 2002.

Homeland Security: Intergovernmental Coordination and Partnership Will Be Critical to Success. GAO-02-900T. Washington, D.C.: July 2, 2002.

Homeland Security: Intergovernmental Coordination and Partnership Will Be Critical to Success. GAO-02-899T. Washington, D.C.: July 1, 2002.

Homeland Security: New Department Could Improve Coordination but May Complicate Priority Setting. GAO-02-893T. Washington, D.C.: June 28, 2002.

Homeland Security: Proposal for Cabinet Agency Has Merit, but Implementation Will Be Pivotal to Success. GAO-02-886T. Washington, D.C.: June 25, 2002.

Homeland Security: New Department Could Improve Coordination but May Complicate Public Health Priority Setting. GAO-02-883T. Washington, D.C.: June 25, 2002.

Homeland Security: Key Elements to Unify Efforts Are Underway but Uncertainty Remains. GAO-02-610. Washington, D.C.: June 7, 2002.

Homeland Security: Responsibility and Accountability for Achieving National Goals. GAO-02-627T. Washington, D.C.: April 11, 2002.

Homeland Security: Progress Made; More Direction and Partnership Sought. GAO-02-490T. Washington, D.C.: March 12, 2002.

Homeland Security: Challenges and Strategies in Addressing Short- and Long-Term National Needs. GAO-02-160T. Washington, D.C.: November 7, 2001.

Homeland Security: A Risk Management Approach Can Guide Preparedness Efforts. GAO-02-208T. Washington, D.C.: October 31, 2001.

Homeland Security: Need to Consider VA's Role in Strengthening Federal Preparedness. GAO-02-145T. Washington, D.C.: October 15, 2001.

Homeland Security: Key Elements of a Risk Management Approach. GAO-02-150T. Washington, D.C.: October 12, 2001.

Homeland Security: A Framework for Addressing the Nation's Efforts. GAO-01-1158T. Washington, D.C.: September 21, 2001.

Public Health

Bioterrorism: The Centers for Disease Control and Prevention's Role in Public Health Protection. GAO-02-235T. Washington, D.C.: November 15, 2001.

Bioterrorism: Review of Public Health Preparedness Programs. GAO-02-149T. Washington, D.C.: October 10, 2001.

Bioterrorism: Public Health and Medical Preparedness. GAO-02-141T. Washington, D.C.: October 9, 2001.

Bioterrorism: Coordination and Preparedness. GAO-02-129T. Washington, D.C.: October 5, 2001.

Bioterrorism: Federal Research and Preparedness Activities. GAO-01-915. Washington, D.C.: September 28, 2001.

Chemical and Biological Defense: Improved Risk Assessment and Inventory Management Are Needed. GAO-01-667. Washington, D.C.: September 28, 2001.

West Nile Virus Outbreak: Lessons for Public Health Preparedness. GAO/HEHS-00-180. Washington, D.C.: September 11, 2000.

Chemical and Biological Defense: Program Planning and Evaluation Should Follow Results Act Framework. GAO/NSIAD-99-159. Washington, D.C.: August 16, 1999.

Combating Terrorism: Observations on Biological Terrorism and Public Health Initiatives. GAO/T-NSIAD-99-112. Washington, D.C.: March 16, 1999.

Combating Terrorism

National Preparedness: Technologies to Secure Federal Buildings. GAO-02-687T. Washington, D.C.: April 25, 2002.

National Preparedness: Integration of Federal, State, Local, and Private Sector Efforts Is Critical to an Effective National Strategy for Homeland Security. GAO-02-621T. Washington, D.C.: April 11, 2002.

Combating Terrorism: Intergovernmental Cooperation in the Development of a National Strategy to Enhance State and Local Preparedness. GAO-02-550T. Washington, D.C.: April 2, 2002.

Combating Terrorism: Enhancing Partnerships Through a National Preparedness Strategy. GAO-02-549T. Washington, D.C.: March 28, 2002.

Combating Terrorism: Critical Components of a National Strategy to Enhance State and Local Preparedness. GAO-02-548T. Washington, D.C.: March 25, 2002.

Combating Terrorism: Intergovernmental Partnership in a National Strategy to Enhance State and Local Preparedness. GAO-02-547T. Washington, D.C.: March 22, 2002.

Combating Terrorism: Key Aspects of a National Strategy to Enhance State and Local Preparedness. GAO-02-473T. Washington, D.C.: March 1, 2002.

Chemical and Biological Defense: DOD Should Clarify Expectations for Medical Readiness. GAO-02-219T. Washington, D.C.: November 7, 2001.

Anthrax Vaccine: Changes to the Manufacturing Process. GAO-02-181T. Washington, D.C.: October 23, 2001.

Chemical and Biological Defense: DOD Needs to Clarify Expectations for Medical Readiness. GAO-02-38. Washington, D.C.: October 19, 2001.

Combating Terrorism: Considerations for Investing Resources in Chemical and Biological Preparedness. GAO-02-162T. Washington, D.C.: October 17, 2001.

Combating Terrorism: Selected Challenges and Related Recommendations. GAO-01-822. Washington, D.C.: September 20, 2001.

Combating Terrorism: Actions Needed to Improve DOD Antiterrorism Program Implementation and Management. GAO-01-909. Washington, D.C.: September 19, 2001.

Combating Terrorism: Comments on H.R. 525 to Create a President's Council on Domestic Terrorism Preparedness. GAO-01-555T. Washington, D.C.: May 9, 2001.

Combating Terrorism: Accountability Over Medical Supplies Needs Further Improvement. GAO-01-666T. Washington, D.C.: May 1, 2001.

Combating Terrorism: Observations on Options to Improve the Federal Response. GAO-01-660T. Washington, DC: April 24, 2001.

Combating Terrorism: Accountability Over Medical Supplies Needs Further Improvement. GAO-01-463. Washington, D.C.: March 30, 2001.

Combating Terrorism: Comments on Counterterrorism Leadership and National Strategy. GAO-01-556T. Washington, D.C.: March 27, 2001.

Combating Terrorism: FEMA Continues to Make Progress in Coordinating Preparedness and Response. GAO-01-15. Washington, D.C.: March 20, 2001.

Combating Terrorism: Federal Response Teams Provide Varied Capabilities; Opportunities Remain to Improve Coordination. GAO-01-14. Washington, D.C.: November 30, 2000.

Combating Terrorism: Need to Eliminate Duplicate Federal Weapons of Mass Destruction Training. GAO/NSIAD-00-64. Washington, D.C.: March 21, 2000.

Combating Terrorism: Chemical and Biological Medical Supplies Are Poorly Managed. GAO/T-HEHS/AIMD-00-59. Washington, D.C.: March 8, 2000.

Combating Terrorism: Chemical and Biological Medical Supplies Are Poorly Managed. GAO/HEHS/AIMD-00-36. Washington, D.C.: October 29, 1999.

Combating Terrorism: Observations on the Threat of Chemical and Biological Terrorism. GAO/T-NSIAD-00-50. Washington, D.C.: October 20, 1999.

Combating Terrorism: Need for Comprehensive Threat and Risk Assessments of Chemical and Biological Attacks. GAO/NSIAD-99-163. Washington, D.C.: September 14, 1999.

Chemical and Biological Defense: Coordination of Nonmedical Chemical and Biological R&D Programs. GAO/NSIAD-99-160. Washington, D.C.: August 16, 1999.

Combating Terrorism: Use of National Guard Response Teams Is Unclear. GAO/T-NSIAD-99-184. Washington, D.C.: June 23, 1999.

Combating Terrorism: Observations on Growth in Federal Programs. GAO/T-NSIAD-99-181. Washington, D.C.: June 9, 1999.

Combating Terrorism: Analysis of Potential Emergency Response Equipment and Sustainment Costs. GAO/NSIAD-99-151. Washington, D.C.: June 9, 1999.

Combating Terrorism: Use of National Guard Response Teams Is Unclear. GAO/NSIAD-99-110. Washington, D.C.: May 21, 1999.

Combating Terrorism: Observations on Federal Spending to Combat Terrorism. GAO/T-NSIAD/GGD-99-107. Washington, D.C.: March 11, 1999.

Combating Terrorism: Opportunities to Improve Domestic Preparedness Program Focus and Efficiency. GAO/NSIAD-99-3. Washington, D.C.: November 12, 1998.

Combating Terrorism: Observations on the Nunn-Lugar-Domenici Domestic Preparedness Program. GAO/T-NSIAD-99-16. Washington, D.C.: October 2, 1998.

Combating Terrorism: Observations on Crosscutting Issues. GAO/T-NSIAD-98-164. Washington, D.C.: April 23, 1998.

Combating Terrorism: Threat and Risk Assessments Can Help Prioritize and Target Program Investments. GAO/NSIAD-98-74. Washington, D.C.: April 9, 1998.

Combating Terrorism: Spending on Governmentwide Programs Requires Better Management and Coordination. GAO/NSIAD-98-39. Washington, D.C.: December 1, 1997.

Disaster Assistance

Disaster Assistance: Improvement Needed in Disaster Declaration Criteria and Eligibility Assurance Procedures. GAO-01-837. Washington, D.C.: August 31, 2001.

Chemical Weapons: FEMA and Army Must Be Proactive in Preparing States for Emergencies. GAO-01-850. Washington, D.C.: August 13, 2001.

Federal Emergency Management Agency: Status of Achieving Key Outcomes and Addressing Major Management Challenges. GAO-01-832. Washington, D.C.: July 9, 2001.

Budget and Management

Budget Issues: Long-Term Fiscal Challenges. GAO-02-467T. Washington, D.C.: February 27, 2002.

Results-Oriented Budget Practices in Federal Agencies. GAO-01-1084SP. Washington, D.C.: August 2001.

Managing for Results: Federal Managers' Views on Key Management Issues Vary Widely Across Agencies. GAO-01-592. Washington, D.C.: May 25, 2001.

Determining Performance and Accountability Challenges and High Risks. GAO-01-159SP. Washington, D.C.: November 2000.

Managing for Results: Using the Results Act to Address Mission Fragmentation and Program Overlap. GAO-AIMD-97-146. Washington, D.C.: August 29, 1997.

Government Restructuring: Identifying Potential Duplication in Federal Missions and Approaches. GAO/T-AIMD-95-161. Washington, D.C.: June 7, 1995.

Government Reorganization: Issues and Principles. GAO/T-GGD/AIMD-95-166. Washington, D.C.: May 17, 1995.

Grant Design

Grant Programs: Design Features Shape Flexibility, Accountability, and Performance Information. GAO/GGD-98-137. Washington, D.C.: June 22, 1998.

Federal Grants: Design Improvements Could Help Federal Resources Go Further. GAO/AIMD-97-7. Washington, D.C.: December 18, 1996.

Block Grants: Issues in Designing Accountability Provisions. GAO/AIMD-95-226. Washington, D.C.: September 1, 1995.

United States General Accounting Office

GAO

Testimony

Before the Subcommittee on Oversight and Investigations, Committee on Energy and Commerce, House of Representatives

For Release on Delivery
Expected at 9:00 a.m. EDT
Tuesday, July 9, 2002

CRITICAL INFRASTRUCTURE PROTECTION

Significant Homeland Security Challenges Need to Be Addressed

Statement of Robert F. Dacey
Director, Information Security Issues

GAO
Accountability * Integrity * Reliability

GAO-02-918T

United States General Accounting Office July 9, 2002

GAO
Accountability * Integrity * Reliability

Highlights

CRITICAL INFRASTRUCTURE PROTECTION

Significant Homeland Security Challenges Need to Be Addressed

Highlights of GAO-02-918T, testimony before the Subcommittee on Oversight and Investigations, House Committee on Energy and Commerce.

Why GAO Did This Study

Since the terrorist attacks of last September 11, the President and the Congress have taken important, aggressive action to protect the nation. Last month, the President proposed elevating homeland security to department status and, at the same time, merging into it several federal organizations. It would comprise four divisions (see graphic).

The six organizations to be moved into the new department's Information Analysis and Infrastructure Protection division (and their current parent organizations) are the National Infrastructure Protection Center (FBI), National Communications System (Defense), Critical Infrastructure Assurance Office (Commerce), Computer Security Division (National Institute of Standards and Technology), National Infrastructure Simulation and Analysis Center (Defense, Energy), and the Federal Computer Incident Response Center (General Services Administration).

At the Subcommittee's request, GAO discussed the functions to be transferred to this new division, along with the potential benefits to be achieved, and the challenges that it will likely face.

What GAO Found

As proposed, the functions of the Information Analysis and Infrastructure Protection division would include receiving and analyzing law enforcement and intelligence information, assessing cyber and physical vulnerabilities of critical infrastructures, and taking measures to protect them.

The consolidation of these six organizations into a single division, if properly implemented, could result in combining similar functions, thereby avoiding duplication and possibly creating more robust capabilities. For example, analysis and warning of cyber incidents is currently performed by both the National Infrastructure Protection Center and the Federal Computer Incident Response Center.

However, prior GAO work has identified and made recommendations concerning several critical infrastructure protection challenges that need to be addressed, which would face the new department. Specifically, they are:

- *Developing a national critical infrastructure protection strategy.*
- *Improving analytical and warning capabilities.*
- *Improving information sharing.*
- *Addressing pervasive weaknesses in federal information security.*

Organization of the Department of Homeland Security

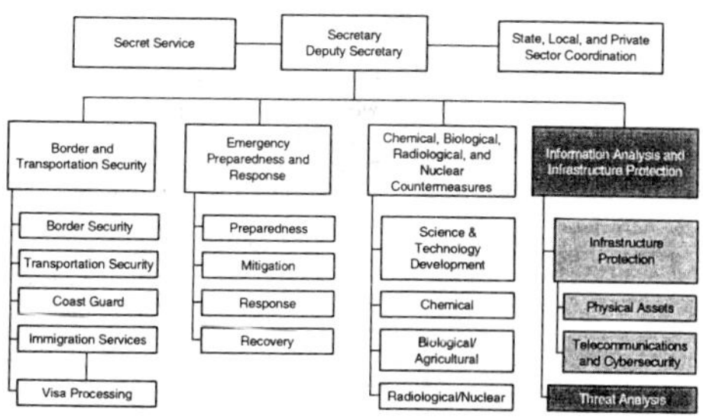

This is a test for developing highlights for a GAO report. The full testimony, including GAO's objectives, scope, methodology, and analysis, is available without charge at *www.gao.gov/cgi-bin/getrpt?GAO-02-918T*. For additional information about this testimony, contact Robert F. Dacey (202-512-3317). To provide comments on this test highlights, contact Keith Fultz (202-512-3200) or E-mail *HighlightsTest@gao.gov*.

Mr. Chairman and Members of the Subcommittee:

I am pleased to be here today to discuss the proposed reorganization of government agencies and the reorientation of their missions to improve our nation's ability to better protect our homeland. This historical transition is clearly one of the most important issues of our time and is already being compared to other large-scale government reorganizations, including the creation of the Department of Defense, the Central Intelligence Agency, and the National Security Council as part of the National Security Act of 1947.

In the months since the events of September 11, the President and the Congress have responded with important and aggressive actions to protect the nation—creating the Office of Homeland Security and the Critical Infrastructure Protection Board, passing new laws such as the USA Patriot Act and an emergency supplemental spending bill, establishing a new agency to improve transportation security, and working in collaboration with federal, state, and local governments and private sector entities to prevent future terrorist acts. More recently, the Congress and the President have sought to remedy long-standing issues and concerns in the government's homeland security functions by proposing greater consolidation and coordination of various agencies and activities. Recent proposals include restructuring the Federal Bureau of Investigation (FBI) and splitting the enforcement and service sections of the Immigration and Naturalization Service (INS). Additionally, Senator Joseph I. Lieberman and Representative William M. "Mac" Thornberry have authored legislation designed to consolidate many homeland security functions.

On June 18, the President transmitted draft legislation to the Congress for the creation of a new Department of Homeland Security whose mission would be preventing terrorist attacks within the United States, reducing America's vulnerability to terrorism, and minimizing the damage and recovering from attacks that do occur. The Comptroller General recently testified on issues that Congress should review in its deliberations on creating the new cabinet department.[1] Specifically, the Comptroller General discussed (1) the need for reorganization and the principles and criteria to help evaluate what agencies and missions should be included or excluded from the new department, and (2) issues related to transition, cost, and implementation challenges.

[1] U.S. General Accounting Office, *Homeland Security: Proposal for Cabinet Agency Has Merit, But Implementation Will be Pivotal to Success*, GAO-02-886T (Washington, D.C.: June 25, 2002).

The new cabinet department would incorporate several federal organizations, including the U.S. Secret Service and the U.S. Coast Guard, and would be organized into four divisions: (1) Information Analysis and Infrastructure Protection; (2) Chemical, Biological, Radiological and Nuclear Countermeasures; (3) Border and Transportation Security; and (4) Emergency Preparedness and Response. In particular, the Information Analysis and Infrastructure Protection division will perform one of the department's most critical missions: analyzing information and intelligence to better foresee terrorist threats to the United States.

Today, as requested, I will discuss the specific functions that would be performed by the department's Information Analysis and Infrastructure Protection division and the organizations that would be transferred to this division. I will also discuss the potential benefits and challenges for this division and, as indicated by our past reports on critical infrastructure protection (CIP) and federal information security, other major challenges that the new department would face. CIP involves activities that enhance the security of our nation's cyber and physical public and private infrastructure that are essential to national security, national economic security, and/or national public health and safety.

In preparing this testimony, we relied on prior GAO reports and testimonies on critical infrastructure protection, information security, and national preparedness, among others. We reviewed and analyzed the President's proposal to establish the Department of Homeland Security and the draft legislation. We also met with officials at the Department of Commerce's Critical Infrastructure Assurance Office and the Federal Bureau of Investigation's (FBI) National Infrastructure Protection Center to follow up on prior recommendations and to discuss their proposed move to the new department. Our work was performed in accordance with generally accepted government auditing standards.

Results in Brief

As proposed, functions of the Homeland Security Department's Information Analysis and Infrastructure Protection Division would include (1) receiving and analyzing law enforcement information, intelligence, and other information to detect and identify potential threats of terrorism within the United States; (2) assessing the vulnerabilities of the key resources and critical infrastructures in the United States; (3) developing a comprehensive national plan for securing these resources and infrastructures; and (4) taking necessary measures to protect these resources and infrastructures, in coordination with other executive agencies and in cooperation with state and local government personnel, agencies, and authorities, the private sector, and other entities. To create this division, six federal organizations that currently play a pivotal role in the protection of our national critical infrastructures would be transferred to this division in the new department. These organizations and their current parent organizations are shown in table 1.

Table 1: Organizations to Be Moved to Information Analysis and Infrastructure Protection Division

Organization to be moved	Current parent organization
National Infrastructure Protection Center (NIPC)[a]	FBI
National Communications System (NCS)	Department of Defense (DOD)[b]
Critical Infrastructure Assurance Office (CIAO)	Department of Commerce
Computer Security Division	National Institute of Standards and Technology (NIST)
National Infrastructure Simulation and Analysis Center	DOD/Department of Energy (DOE)
Federal Computer Incident Response Center (FedCIRC)	General Services Administration (GSA)

[a]The Computer Investigations and Operations Section currently within NIPC would remain at the FBI.

[b]DOD is the executive agent for the NCS, which reports to multiple Executive Office of the President organizations.

The consolidation of essential CIP functions and organizations may, if properly organized and implemented, lead over time to more efficient, effective, and coordinated programs. For example, two of the organizations proposed for consolidation—the GSA's FedCIRC and the FBI's NIPC—conduct incident reporting, analysis, and warning functions. Combining such efforts could not only eliminate possible duplicative efforts, but might also result in stronger and more coordinated capabilities. Other potential benefits include better control of funding through a single appropriation process for the new department and through establishing budget priorities for transferred functions based on their homeland security mission, and the consolidation of points of contact for federal agencies, state and local governments, and the private sector in coordinating activities to protect our homeland.

The Information Analysis and Infrastructure Protection Division will also face implementation challenges. For example, the new department will face tremendous information management and technology challenges, not the least of which will be integrating the diverse communications and information systems of the programs and agencies being brought together and securing the sensitive information these networks and systems process.

Further, through our past work, we have identified other significant challenges for many aspects of the functions to be transferred to the Information Analysis and Infrastructure Protection Division, and have recommended numerous changes to improve information analysis and protect our critical infrastructures. These challenges, which would face the new department, include the following:

- *Developing a national CIP strategy.* Although steps have been taken in this direction, a more complete strategy is needed that will address specific CIP roles and responsibilities for entities both within and outside of the new department, clearly define interim objectives and milestones, set time frames for achieving objectives, establish performance measures, and clarify how CIP entities will coordinate their activities.
- *Improving analytical and warning capabilities.* Although improvement efforts have been initiated, more robust analysis and warning capabilities, including a methodology for strategic analysis and a framework for collecting needed threat and vulnerability information, are still needed to identify threats and provide timely warnings. Such capabilities need to include both cyber and physical threats.
- *Improving information sharing on threats and vulnerabilities.* Information sharing needs to be improved both within the government and between the federal government and the private sector and state and local governments.
- *Addressing pervasive weaknesses in federal information security.* A comprehensive strategy for improving federal information security is needed, in which roles and responsibilities are clearly delineated, appropriate guidance is given, regular monitoring is undertaken, and security information and expertise are shared to maximize their value.

Critical Infrastructure Protection Policy Has Been Evolving Since the Mid-1990's

Federal awareness of the importance of securing our nation's critical infrastructures, which underpin our society, economy, and national security, has been evolving since the mid-1990's. Over the years, a variety of working groups have been formed, special reports written, federal policies issued, and organizations created to address the issues that have been raised. In October 1997, the President's Commission on Critical Infrastructure Protection issued its report,[2] which described the potentially devastating implications of poor information security from a national perspective. The report recommended several measures to achieve a higher level of critical infrastructure protection, including infrastructure protection through industry cooperation and information sharing, a national organization structure, a revised program of research and development, a broad program of awareness and education, and reconsideration of laws related to infrastructure protection. The report stated that a comprehensive effort would need to "include a system of surveillance, assessment, early warning, and response mechanisms to mitigate the potential for cyberthreats." It said that the FBI had already begun to develop warning and threat analysis capabilities and urged it to continue in these efforts. In addition, the report noted that the FBI could serve as the preliminary national warning center for infrastructure attacks and provide law enforcement, intelligence, and other information needed to ensure the highest quality analysis possible.

In 1998, the President issued Presidential Decision Directive (PDD) 63, which describes a strategy for cooperative efforts by government and the private sector to protect the physical and cyber-based systems essential to the minimum operations of the economy and the government. PDD 63 called for a range of actions intended to improve federal agency security programs, improve the nation's ability to detect and respond to serious computer-based and physical attacks, and establish a partnership between the government and the private sector. The directive called on the federal government to serve as a model of how infrastructure assurance is best achieved and designated lead agencies to work with private-sector and government organizations. Further, it established critical infrastructure protection as a national goal, and stated that, by the close of 2000, the United States was to have achieved an initial operating capability to protect the nation's critical infrastructures from intentional destructive acts and, no later than 2003, an enhanced capability.

To accomplish its goals, PDD 63 designated and established organizations to provide central coordination and support, including

---

[2] *Critical Foundations: Protecting America's Infrastructures*, Report of the President's Commission on Critical Infrastructure Protection (October 1997).

- the Critical Infrastructure Assurance Office (CIAO), an interagency office housed in the Department of Commerce, which was established to develop a national plan for CIP on the basis of infrastructure plans developed by the private sector and federal agencies;
- the National Infrastructure Protection Center (NIPC), an organization within the FBI, which was expanded to address national-level threat assessment, warning, vulnerability, and law enforcement investigation and response; and
- the National Infrastructure Assurance Council, which was established to enhance the partnership of the public and private sectors in protecting our critical infrastructures.[3]

To ensure coverage of critical sectors, PDD 63 also identified eight private-sector infrastructures and five special functions. The infrastructures are (1) information and communications; (2) banking and finance; (3) water supply; (4) aviation, highway, mass transit, pipelines, rail, and waterborne commerce; (5) emergency law enforcement; (6) emergency fire services and continuity of government; (7) electric power and oil and gas production and storage; and (8) public health services. The special functions are (1) law enforcement and internal security, (2) intelligence, (3) foreign affairs, (4) national defense, and (5) research and development. For each of the infrastuctures and functions, the directive designated lead federal agencies to work with their counterparts in the private-sector. For example, the Department of the Treasury is responsible for working with the banking and finance sector, and the Department of Energy is responsible for working with the electrical power industry. Similarly, regarding special function areas, DOD is responsible for national defense, and the Department of State is responsible for foreign affairs.

To facilitate private-sector participation, PDD 63 also encouraged the creation of information sharing and analysis centers (ISACs) that could serve as mechanisms for gathering, analyzing, and appropriately sanitizing and disseminating information to and from infrastructure sectors and the federal government through the NIPC. Figure 1 displays a high-level overview of the organizations with CIP responsibilities as outlined by PDD 63.

[3]Executive Order 13231 replaces this council with the National Infrastructure Advisory Council.

Figure 1: Organizations with CIP Responsibilities as Outlined by PDD 63

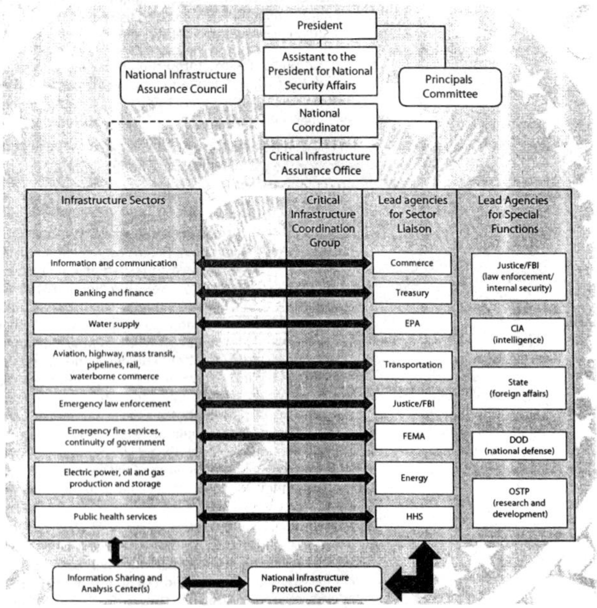

Note: In February 2001, the Critical Infrastructure Coordination Group was replaced by the Information Infrastructure Protection and Assurance Group under the Policy Coordinating Committee on Counter-terrorism and National Preparedness. In October 2001, the National Infrastructure Assurance Council was replaced by the National Infrastructure Advisory Council, and cyber CIP functions performed by the national coordinator were assigned to the chair of the President's Critical Infrastructure Protection Board.

Source: CIAO.

In response to PDD 63, in January 2000 the White House issued its "National Plan for Information Systems Protection."[4] The national plan provided a vision and framework for the federal government to prevent, detect, respond to, and protect the nation's critical cyber-based infrastructure from attack and reduce existing vulnerabilities by complementing and focusing existing federal computer security and information technology requirements. Subsequent versions of the plan were expected to (1) define the roles of industry and state and local governments working in partnership with the federal government to protect physical and cyber-based infrastructures from deliberate attack and (2) examine the international aspects of CIP.

The most recent federal CIP guidance was issued in October 2001, when President Bush signed Executive Order 13231, establishing the President's Critical Infrastructure Protection Board to coordinate cyber-related federal efforts and programs associated with protecting our nation's critical infrastructures. The Special Advisor to the President for Cyberspace Security chairs the board. Executive Order 13231 tasks the board with recommending policies and coordinating programs for protecting CIP-related information systems. The executive order also established 10 standing committees to support the board's work on a wide range of critical information infrastructure efforts. The board is intended to coordinate with the Office of Homeland Security in activities relating to the protection of and recovery from attacks against information systems for critical infrastructure, including emergency preparedness communications that were assigned to the Office of Homeland Security by Executive Order 13228, dated October 8, 2001. The board recommends policies and coordinates programs for protecting information systems for critical infrastructure, including emergency preparedness communications, and the physical assets that support such systems. In addition, the chair coordinates with the Assistant to the President for Economic Policy on issues relating to private-sector systems and economic effects and with the Director of OMB on issues relating to budgets and the security of federal computer systems. Further, the Special Advisor reports to the Assistant to the President for National Security Affairs and to the Assistant to the President for Homeland Security.

Implementing PDD 63 Has Not Been Completely Successful

Both GAO and the inspectors general have issued reports highlighting concerns about PDD 63 implementation. As we reported in September 2001, efforts to perform substantive, comprehensive analyses of infrastructure sector vulnerabilities and development of related remedial plans had been limited. Further, a March 2001 report by the President's

[4] The White House, *Defending America's Cyberspace: National Plan for Information Systems Protection: Version 1.0: An Invitation to a Dialogue* (Washington, D.C.: 2000).

Council on Integrity and Efficiency and the Executive Council on Integrity and Efficiency (PCIE/ECIE) identified significant deficiencies in federal agencies' implementation of PDD 63 requirements to (1) establish plans for protecting their own critical infrastructure that were to be implemented within 2 years, or by May 2000, and (2) develop procedures and conduct vulnerability assessments.[5]

Specifically,

- many agency critical infrastructure protection plans were incomplete and some agencies had not developed such plans,
- most agencies had not completely identified their mission-essential infrastructure assets, and
- few agencies had completed vulnerability assessments of their minimum essential infrastructure assets or developed remediation plans.

Our subsequent review of PDD 63-related activities at eight lead agencies found similar problems, although some agencies had made progress since their respective inspectors general reviews.[6] Further, OMB reported in February 2002 that it planned to direct all large agencies to undertake a Project Matrix review to identify critical infrastructure assets and their interdependencies with other agencies and the private sector.[7]

We identified several other factors that had impeded federal agency efforts to comply with PDD 63. First, no clear definitions had been developed to guide development and implementation of agency plans and measure performance. For example, PDD 63 established December 2000 as the deadline for achieving an initial operating capability and May 2003 for achieving full operational capability of key functions. However, the specific capabilities to be achieved at each milestone had not been defined. The PCIE/ECIE report noted that agencies had used various interpretations of initial operating capability and stated that, without a definition, there is no consistent measure of progress toward achieving full security preparedness. In addition, several agency officials said that funding and staffing constraints contributed to their delays in

[5]The PCIE primarily is comprised of the presidentially appointed inspectors general and the ECIE is primarily comprised of the agency head-appointed inspectors general. In November 1999, PCIE and ECIE formed a working group to review the adequacy of federal agencies' implementation of PDD 63. The March 2001 report is based on reviews by 21 inspectors general of their respective agencies' PDD 63 planning and assessment activities.

[6]GAO-01-822, September 20, 2001.

[7]Project Matrix is a CIAO methodology that identifies all critical assets, nodes, networks, and associated infrastructure dependencies and interdependencies.

implementing PDD 63 requirements. Further, the availability of adequate technical expertise to provide information security has been a continuing concern to agencies.

Cyber Threats Are Increasing

Dramatic increases in computer interconnectivity, especially in the use of the Internet, are revolutionizing the way our government, our nation, and much of the world communicate and conduct business. The benefits have been enormous. Vast amounts of information are now literally at our fingertips, facilitating research on virtually every topic imaginable; financial and other business transactions can be executed almost instantaneously, often on a 24-hour-a-day basis; and electronic mail, Internet web sites, and computer bulletin boards allow us to communicate quickly and easily with a virtually unlimited number of individuals and groups.

In addition to such benefits, however, this widespread interconnectivity poses significant risks to our computer systems and, more important, to the critical operations and infrastructures they support. For example, telecommunications, power distribution, water supply, public health services, and national defense (including the military's warfighting capability), law enforcement, government services, and emergency services all depend on the security of their computer operations. The speed and accessibility that create the enormous benefits of the computer age likewise, if not properly controlled, allow individuals and organizations to inexpensively eavesdrop on or interfere with these operations from remote locations for mischievous or malicious purposes, including fraud or sabotage.

Government officials are increasingly concerned about attacks from individuals and groups with malicious intent, such as crime, terrorism, foreign intelligence gathering, and acts of war. According to the FBI, terrorists, transnational criminals, and intelligence services are quickly becoming aware of and using information exploitation tools such as computer viruses, Trojan horses, worms, logic bombs, and eavesdropping sniffers that can destroy, intercept, degrade the integrity of, or deny access to data. As greater amounts of money are transferred through computer systems, as more sensitive economic and commercial information is exchanged electronically, and as the nation's defense and intelligence communities increasingly rely on commercially available information technology, the likelihood increases that information attacks will threaten vital national interests. In addition, the disgruntled organization insider is a significant threat, since such individuals often have knowledge that allows them to gain unrestricted access and inflict damage or steal assets without possessing a great deal of knowledge about computer intrusions.

Reports of attacks and disruptions abound. The 2002 report of the "Computer Crime and Security Survey," conducted by the Computer Security Institute and the FBI's San Francisco Computer Intrusion Squad, showed that 90 percent of respondents (primarily large corporations and government agencies) had detected computer security breaches within the last 12 months. In addition, the number of computer security incidents reported to the CERT® Coordination Center rose from 9,859 in 1999 to 52,658 in 2001 and 26,829 for just the first quarter of 2002. And these are only the reported attacks.[8] The CERT® Coordination Center estimates that as much as 80 percent of actual security incidents go unreported, in most cases because the organization was unable to recognize that its systems had been penetrated or because there were no indications of penetration or attack.

Since the September 11 attacks, warnings of the potential for terrorist cyber attacks against our critical infrastructures have also increased. For example, earlier this year, the Special Advisor to the President for Cyberspace Security stated in a Senate briefing that although to date none of the traditional terrorist groups such as al Qaeda have used the Internet to launch a known attack on the United States infrastructure, information on computerized water systems was recently discovered on computers found in al Qaeda camps in Afghanistan. Further, in his October congressional testimony, Governor James Gilmore, Governor of the Commonwealth of Virginia and Chairman of the Advisory Panel to Assess Domestic Response Capabilities for Terrorism Involving Weapons of Mass Destruction (commonly known as the "Gilmore Commission"), warned that systems and services critical to the American economy and the health of our citizens—such as banking and finance, "just-in-time" delivery systems for goods, hospitals, and state and local emergency services—could all be shut down or severely handicapped by a cyber attack or a physical attack against computer hardware.[9]

[8]CERT® Coordination Center (CERT-CC) is a center of Internet security expertise located at the Software Engineering Institute, a federally funded research and development center operated by Carnegie Mellon University.

[9]Testimony of Governor James S. Gilmore III, Governor of the Commonwealth of Virginia and Chairman of the Advisory Panel to Assess the Capabilities for Domestic Response to Terrorism Involving Weapons of Mass Destruction before the House Science Committee, October 17, 2001.

Information Analysis and Infrastructure Protection Division Consolidates Several CIP Functions

On June 6, President Bush announced a new proposal to create a Department of Homeland Security and submitted draft legislation to Congress on June 18. Like the congressional approaches to create a new department, the President's plan also reflected many of the recent commissions' suggestions and our recommendations for improved coordination and consolidation of homeland security functions. As indicated by Governor Ridge is his recent testimony before Congress, the creation of this department would empower a single cabinet official whose primary mission is to protect the American homeland from terrorism, including: (1) preventing terrorist attacks within the United States; (2) reducing America's vulnerability to terrorism; and (3) minimizing the damage and recovering from attacks that do occur.[10]

In our initial review of the proposed department, we have used the President's draft bill of June 18 as the basis of our comments. Nevertheless, we recognize that the proposal has already—and will continue—to evolve in the coming days and weeks ahead. The President's proposal creates a cabinet department with four divisions, including:

- Information Analysis and Infrastructure Protection;
- Chemical, Biological, Radiological and Nuclear Countermeasures;
- Border and Transportation Security; and
- Emergency Preparedness and Response.

One of the most critical functions that the new department will have is the analysis of information and intelligence to better foresee terrorist threats to the United States—a function that would be performed by the Information Analysis and Infrastructure Protection Division. The primary responsibilities of this division would be

- receiving and analyzing law enforcement information, intelligence, and other information in order to understand the nature and scope of the terrorist threat to the American homeland and to detect and identify potential threats of terrorism within the U.S;
- assessing the vulnerabilities of the key resources and critical infrastructures in the United States including food and water systems, agriculture, health systems, emergency services, banking and finance, communications and information systems, energy (including electric,

[10] *The Department of Homeland Security: Making Americans Safer*, Written Statement of Governor Tom Ridge before the Committee on Governmental Affairs, U.S. Senate, June 20, 2002.

nuclear, gas and oil and hydropower), transportation systems, and national monuments;

- integrating relevant information, intelligence analyses, and vulnerability assessments to identify protective priorities and support protective measures by the Department, by other executive agencies, by state and local government personnel, agencies, and authorities, by the private sector, and by other entities;
- developing a comprehensive national plan for securing the key resources and critical infrastructures in the United States;
- taking or seeking to effect necessary measures to protect the key resources and critical infrastructures in the United States, in coordination with other executive agencies and in cooperation with state and local government personnel, agencies, and authorities, the private sector, and other entities;
- administering the Homeland Security Advisory System, exercising primary responsibility for public threat advisories, and (in coordination with other executive agencies) providing specific warning information to state and local government personnel, agencies, and authorities, the private sector, other entities, and the public, as well as advice about appropriate protective actions and countermeasures; and
- reviewing, analyzing, and making recommendations for improvements in the policies and procedures governing the sharing of law enforcement, intelligence, and other information relating to homeland security within the federal government and between such government and state and local government personnel, agencies, and authorities.

To create this division, the proposed reorganization would transfer six federal organizations that currently play a pivotal role in the protection of our national critical infrastructures—the FBI's National Infrastructure Protection Center (other than the computer investigations and operations center), DOD's National Communications System, the Commerce Department's Critical Infrastructure Assurance Office, the Computer Security Division of Commerce's NIST, the National Infrastructure Simulation and Analysis Center of DOD/DOE, and GSA's FedCIRC. (See the appendix for a description of the principal activities of these six organizations.)

Potential Benefits Could Be Achieved By Consolidating Similar Activities

The administration has indicated that this new division would for the first time merge under one roof the capability to identify and assess threats to the homeland, map those threats against our vulnerabilities, issue timely warnings, and organize preventive or protective action to secure the homeland. The agencies and programs included in the Administration's proposal to consolidate information analysis functions are clear contributors to the homeland security mission and, if well coordinated or consolidated, could provide greater benefits by avoiding duplication and more closely coordinating activities.

Three areas are clearly opportunities for synergy: outreach and education; the identification of critical assets; and incident reporting, analysis, and warning. Currently the NIPC and CIAO both provide outreach to educate groups regarding the importance of protecting our critical infrastructures. These two organizations are also involved in the identification of critical assets. For instance, the NIPC is responsible for the Key Asset Initiative—a database of the most important components of the nation's critical infrastructures—while the CIAO is responsible for project matrix—a methodology that identifies all critical assets, nodes, networks, and associated infrastructure dependencies and interdependencies. Further, both the NIPC and FedCIRC have threat identification, incident reporting, analysis, and warning responsibilities. The CIAO Director recently testified that the new division will combine functions that are currently fragmented and inefficient, minimize duplication or redundancy of efforts, and ensure that critical infrastructure and cyber security activities can be more closely coordinated.

Several other potential benefits could be realized with the consolidation of related organizations and responsibilities within a single department. First, funding for critical infrastructure protection activities of the transferred organizations such as the NIPC and the CIAO could be better controlled through a single appropriation process rather than through separate processes for different departments. For example, as we reported in April 2001, NIPC's budget requests—including staffing and other financial resources—are controlled by the FBI and the Department of Justice, raising concern at that time among NIPC officials that its priorities, which are intended to reflect the interests of national critical infrastructure protection, may be subordinated to the FBI's law enforcement priorities. NIPC officials told us that the FBI had not approved their repeated requests for additional resources as part of the budget process. Another potential benefit is the consolidation of points of contact for use by other federal agencies, state and local governments, the private sector, and other entities so that those within and external to the federal government have clear understanding of whom to coordinate with on homeland security issues.

New Department Needs to Focus on Critical Success Factors

In his June 2002 testimony, the Comptroller General noted key factors that should be considered for successfully implementing the new department.[11] These key factors include strategic planning, organizational alignment, communication and building partnerships, performance management, human capital strategy, information management and technology, knowledge management, financial management, acquisition management, and risk management. Given the transfer of organizations and responsibilities, the analysis and assessment functions to be performed, and the sensitivity of information to be collected, several of these factors will also be particularly important for the proposed Information Analysis and Infrastructure Protection Division. Specifically:

Human capital strategy. An organization's people are its most important asset. People define an organization, affect its capacity to perform, and represent the knowledge base of the organization. In an effort to help agency leaders integrate human capital considerations into daily decision-making and in the program results they seek to achieve, we have recently released an exposure draft of a model of strategic human capital management that highlights the kinds of thinking that agencies should apply and steps they can take to manage their human capital more strategically.[12] The model focuses on four cornerstones for effective human capital management—leadership; strategic human capital planning; acquiring, developing, and retaining talent; and results-oriented organization culture—and both the new department and the new division may find this model useful in helping guide its efforts. Hiring and retaining personnel with appropriate technology and analytical skills will also be critical to the new division.

Information management and technology. The new department will face significant information management and technology challenges. Programs and agencies will be brought together in the new department from throughout the government, and each will bring their own communications and information systems. It will be a tremendous undertaking to integrate these diverse systems and enable effective communication and share information among themselves, as well as those outside the department.

[11] GAO-02-886T, June 25, 2002.

[12] U.S. General Accounting Office, *A Model of Strategic Human Capital Management*, GAO-02-373SP (Washington, D.C.: Mar. 15, 2002).

To address the challenge, it will be critical that an enterprise architecture be developed to guide the integration and modernization of information systems. Such architecture, required by the Clinger-Cohen Act, consist of models that describe how the enterprise operates now and how it needs to operate in the future. Without an enterprise architecture to guide and constrain information technology investments, stovepipe operations and systems can emerge, which in turn lead to needless duplication, incompatibilities, and additional costs. This will be quite a challenge given that, as we reported earlier this year, few federal departments and agencies have the management practices necessary to develop and leverage enterprise architectures.[13] It will be particularly important for the new division to leverage technology to enhance its ability to transform capabilities and capacities to share and act upon timely, quality information about terrorist threats.

Further, as discussed later, since 1996, we have reported that poor information security is a widespread federal government problem with potentially devastating consequences. Considering the sensitivity of the data at the proposed department, securing its information systems and networks will be of utmost importance.

Proposed Homeland Security Department Faces Ongoing Challenges

We have reported for years on many aspects of the functions that are to transferred to the Information Analysis and Infrastructure Protection division and have made numerous recommendations to improve information analysis and to protect our critical infrastructures. Specific challenges, which would face the new department, include developing a national CIP strategy, improving analytical and warning capabilities, improving information sharing, and addressing pervasive weaknesses in federal information security.

National CIP Strategy Needs to Be Developed

A clearly defined strategy is essential for defining the relationships among all CIP organizations, both internal as well as external to the proposed Department of Homeland Security, to ensure that the approach is comprehensive and well coordinated. The President's proposal states that one of the primary responsibilities of the new Information Analysis and Infrastructure Protection division is to develop such a strategy.

An underlying issue in the implementation of PDD 63, and a major challenge for the new department, is that no national strategy yet exists that clearly delineates the roles and responsibilities of federal and

[13] U.S. General Accounting Office, *Information Technology: Enterprise Architecture Use Across the Federal Government Can Be Improved*, GAO-02-6 (Washington, D.C.: Feb. 19, 2002).

80-680 - 02 - 12

nonfederal CIP entities and defines interim objectives.[14] We first identified the need for a detailed plan in September 1998, when we reported that developing a governmentwide strategy that clearly defined and coordinated the roles of new and existing federal entities was important to ensure governmentwide cooperation and support for PDD 63.[15] At that time, we recommended that OMB and the Assistant to the President for National Security Affairs ensure such coordination.

In January 2000, the President issued Defending America's Cyberspace: National Plan for Information Systems Protection: Version 1.0: An Invitation to a Dialogue as a first major element of a more comprehensive effort to protect the nation's information systems and critical assets from future attacks. The plan proposed achieving the twin goals of making the U.S. government a model of information security and developing a public/private partnership to defend our national infrastructures by achieving three crosscutting infrastructure protection objectives:

- minimize the possibility of significant and successful attacks;
- identify, assess, contain, and quickly recover from an attack; and
- create and build strong foundations, including people, organizations, and laws, for preparing, preventing, detecting and responding to attacks.

However, this plan focused largely on federal cyber CIP efforts, saying little about the private-sector role. Subsequently, in July 2000, we reiterated the importance of defining and clarifying organizational roles and responsibilities, noting that numerous federal entities were collecting, analyzing, and disseminating data or guidance on computer security vulnerabilities and incidents and that clarification would help ensure a common understanding of (1) how the activities of these many organizations interrelate, (2) who should be held accountable for their success or failure, and (3) whether such activities will effectively and efficiently support national goals.[16]

A May 2001 White House press statement announced that the administration was reviewing how it was organized to deal with information security issues and that recommendations would be made on how to structure an integrated approach to cyber security and critical

[14] GAO-01-822, September 20, 2001.

[15] U.S. General Accounting Office, *Information Security: Serious Weaknesses Place Critical Federal Operations and Assets at Risk*; GAO/AIMD-98-92 (Washington, D.C.: Sep. 23, 1998).

[16] U.S. General Accounting Office, *Critical Infrastructure Protection: Challenges to Building a Comprehensive Strategy for Information Sharing and Cooperation*; GAO/T-AIMD-00-268 (Washington, D.C.: July 26, 2000).

infrastructure protection. Specifically, the announcement stated that the White House, federal agencies, and private industry had begun to collaboratively prepare a new version of a "national plan for cyberspace security and critical infrastructure protection" and reviewing how the government is organized to deal with information security issues.

In September 2001, we reported that agency questions had surfaced regarding specific roles and responsibilities of entities involved in cyber CIP and the timeframes within which CIP objectives are to be met, as well as guidelines for measuring progress.[17] Accordingly, we made several recommendations to supplement those we had made in the past, including those regarding the NIPC. Specifically, we recommended that the Assistant to the President for National Security Affairs ensure that the federal government's strategy to address computer-based threats define

- specific roles and responsibilities of organizations involved in critical infrastructure protection and related information security activities;
- interim objectives and milestones for achieving critical infrastructure protection goals and a specific action plan for achieving these objectives, including implementation of vulnerability assessments and related remedial plans; and
- performance measures for which entities can be held accountable.
- The national strategy for cyber CIP is still being developed and is now planned to be issued in September 2002.

Further, an important aspect of this strategy will be the inclusion of all CIP-related federal activities. For example, it should include additional sectors not included in PDD 63. This was acknowledged by the chair of the President's Critical Infrastructure Protection Board recently, when he told a Senate subcommittee that the critical infrastructure sectors were being reviewed after the September 11 attacks and the subsequent anthrax attacks on the U.S. Capitol. In addition, the proposal to create a Department of Homeland Security refers to the need to consider additional sectors. According to the proposal, "the Department would be responsible for comprehensively evaluating the vulnerabilities of America's critical infrastructure, including food and water systems, agriculture, health systems and emergency services, information and telecommunications, banking and finance, energy (electrical, nuclear, gas and oil, dams), transportation (air, road, rail, ports, waterways), the chemical and defense industries, postal and shipping entities, and national

[17] GAO-01-822, September 20, 2001.

monuments and icons." It is also important that any CIP-related efforts or proposals outside the current scope of PDD 63 be coordinated with other CIP efforts. For example, we understand that EPA is considering a proposal that would require the 15,000 industrial facilities using hazardous chemicals to submit detailed vulnerability assessments.

A clearly defined strategy is also essential for clarifying how CIP entities will coordinate their activities with each other, both those that are to be included in the proposed department and those external to it. For example, Information Analysis and Infrastructure Protection division's responsibilities include receiving and analyzing law enforcement information, intelligence, and other information. Similar functions are also performed by the recently created Transportation Security Agency, which the bill transfers to another division of the new department. Coordinating these similar activities within the new department will be critical to avoiding unnecessarily duplicative efforts and ensuring the effective flow of appropriate law enforcement, intelligence, and other information to the entities that need it. In addition, the numerous federal CIP organizations that will remain in place, such as the President's Critical Infrastructure Protection Board, NIPC's Computer Investigations and Operations Section that is to remain with the FBI, and the Joint Task Force for Computer Network Operations within the Department of Defense will need to be closely coordinated with the other CIP players. Coordination will be especially critical between the department and the other federal entities that are to provide it with intelligence and other threat information, such as the FBI and the CIA.

A national strategy that covers both cyber and physical CIP could greatly facilitate such organizational coordination and the success of the new department. CIAO officials told us that separate cyber and physical strategies are now planned to be issued. Without a comprehensive and coordinated strategy that identifies roles and responsibilities for all CIP efforts, our nation risks not having a consistent and appropriate structure to deal with the growing threat of computer-based attacks on its critical infrastructure.

Analytical and Warning Capabilities Need to Be Improved

Another key challenge for the new department is to develop the analysis and warning capabilities called for in the President's proposal. NIPC was established in PDD 63 as "a national focal point" for gathering information on threats and facilitating the federal government's response to computer-based incidents. Specifically, the directive assigned the NIPC the responsibility for providing comprehensive analyses on threats, vulnerabilities, and attacks; issuing timely warnings on threats and attacks; facilitating and coordinating the government's response to computer-based incidents; providing law enforcement investigation and response, monitoring reconstitution of minimum required capabilities after an infrastructure attack; and promoting outreach and information sharing. This responsibility requires obtaining and analyzing intelligence, law enforcement, and other information to identify patterns that may signal that an attack is underway or imminent. Similar activities are also called for in the President's proposal for the Information Analysis and Infrastructure Protection division.

In April 2001, we reported on NIPC's progress in developing national capabilities for analyzing threat and vulnerability data and issuing warnings, responding to attacks, among others.[18] Overall, we found that while progress in developing these capabilities was mixed, the NIPC had initiated a variety of critical infrastructure protection efforts that had laid a foundation for future governmentwide efforts. In addition, the NIPC had provided valuable support and coordination related to investigating and otherwise responding to attacks on computers. However, at the close of our review, the analytical capabilities that PDD 63 asserted are needed to protect the nation's critical infrastructures had not yet been achieved, an the NIPC had developed only limited warning capabilities. Developing such capabilities is a formidable task that experts say will take an intense interagency effort.

At the time of our review, the NIPC had issued a variety of analytical products, most of which have been tactical analyses pertaining to individual incidents. In addition, it had issued a variety of publications, most of which were compilations of information previously reported by others with some NIPC analysis.

We reported that the use of strategic analysis to determine the potential broader implications of individual incidents had been limited. Such analysis looks beyond one specific incident to consider a broader set of incidents or implications that may indicate a potential threat of national importance. Identifying such threats assists in proactively managing risk, including evaluating the risks associated with possible future incidents and effectively mitigating the impact of such incidents.

We reported last year that three factors hindered NIPC's ability to develop strategic analytical capabilities:

[18] U.S. General Accounting Office, *Critical Infrastructure Protection: Significant Challenges in Developing National Capabilities*; GAO-01-323 (Washington, D.C.: Apr. 25, 2001).

- First, there was no generally accepted methodology for analyzing strategic cyber-based threats. For example, there was no standard terminology, no standard set of factors to consider, and no established thresholds for determining the sophistication of attack techniques. According to officials in the intelligence and national security community, developing such a methodology would require an intense interagency effort and dedication of resources.
- Second, the NIPC had sustained prolonged leadership vacancies and did not have adequate staff expertise, in part because other federal agencies had not provided the originally anticipated number of detailees. For example, at the close of our review in February, the position of Chief of the Analysis and Warning Section, which was to be filled by the Central Intelligence Agency, had been vacant for about half of NIPC's 3-year existence. In addition, the NIPC had been operating with only 13 of the 24 analysts that NIPC officials estimate are needed to develop analytical capabilities.
- Third, the NIPC did not have industry-specific data on factors such as critical system components, known vulnerabilities, and interdependencies. Under PDD 63, such information is to be developed for each of eight industry segments by industry representatives and the designated federal lead agencies. However, at the close of our work, only three industry assessments had been partially completed, and none had been provided to the NIPC.

To provide a warning capability, the NIPC established a Watch and Warning Unit that monitors the Internet and other media 24 hours a day to identify reports of computer-based attacks. While some warnings were issued in time to avert damage, most of the warnings, especially those related to viruses, pertained to attacks underway. We reported that NIPC's ability to issue warnings promptly was impeded because of (1) a lack of a comprehensive governmentwide or nationwide framework for promptly obtaining and analyzing information on imminent attacks, (2) a shortage of skilled staff, (3) the need to ensure that the NIPC does not raise undue alarm for insignificant incidents, and (4) the need to ensure that sensitive information is protected, especially when such information pertains to law enforcement investigations underway.

Further, the relationships between the Center, the FBI, and the National Coordinator for Security, Infrastructure Protection, and Counter-Terrorism at the National Security Council were unclear regarding who had direct authority for setting NIPC priorities and procedures and providing NIPC oversight. In addition, NIPC's own plans for further developing its analytical and warning capabilities were fragmented and incomplete. As a result, no specific priorities, milestones, or program performance measures existed to guide NIPC's actions or provide a basis for evaluating its progress.

In our report, we recognized that the administration was reviewing the government's infrastructure protection strategy and recommended that, as the administration proceeds, the Assistant to the President for National Security Affairs, in coordination with pertinent executive agencies,

- establish a capability for strategically analyzing computer-based threats, including developing related methodology, acquiring staff expertise, and obtaining infrastructure data;
- require development of a comprehensive data collection and analysis framework and ensure that national watch and warning operations for computer-based attacks are supported by sufficient staff and resources, and
- clearly define the role of the NIPC in relation to other government and private-sector entities.

In response to our report recommendations, the NIPC Director recently told us that NIPC had developed a plan with goals and objectives to improve their analytical and warning capabilities and that NIPC has made considerable progress in this area. For example, the Director told us that the analysis and warning section has created two additional teams to bolster its analytical capabilities—(1) the critical infrastructure assessment team to focus efforts on learning about particular infrastructures and coordinating with respective infrastructure efforts and (2) the collection operations intelligence liaison team to coordinate with various entities within the intelligence community. The Director added that NIPC (1) now holds a quarterly meeting with senior government leaders of entities that it regularly works with to better coordinate their analytical and warning capabilities, (2) has developed close working relationships with other CIP entities involved in analysis and warning activities, such as FedCirc, DOD's Joint Task Force for Computer Network Operations, the Carnegie Mellon's Computer Emergency Response Team (CERT) Coordination Center, and the intelligence and anti-virus communities, and (3) had developed and implemented procedures to more quickly share relevant CIP information, while separately continuing any related law enforcement investigation. The Director also stated that NIPC has received sustained leadership commitment from key entities, such as CIA and NSA, and that it continues to increase its staff primarily through reservists and contractors. The Director acknowledged that our recommendations are not fully implemented and that despite the accomplishments to date, much more work remains to create the robust analysis and warning capabilities needed to adequately address cyberthreats.

Another challenge confronting the analysis and warning capabilities of the new department is that the functions proposed to be transferred to the new department for Information Analysis and Infrastructure Protection have historically focused their attention and efforts on cyber threats. In April 2001, we reported that while PDD 63 covers both physical and computer-based threats, federal efforts to meet the directive's requirements have pertained primarily to computer-based threats, since this was an area that the leaders of the administration's critical infrastructure protection strategy viewed as needing attention. Not only is physical protection of our critical infrastructures important in and of itself, but a physical attack in conjunction with a cyber attack has recently been highlighted as a major concern. Also, exploiting cyber vulnerabilities can be used as a means to attack our nation's critical physical infrastructures. The Director told us that NIPC had begun to develop some capabilities for the identification of physical CIP threats. For example, NIPC has developed thresholds with several ISACs for reporting physical incidents and has, since January 2002, issued several information bulletins concerning physical CIP threats. However, NIPC Director acknowledged that fully developing this capability will be a significant challenge. It is important that the national CIP strategy adequately addresses physical threats.

Another critical issue in developing effective analysis and warning capabilities is to ensure that appropriate intelligence and other threat information, both cyber and physical, is received from the intelligence and law enforcement communities. For example, considerable debate has ensued in recent weeks with respect to the quality and timeliness of intelligence data shared between and among relevant intelligence, law enforcement, and other agencies. The proposal would provide for the new department to receive all reports and analysis related to threats of terrorism and vulnerabilities to our infrastructure and, if the President directs, information in the "raw" state that has not been analyzed. Also, with the proposed separation of NIPC from the FBI's law enforcement activities, including the Counterterrorism Division and NIPC field agents, it will be critical to establish mechanisms for continued communication to occur. Further, it will be important that the relationships between the law enforcement and intelligence communities and the new department are effective and that appropriate information is exchanged on a timely basis.

Further, according to the NIPC Director, a significant challenge in developing a robust analysis and warning function is the development of the technology and human capital capacities to collect and analyze substantial amounts of information. Similarly, the Director of the FBI recently testified that implementing a more proactive approach to preventing terrorist acts and denying terrorist groups the ability to operate and raise funds requires a centralized and robust analytical capacity that does not currently exist in the FBI's Counterterrorism Division. He also stated that processing and exploiting information gathered domestically and abroad during the course of investigations requires an enhanced analytical and data mining capacity that is not presently available. Also, the NIPC Director stated that multi-agency staffing, similar to NIPC, is a critical success factor in establishing an effective analysis and warning function and that appropriate funding for such staff was important.

Government Faces Information Sharing Challenges

Information sharing is a key element in developing comprehensive and practical approaches to defending against cyber attacks, which could threaten the national welfare. Information on threats and incidents experienced by others can help identify trends, better understand the risks faced, and determine what preventive measures should be implemented. However, as we testified in July 2000,[19] establishing the trusted relationships and information-sharing protocols necessary to support such coordination can be difficult. Last October we reported on information sharing practices that could benefit critical infrastructure protection.[20] These practices include

- establishing trust relationships with a wide variety of federal and nonfederal entities that may be in a position to provide potentially useful information and advice on vulnerabilities and incidents,
- developing standards and agreements on how shared information will be used and protected,
- establishing effective and appropriately secure communications mechanisms, and
- taking steps to ensure that sensitive information is not inappropriately disseminated, which may require statutory changes.

[19] GAO/T-AIMD-00-268, July 26, 2000.

[20] U.S. General Accounting Office, *Information Sharing: Practices That Can Benefit Critical Infrastructure Protection*, GAO-02-24 (Washington, D.C.: Oct. 15, 2001).

In June of this year, we also reported on the information sharing barriers confronting homeland security, both within the federal government and with the private sector.[21]

A number of activities have been undertaken to build relationships between the federal government and the private sector, such as InfraGard, the Partnership for Critical Infrastructure Security, efforts by the CIAO, and efforts by lead agencies to establish information sharing and analysis centers (ISACs). For example, the InfraGard Program, which provides the FBI and the NIPC with a means of securely sharing information with individual companies, has expanded substantially. By early January 2001, 518 entities were InfraGard members—up from 277 members in October 2000. Members included representatives from private industry, other government agencies, state and local law enforcement, and the academic community. Currently, NIPC reports over 5,000 InfraGard members. Although each of these efforts is commendable, more needs to be done.

PDD 63 encouraged the voluntary creation of ISACs that could serve as the mechanism for gathering, analyzing, and appropriately sanitizing and disseminating information between the private sector and the federal government through NIPC. Such centers are critical since the private sector entities control over 80 percent of our nation's critical infrastructures. In September 2001, we reported that although outreach efforts had raised awareness and improved information sharing, substantive, comprehensive analysis of infrastructure sector interdependenices, vulnerabilities and related remedial plans had been limited.

In April 2001, we reported that NIPC had undertaken a range of initiatives to foster information sharing relationships with ISACs, as well as government and international entities. We recommended that NIPC formalize relationships with ISACs and develop a plan to foster a two-way exchange of information between them. In response to our recommendations, NIPC officials told us that a new ISAC development and support unit had been created, whose mission is to enhance private sector cooperation and trust, resulting in a two-way sharing of information. NIPC now reports that 11 ISACs have been established, including those for the chemical industry, surface transportation, electric power, telecommunications, information technology, financial services, water supply, oil and gas, emergency fire services, food, and emergency law enforcement. Officials informed us that the Center has signed

[21] U.S. General Accounting Office, *National Preparedness: Integrating New and Existing Technology and Information Sharing Into an Effective Homeland Security Strategy*, GAO-02-811T (Washington, D.C.: June 7, 2002).

information sharing agreements with most ISACs, including those representing telecommunications, information technology, air transportation, water supply, food, emergency fire services, banking and finance, and chemical sectors. NIPC officials added that these agreements contained industry specific cyber and physical incident reporting thresholds. Further, officials told us that it has developed a program with the electric power ISAC whereby members transmit incident reports directly to NIPC.

Our ongoing work for this Subcommittee on five of these ISACs has shown that while progress has been made, each sector does not have a fully established ISAC, those that do have varied participation, and the amount of information being shared between the federal government and private sector organizations also varies.[22] In the Comptroller General's testimony several weeks ago, he stated that intelligence and information sharing challenges highlight the need for strong partnerships with those outside the federal government and that the new department will need to design and manage tools of public policy (e.g., grants to non-federal entities) to engage and work constructively with third parties.[23]

Some in the private sector have expressed concerns about voluntarily sharing information with the government. For example, concerns have been raised that industry could potentially face antitrust violations for sharing information with other industry partners, have their information be subject to the Freedom of Information Act (FOIA), or face potential liability concerns for information shared in good faith. Many suggest that the government should model the Year 2000 Information and Readiness Disclosure Act, which provided limited exemptions and protections for the private sector in order to facilitate the sharing of information on Year 2000 readiness.

In addition, other actions have been taken by the Congress and the administration to strengthen information sharing. The USA Patriot Act, for example, enhances or promotes information sharing among federal agencies, and numerous terrorism task forces have been established to coordinate investigations and improve communications among federal and local law enforcement. There will be continuing debate as to whether adequate protection is being provided to the private sector as these entities are encouraged to disclose and exchange information on both physical and cyber security problems and solutions that are essential to protecting our nation's critical infrastructures.

[22] The five ISACs are information technology, telecommunications, energy, electricity, and water.

[23] GAO-02-866T, June 25, 2002.

Information sharing within the government also remains a challenge. In April of last year, we reported that the NIPC and other government entities had not developed fully productive information sharing and cooperative relationships. For example, federal agencies had not routinely reported incident information to the NIPC, at least in part because guidance provided by the federal Chief Information Officers Council, which is chaired by the Office of Management and Budget, directs agencies to report such information to the General Services Administration's Federal Computer Incident Response Center. Further, NIPC and Defense officials agreed that their information-sharing procedures needed improvement, noting that protocols for reciprocal exchanges of information had not been established. In addition, the expertise of the U.S. Secret Service regarding computer crime had not been integrated into NIPC efforts. According to the NIPC director, the relationship between the NIPC and other government entities has significantly improved since our review, and that the quarterly meetings with senior government leaders have been instrumental in improving information sharing. In addition, officials from the Federal Computer Incident Response Center and the U.S. Secret Service in testimony have discussed the collaborative and cooperative relationships that now exist between their agencies and the NIPC.

Pervasive Federal Information Security Weaknesses Need to Be Addressed

At the federal level, cyber CIP activities are a component, perhaps the most critical, of a federal department or agency's overall information security program. Federal agencies have significant critical infrastructures that need effective information security to adequately protect them. However, since September 1996, we have reported that poor information security is a widespread federal problem with potentially devastating consequences.[24] Our analyses of information security at major federal agencies have shown that federal systems were not being adequately protected from computer-based threats, even though these systems process, store, and transmit enormous amounts of sensitive data and are indispensable to many federal agency operations. In addition, in both 1998 and in 2000, we analyzed audit results for 24 of the largest federal agencies and found that all 24 agencies had significant information security weaknesses.[25] As a result of these analyses, we have identified information security as a governmentwide high-risk issue in reports to the Congress

[24]U.S. General Accounting Office, *Information Security: Opportunities for Improved OMB Oversight of Agency Practices*; GAO/AIMD-96-110 (Washington, D.C.: Sep. 24, 1996).

[25]U.S. General Accounting Office, *Information Security: Serious Weaknesses Place Critical Federal Operations and Assets at Risk*; GAO/AIMD-98-92 (Washington, D.C.: Sep. 23, 1998); *Information Security: Serious and Widespread Weaknesses Persist at Federal Agencies*; GAO/AIMD-00-295 (Washington, D.C.: Sep. 6, 2000).

since 1997—most recently in January 2001.[26] More current analyses of audit results, as well as of the agencies' own reviews of their information security programs continue to show significant weaknesses that put critical federal operations and assets at risk.

Weaknesses Remain Pervasive

Our November 2001 analyses of audit results for 24 of the largest federal agencies showed that weaknesses continued to be reported in each of the 24 agencies.[27] These analyses considered GAO and inspector general (IG) reports published from July 2000 through September 2001, which included the first annual independent IG evaluations of agencies' information security programs required by government information security reform legislation (commonly referred to as "GISRA").[28]

Our analyses showed that the weaknesses reported for the 24 agencies covered all six major areas of general controls, that is, the policies, procedures, and technical controls that apply to all or a large segment of an entity's information systems and help ensure their proper operation. These six areas are (1) security program management, which provides the framework for ensuring that risks are understood and that effective controls are selected and properly implemented; (2) access controls, which ensure that only authorized individuals can read, alter, or delete data; (3) software development and change controls, which ensure that only authorized software programs are implemented; (4) segregation of duties, which reduces the risk that one individual can independently perform inappropriate actions without detection; (5) operating systems controls, which protect sensitive programs that support multiple applications from tampering and misuse; and (6) service continuity, which ensures that computer-dependent operations experience no significant disruptions. Figure 2 illustrates the distribution of weaknesses for the six general control areas across the 24 agencies.

[26] U.S. General Accounting Office, *High-Risk Series: Information Management and Technology*; GAO/HR-97-9 (Washington, D.C.: Feb. 1, 1997); *High-Risk Series: An Update*; GAO/HR-99-1 (Washington, D.C.: Jan. 1999); *High-Risk Series: An Update*, GAO-01-263 (Washington, D.C.: Jan. 2001).

[27] U.S. General Accounting Office, *Computer Security: Improvements Needed to Reduce Risk to Critical Federal Operations and Assets*, GAO-02-231T (Washington, D.C.: Nov. 9, 2001).

[28] Title X, Subtitle G—Government Information Security Reform, Floyd D. Spence National Defense Authorization Act for Fiscal Year 2001, P.L. 106-398, October 30, 2000. Congress enacted "GISRA" to supplement information security requirements established in the Computer Security Act of 1987, the Paperwork Reduction Act of 1995, and the Clinger-Cohen Act of 1996 and is consistent with existing information security guidance issued by OMB and the National Institute of Standards and Technology, as well as audit and best practice guidance issued by GAO. Most importantly, however, GISRA consolidates these separate requiremen' and guidance into an overall framework for managing information security and establishes new annual review independent evaluation, and reporting requirements to help ensure agency implementation and both OMB and congressional oversight. Effective November 29, 2000, GISRA is in effect for 2 years after this date.

Figure 2: Information Security Weaknesses at 24 Major Agencies

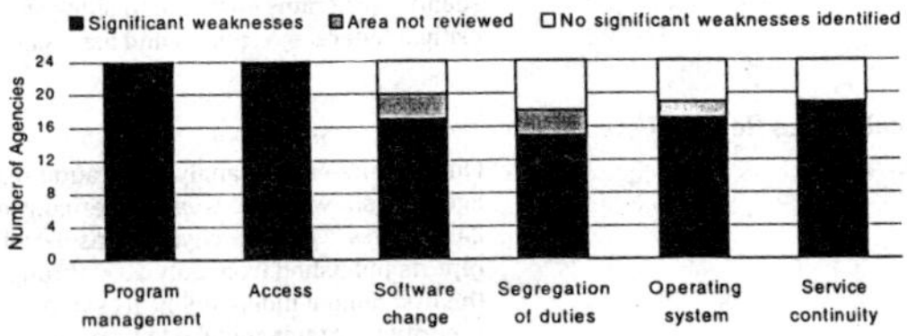

Source: Audit reports issued July 2000 through September 2001.

As in 2000, our current analysis shows that weaknesses were most often identified for security program management and access controls. For security program management, we found weaknesses for all 24 agencies in 2001 as compared to 21 of the 24 agencies (88 percent) in 2000. Security program management, which is fundamental to the appropriate selection and effectiveness of the other categories of controls, covers a range of activities related to understanding information security risks; selecting and implementing controls commensurate with risk; and ensuring that controls, once implemented, continue to operate effectively. For access controls, we also found weaknesses for all 24 agencies in 2001—the same condition we found in 2000. Weak access controls for sensitive data and systems make it possible for an individual or group to inappropriately modify, destroy, or disclose sensitive data or computer programs for purposes such as personal gain or sabotage. In today's increasingly interconnected computing environment, poor access controls can expose an agency's information and operations to attacks from remote locations all over the world by individuals with only minimal computer and telecommunications resources and expertise. In 2001, we also found that 19 of the 24 agencies (79 percent) had weaknesses in service continuity controls (compared to 20 agencies or 83 percent in 2000). These controls are particularly important because they ensure that when unexpected events occur, critical operations will continue without undue interruption and that crucial, sensitive data are protected. If service continuity controls are inadequate, an agency can lose the capability to process, retrieve, and protect electronically maintained information, which can significantly affect an agency's ability to accomplish its mission.

Our current analyses of information security at federal agencies also showed that the scope of audit work performed has continued to expand to more fully cover all six major areas of general controls at each agency. Not surprisingly, this has led to the identification of additional areas of weakness at some agencies. These increases in reported weaknesses do not necessarily mean that information security at federal agencies is getting worse. They more likely indicate that information security weaknesses are becoming more fully understood—an important step toward addressing the overall problem. Nevertheless, the results leave no doubt that serious, pervasive weaknesses persist. As auditors increase their proficiency and the body of audit evidence expands, it is probable that additional significant deficiencies will be identified.

Most of the audits represented in figure 2 were performed as part of financial statement audits. At some agencies with primarily financial missions, such as the Department of the Treasury and the Social Security Administration, these audits covered the bulk of mission-related operations. However, at agencies whose missions are primarily nonfinancial, such as the Departments of Defense and Justice, the audits may provide a less complete picture of the agency's overall security posture because the audit objectives focused on the financial statements and did not include evaluations of individual systems supporting nonfinancial operations. In response to congressional interest, beginning in fiscal year 1999, we expanded our audit focus to cover a wider range of nonfinancial operations—a trend we expect to continue. Audit coverage for nonfinancial systems is also likely to increase as agencies review and evaluate their information security programs as required by GISRA.

Weaknesses Pose Substantial Risks for Federal Operations, Assets, and Confidentiality

To fully understand the significance of the weaknesses we identified, it is necessary to link them to the risks they present to federal operations and assets. Virtually all federal operations are supported by automated systems and electronic data, and agencies would find it difficult, if not impossible, to carry out their missions and account for their resources without these information assets. Hence, the degree of risk caused by security weaknesses is extremely high.

The weaknesses identified place a broad array of federal operations and assets at risk. For example,

- resources, such as federal payments and collections, could be lost or stolen;
- computer resources could be used for unauthorized purposes or to launc attacks on others;

- sensitive information, such as taxpayer data, social security records, medical records, and proprietary business information, could be inappropriately disclosed or browsed or copied for purposes of espionage or other types of crime;
- critical operations, such as those supporting national defense and emergency services, could be disrupted;
- data could be modified or destroyed for purposes of fraud or disruption; and
- agency missions could be undermined by embarrassing incidents that result in diminished confidence in their ability to conduct operations and fulfill their fiduciary responsibilities.

Examples from recent audit reports issued in 2001 illustrate the serious weaknesses found in the agencies that continue to place critical federal operations and assets at risk:

- In August, we reported that significant and pervasive weaknesses placed Commerce's systems at risk. Many of these systems are considered critical to national security, national economic security, and public health and safety. Nevertheless, we demonstrated that individuals, both within and outside of Commerce, could gain unauthorized access to Commerce systems and thereby read, copy, modify, and delete sensitive economic, financial, personnel, and confidential business data. Moreover, intruders could disrupt the operations of systems that are critical to the mission of the department.[29] Commerce's IG has also reported significant computer security weaknesses in several of the department's bureaus and, in February 2001, reported multiple material information security weaknesses affecting the department's ability to produce accurate data for financial statements.[30]
- In July, we reported serious weaknesses in systems maintained by the Department of Interior's National Business Center, a facility processing more than $12 billion annually in payments that place sensitive financial and personnel information at risk of unauthorized disclosure, critical operations at risk of disruption, and assets at risk of loss. While Interior has made progress in correcting previously identified weaknesses, the newly identified weaknesses impeded the center's ability to (1) prevent

[29] U.S. General Accounting Office, *Information Security: Weaknesses Place Commerce Data and Operations at Serious Risk*; GAO-01-751 (Washington, D.C.: Aug. 13, 2001).

[30] Department of Commerce's Fiscal Year 2000 Consolidated Financial Statements, Inspector General Audit Report No. FSD-12849-1-0001.

and detect unauthorized changes, (2) control electronic access to sensitive information, and (3) restrict physical access to sensitive computing areas.[31]

- In March, we reported that although DOD's Departmentwide Information Assurance Program made progress, it had not yet met its goals of integrating information assurance with mission-readiness criteria, enhancing information assurance capabilities and awareness of department personnel, improving monitoring and management of information assurance operations, and establishing a security management infrastructure. As a result, DOD was unable to accurately determine the status of information security across the department, the progress of its improvement efforts, or the effectiveness of its information security initiatives.[32]
- In February, the Department of Health and Human Services' IG again reported serious control weaknesses affecting the integrity, confidentiality, and availability of data maintained by the department.[33] Most significant were weaknesses associated with the department's Centers for Medicare and Medicaid Services (CMS), formerly known as the Health Care Financing Administration, which, during fiscal year 2000, wa responsible for processing more than $200 billion in Medicare expenditures. CMS relies on extensive data processing operations at its central office to maintain administrative data (such as Medicare enrollment, eligibility, and paid claims data) and to process all payments for managed care. Significant weaknesses were also reported for the Food and Drug Administration and the department's Division of Financial Operations.

To correct reported weaknesses, several agencies took significant steps to redesign and strengthen their information security programs. For example, the Environmental Protection Agency has moved aggressively to reduce the exposure of its systems and data and to correct weaknesses we identified in February 2000.[34] While we have not tested their effectiveness, these actions show that the agency is taking a comprehensive and systematic approach that should help ensure that its efforts are effective.

---

[31]U.S. General Accounting Office, *Information Security: Weak Controls Place Interior's Financial and Other Data at Risk*; GAO-01-615 (Washington, D.C.: July 3, 2001).

[32]U.S. General Accounting Office, *Information Security: Progress and Challenges to an Effective Defense-wide Information Assurance Program*; GAO-01-307 (Washington, D.C.: Mar. 30, 2001).

[33]Report on the Financial Statement Audit of the Department of Health and Human Services for Fiscal Year 2000, A-17-00-00014, Feb. 26, 2001.

[34]U.S. General Accounting Office, *Information Security: Fundamental Weaknesses Place EPA Data and Operations at Risk*; GAO/AIMD-00-215 (Washington, D.C.: July 6, 2000).

Agencies' GISRA Results Also Highlight Weaknesses

As required by GISRA, agencies reviewed their information security programs, reported the results of these reviews and the IGs' independent evaluations to OMB, and developed plans to correct identified weaknesses. These reviews and evaluations showed that agencies have not established information security programs consistent with GISRA requirements and that significant weaknesses exist. Although agency actions are now underway to strengthen information security and implement these requirements, significant improvement will require sustained management attention and OMB and congressional oversight.

In its fiscal year 2001 report to Congress on GISRA, OMB notes that although examples of good security exist in many agencies, and others are working very hard to improve their performance, many agencies have significant deficiencies in every important area of security.[35] In particular, the report highlights six common security weaknesses: (1) a lack of senior management attention to information security; (2) inadequate accountability for job and program performance related to information technology security; (3) limited security training for general users, information technology professionals, and security professionals; (4) inadequate integration of security into the capital planning and investment control process; (5) poor security for contractor-provided services; and (6) limited capability to detect, report, and share information on vulnerabilities or to detect intrusions, suspected intrusions, or virus infections.

In general, our analyses of the results of agencies' GISRA reviews and evaluations also showed that agencies are making progress in addressing information security, but that none of the agencies had fully implemented the information security requirements of GISRA and all continue to have significant weaknesses. In particular, our review of 24 of the largest federal agencies showed that agencies had not fully implemented requirements to:

- conduct risk assessments for all their systems;
- establish information security policies and procedures that are commensurate with risk and that comprehensively address the other reform provisions;

[35] Office of Management and Budget, *FY 2001 Report to Congress on Federal Government Information Security Reform* (February, 2002).

- provide adequate computer security training to their employees including contractor staff;
- test and evaluate controls as part of their management assessments;
- implement documented incident handling procedures agencywide;
- identify and prioritize their critical operations and assets, and determine the priority for restoring these assets should a disruption in critical operations occur; or
- have a process to ensure the security of services provided by a contractor or another agency.

Improvement Efforts are Underway, but Challenges to Federal Information Security Remain

Information security improvement efforts have been undertaken in the past few years both at an agency and governmentwide level. However, given recent events and reports that critical operations and assets continue to be highly vulnerable to computer-based attacks, the government still faces a challenge in ensuring that risks from cyber threats are appropriately addressed. Accordingly, it is important that federal information security efforts be guided by a comprehensive strategy for improvement.

First, it is important that the federal strategy delineate the roles and responsibilities of the numerous entities involved in federal information security. This strategy should also consider other organizations with information security responsibilities, including OMB, which oversees and coordinates federal agency security, and interagency bodies like the CIO Council, which are attempting to coordinate agency initiatives. It should also describe how the activities of these many organizations interrelate, who should be held accountable for their success or failure, and whether they will effectively and efficiently support national goals.

Second, more specific guidance to agencies on the controls that they need to implement could help ensure adequate protection. Currently, agencies have wide discretion in deciding what computer security controls to implement and the level of rigor with which to enforce these controls. In theory, this discretion is appropriate since, as OMB and NIST guidance states, the level of protection that agencies provide should be commensurate with the risk to agency operations and assets. In essence, one set of specific controls will not be appropriate for all types of systen and data. Nevertheless, our studies of best practices at leading organizations have shown that more specific guidance is important.[36] In particular, specific mandatory standards for varying risk levels can clarify expectations for information protection, including audit criteria; provide a standard framework for assessing information security risk; help ensure that shared data are appropriately protected; and reduce demands for limited resources to independently develop security controls. Implementing such standards for federal agencies would require developing a single set of information classification categories for use by all agencies to define the criticality and sensitivity of the various types of information they maintain. It would also necessitate establishing minimum mandatory requirements for protecting information in each classification category.

Third, ensuring effective implementation of agency information security and critical infrastructure protection plans will require active monitoring by the agencies to determine if milestones are being met and testing to determine if policies and controls are operating as intended. Routine periodic audits, such as those required by GISRA, would allow for more meaningful performance measurement. In addition, the annual evaluation, reporting, and monitoring process established through these provisions, is an important mechanism, previously missing, to hold agencies accountable for implementing effective security and to manage the problem from a governmentwide perspective. Moreover, with GISRA expiring on November 29, 2002, we believe that continued authorization of information security legislation is essential to improving federal information security.

Fourth, the Congress and the executive branch can use audit results to monitor agency performance and take whatever action is deemed advisable to remedy identified problems. Such oversight is essential for holding agencies accountable for their performance, as was demonstrated by the OMB and congressional efforts to oversee the Year 2000 computer challenge.

Fifth, agencies must have the technical expertise they need to select, implement, and maintain controls that protect their information systems. Similarly, the federal government must maximize the value of its technical staff by sharing expertise and information. Highlighted during the Year 2000 challenge, the availability of adequate technical and audit expertise is a continuing concern to agencies.

---

[36]U.S. General Accounting Office, *Information Security Management: Learning from Leading Organizations*, GAO/AIMD-98-68 (Washington, D.C.: May 1998).

Sixth, agencies can allocate resources sufficient to support their information security and infrastructure protection activities. Funding for security is already embedded to some extent in agency budgets for computer system development efforts and routine network and system management and maintenance. However, some additional amounts are likely to be needed to address specific weaknesses and new tasks. OMB and congressional oversight of future spending on information security will be important to ensuring that agencies are not using the funds they receive to continue ad hoc, piecemeal security fixes that are not supported by a strong agency risk management process.

Seventh, expanded research is needed in the area of information systems protection. While a number of research efforts are underway, experts have noted that more is needed to achieve significant advances. As stated by the director of the CERT® Coordination Center in congressional testimony last September, "It is essential to seek fundamental technological solutions and to seek proactive, preventive approaches, not just reactive, curative approaches." In addition, in its December 2001 third annual report, the Gilmore Commission recommended that the Office of Homeland Security develop and implement a comprehensive plan for research, development, test, and evaluation to enhance cyber security.[37]

In conclusion, consolidating the six federal CIP-focused organizations into a single division within the proposed Department of Homeland Security, if properly organized and implemented, could minimize duplication and allow for closer coordination of national CIP approach, especially in the areas of outreach and education, the identification of critical assets, and incident reporting, analysis, and warning. However, prior GAO work has identified and made recommendations concerning several critical infrastructure protection challenges that need to be addressed. The new department should be viewed as a catalyst for addressing these recommendations, which include:

- completing a comprehensive and coordinated CIP strategy to include both cyber and physical aspects,
- improving analytical and warning capabilities,
- improving information sharing both within the federal government and between the federal government and the private sector and state and local governments.

[37] *Third Annual Report to the President and Congress of the Advisory Panel to Assess Domestic Response Capabilities for Terrorism Involving Weapons of Mass Destruction* (Dec. 15, 2001).

- addressing pervasive weaknesses in federal information security.

Mr. Chairman, this concludes my written testimony. I would be pleased to answer any questions that you or other members of the Subcommittee may have at this time.

If you should have any questions about this testimony, please contact me at (202) 512-3317. I can also be reached by e-mail at daceyr@gao.gov.

Appendix

Description of the Six Organizations Proposed to be Moved to the Information Analysis and Infrastructure Protection Division

Below is a description of the current roles and responsibilities for the federal organizations that are proposed to be moved to the new division.

Critical Infrastructure Assurance Office

As established under PDD 63, the Critical Infrastructure Assurance Office (CIAO) performs a variety of CIP functions in three major areas: (1) educating the private sector on the importance of CIP, (2) preparing the national CIP strategy, and (3) assisting federal civilian agencies and departments in determining their dependencies on critical infrastructures.

First, the CIAO works to educate industry representatives that critical infrastructure assurance must be addressed through corporate risk management activities. Its efforts focus on the critical infrastructure industries (e.g., information and communications, banking and finance, transportation, energy, and water supply), particularly the corporate boards and chief executive officers who are responsible for setting policy and allocating resources for risk management. In addition to infrastructure owners and operators, this office's awareness and outreach efforts also target members of the audit, insurance, and investment communities. CIAO's goal is to educate these groups on the importance of assuring effective corporate operations, accountability, and information security.

Second, the CIAO is tasked with working with government and industry to prepare the national strategy for CIP, which is due for completion in 2002. This strategy will serve as the basis for CIP legislative and public policy reforms, where needed. The development of the national strategy for CIP is to also serve as part of an ongoing process in which government and industry will continuously modify and refine their efforts to ensure the safety of critical information systems.

Third, the CIAO is responsible for assisting civilian federal agencies and departments in analyzing their dependencies on critical infrastructures. This mission is conducted under Project Matrix, a program designed to identify and characterize the assets and associated infrastructure dependencies and interdependencies that the government requires to fulfill its most critical responsibilities. Project Matrix involves a three-step process in which each federal civilian agency identifies (1) its critical assets; (2) other federal government assets, systems, and networks on which those critical assets depend to operate; and (3) all associated dependencies on privately owned and operated critical infrastructures.

Additional cyber CIP duties were added to CIAO under Executive Order 13231, including having its director serve as a member of and advisor to the President's Critical Infrastructure Protection Board. The CIAO is also to support the activities of the National Infrastructure Advisory Council, a group of 30 representatives from private industry and state and local government that will advise the President on matters relating to cybersecurity and CIP. In addition, the CIAO is expected to administer a Homeland Security Information Technology and Evaluation Program to study and develop methods to improve information sharing among federal agencies and state and local governments.

Federal Computer Incident Response Center

The Federal Computer Incident Response Center (FedCIRC) is the focal point for dealing with computer-related incidents affecting federal civilian agencies. Originally established in 1996 by the National Institute of Standards and Technology, the center has been administered by the General Services Administration since October 1998.

FedCIRC's primary purposes are to provide a means for federal civilian agencies to work together to handle security incidents, share related information, and solve common security problems. In this regard, FedCIRC

- provides federal civilian agencies with technical information, tools, methods, assistance, and guidance;
- provides coordination and analytical support;
- encourages development of quality security products and services through collaborative relationships with federal agencies, academia, and private industry;
- promotes incident response and handling procedural awareness within the federal government;
- fosters cooperation among federal agencies for effectively preventing, detecting, handling, and recovering from computer security incidents;
- communicates alert and advisory information regarding potential threats and emerging incident situations; and

- augments the incident response capabilities of federal agencies.
- In accomplishing these efforts, FedCIRC draws on expertise from the Department of Defense, the intelligence community, academia, and federal civilian agencies. In addition, FedCIRC collaborates with the Federal Bureau of Investigation's (FBI) National Infrastructure Protection Center in planning for and dealing with criminal activities that pose a threat to the critical information infrastructure.

National Communications System

Created by Executive Order 12472, the National Communications System's (NCS's) CIP mission is to assure the reliability and availability of national security and emergency preparedness (NS/EP) telecommunications. Its mission includes, but it is not necessarily limited to, responsibility for (1) assuring the government's ability to receive priority services for NS/EP purposes in current and future telecommunications networks by conducting research and development and participating in national and international standards bodies and (2) operationally coordinating with industry for protecting and restoring NS/EP services in an all-hazards environment. NCS's mission is externally focused on the reliability and availability of the public telecommunications network. This mission is carried out within government through the NS/EP Coordinating Committee, with industry on a policy level in coordination with the National Security Telecommunications Advisory Committee (NSTAC), and operationally through the National Coordinating Center for Telecommunications and through its participation in national and international standards bodies. Furthermore, in January 2000, National Coordinating Center was designated an ISAC for telecommunications under the provisions of PDD 63.

NCS reports to the Executive Office of the President–NSC for policy, to the Office of Science Technology and Policy for operations, and to OMB for budget through the Secretary of Defense, who is the Executive Agent for NCS. NCS's NS/EP Coordinating Committee is a standing committee under the President's Critical Infrastructure Protection Board. Externally, NCS coordinates with the Office of Cyberspace Security; CIAO; the National Telecommunications and Information Administration; the NIPC; GSA's FedCIRC; the Department of Energy (DOE) (including several of the laboratories); the Department of Transportation, industry members of the National Coordinating Center for Telecommunications; ISACs; and numerous DOD organizations.

National Infrastructure Protection Center

NIPC, a multiagency organization located within the FBI, detects, analyzes, and warns of cyberthreats to and/or attacks on the infrastructure, should they occur. NIPC's mission is based on authorities given in Executive Order 13231 and PDD 63. In addition, the center is responsible for accomplishing the FBI's role as lead agency for sector liaison for the Emergency Law Enforcement Services Sector. As a sector liaison, NIPC provides law enforcement response for cyberthreats and crimes involving or affecting critical infrastructures. NIPC also facilitates and coordinates the federal government's response to cyber incidents, mitigating attacks, and investigating threats, as well as monitoring reconstitution efforts. NIPC regularly coordinates with federal, state, local, and law enforcement and intelligence agencies resident in the NIPC: FBI, DOD, the Central Intelligence Agency (CIA), the National Security Agency (NSA), the United States Secret Service (USSS), Commerce, DOT, DOE, and other federal agencies on the President's Critical Infrastructure Protection Board, as well as Canada and Great Britain.

In addition, NIPC runs the National InfraGard program, which is a cooperative undertaking between the federal government and an association of businesses, academic institutions, state and local law enforcement agencies, and other participants dedicated to increasing the security of critical infrastructures. InfraGard's goal is to enable the flow of information so that the owners and operators of infrastructure assets, the majority of which are from the private sector, can better protect themselves and so that the U.S. government can better discharge its law enforcement and national security responsibilities. InfraGard provides members a forum for education and training on infrastructure vulnerabilities and protection measures and with threat advisories, alerts, and warnings.

NIPC comprises three sections: (1) the Computer Investigations and Operations Section, which is the operational and response arm and is responsible for designing, developing, implementing, and managing automated tools NIPC uses to collect, analyze, share, and distribute information; and coordinating computer investigations conducted by the FBI's 56 field offices and approximately 400 sublocations throughout the country; (2) the Analysis and Warning Section, which is the indication and warning arm, which provides support during computer intrusion investigations; and (3) the Training, Outreach, and Strategy Section, which provides outreach to the private sector and to local law enforcement, and training and exercise programs for cyber and infrastructure protection investigators within the FBI and other agencies.

National Infrastructure Simulation and Analysis Center

The National Infrastructure and Analysis Center (NISAC) exists as a partnership between the Defense Threat Reduction Agency and the Los Alamos and Sandia national laboratories. Its mission is to improve the robustness of the nation's critical infrastructures by providing an advanced modeling and simulation capability that will enable an understanding of how the infrastructure operates; help identify vulnerabilities; determine the consequences of infrastructure outages; and optimize protection and mitigation strategies. NISAC's objectives are to

- leverage the existing capabilities of the NISAC partners to provide leadership in critical infrastructure interdependencies modeling, simulation, and analysis;
- establish a virtual capability that will provide a portal for nation-wide remote access and communications to infrastructure-related modeling, simulation, and analysis capabilities;
- move toward a predictive capability that uses science-based tools to understand the expected performance of interrelated infrastructures under various conditions;
- provide simulation and analysis capabilities to a wide range of users that will enhance the understanding of vulnerabilities of the national infrastructures and establish priorities and potential mitigation strategies for protecting the infrastructures;
- provide decision-makers the ability to assess policy and investment options that address critical infrastructure needs – near and long term;
- provide education and training to public and private decision makers on how to cope effectively with crisis events; and
- provide an integrating function that includes interdependencies; bring disparate users and information providers and individual infrastructure sector leaders together.

National Institute of Standards and Technology—Computer Security Division

Under the Computer Security Act of 1987, NIST's Computer Security Division of the Information Technology Laboratory develops computer security prototypes, tests, standards, and procedures to protect sensitive information from unauthorized access or modification. Specifically, the Computer Security Division's mission is to improve information systems security by

- raising awareness of IT risks, vulnerabilities, and protection requirements, particularly for new and emerging technologies;
- researching, studying, and advising agencies of IT vulnerabilities and devising techniques for the cost-effective security and privacy of sensitive federal systems;
- developing standards, metrics, tests, and validation programs to
  - promote, measure, and validate security in systems and services
  - educate consumers and
  - establish minimum security requirements for federal systems; and
- developing guidance to increase secure IT planning, implementation, management, and operation.

Further, the division's functions are focused on five areas:

- cryptographic standards and applications,
- security research and emerging technologies,
- security management and guidance,
- security testing, and
- outreach, awareness, and education.

(310161)